Lecture Notes in Computer Science 15720

The series Lecture Notes in Computer Science (LNCS), including its subseries Lecture Notes in Artificial Intelligence (LNAI) and Lecture Notes in Bioinformatics (LNBI), has established itself as a medium for the publication of new developments in computer science and information technology research, teaching, and education.

LNCS enjoys close cooperation with the computer science R & D community, the series counts many renowned academics among its volume editors and paper authors, and collaborates with prestigious societies. Its mission is to serve this international community by providing an invaluable service, mainly focused on the publication of conference and workshop proceedings and postproceedings. LNCS commenced publication in 1973.

Jörg Endrullis · Matthias Tichy
Editors

Graph Transformation

18th International Conference, ICGT 2025
Held as Part of STAF 2025
Koblenz, Germany, June 11–12, 2025
Proceedings

Springer

Editors
Jörg Endrullis
Vrije Universiteit Amsterdam
Amsterdam, The Netherlands

Matthias Tichy
Universität Ulm
Ulm, Germany

ISSN 0302-9743 ISSN 1611-3349 (electronic)
Lecture Notes in Computer Science
ISBN 978-3-031-94705-6 ISBN 978-3-031-94706-3 (eBook)
https://doi.org/10.1007/978-3-031-94706-3

This Springer imprint is published by the registered company Springer Nature Switzerland AG
The registered company address is: Gewerbestrasse 11, 6330 Cham, Switzerland

Preface

This volume contains the proceedings of ICGT 2025, the 18th International Conference on Graph Transformation, held on 11–12 June 2025, at the University of Koblenz in Germany. ICGT 2025 was affiliated with STAF (Software Technologies: Applications and Foundations), a federation of leading conferences on software technologies. ICGT 2025 took place under the auspices of the European Association of Theoretical Computer Science (EATCS), the European Association of Software Science and Technology (EASST), and the IFIP Working Group 1.3, Foundations of Systems Specification.

The ICGT series aims to foster exchange of knowledge and collaboration among researchers from different backgrounds working with graphs and graph transformation, either by contributing to their theoretical foundations or by applying established formalisms to classic or novel areas. The series not only serves as a well-established scientific publication outlet but also as a platform to boost inter- and intra-disciplinary research and to stimulate new ideas. The use of graphs and graph-like structures as a formalism for specification and modelling is widespread in all areas of computer science as well as in many fields of computational research and engineering. Relevant examples include software architectures, pointer structures, state-space and control/data flow graphs, UML and other domain-specific models, network layouts, topologies of cyber-physical environments, quantum computing, and molecular structures. Often, these graphs undergo dynamic change, ranging from reconfiguration and evolution to various kinds of behaviour, all of which may be captured by rule-based graph manipulation. Thus, graphs and graph transformation form a fundamental, universal modelling paradigm that serves as a means for formal reasoning and analysis, ranging from the verification of certain properties of interest to the discovery of fundamentally new insights.

ICGT 2025 continued the series of conferences previously held in Barcelona (Spain) in 2002, Rome (Italy) in 2004, Natal (Brazil) in 2006, Leicester (UK) in 2008, Enschede (The Netherlands) in 2010, Bremen (Germany) in 2012, York (UK) in 2014, L'Aquila (Italy) in 2015, Vienna (Austria) in 2016, Marburg (Germany) in 2017, Toulouse (France) in 2018, Eindhoven (The Netherlands) in 2019, online in 2020 and 2021, Nantes (France) in 2022, Leicester (UK) in 2023, and Twente (The Netherlands) in 2024, following a series of six International Workshops on Graph Grammars and Their Application to Computer Science from 1978 to 1998 in Europe and in the USA.

This year, the conference solicited papers across three categories: research papers, describing new and unpublished contributions to the theory and applications of graph transformation; tool presentation papers, demonstrating the main features and functionalities of graph-based tools; and blue skies papers, reporting on new research directions or ideas that are at an early or emerging stage.

ICGT 2025 received 19 submissions, each reviewed by three Programme Committee members and/or additional reviewers. Following an extensive discussion phase, the Programme Committee selected 10 research papers and one blue skies paper for publication in these proceedings.

The topics of the accepted papers cover a wide spectrum, ranging from advancements in the classical theory of graph transformation to the integration of artificial intelligence approaches with graph transformations, the fuzzing of graph databases, and applications of graph transformation in areas such as dialogue management systems and threat analysis.

In addition, we solicited proposals for 'journal-first' talks, allowing authors to present relevant (previously published) work to the community. Finally, we were delighted to host an invited talk by Erik Burger from the Karlsruhe Institute of Technology.

We would like to thank everyone who contributed to the success of ICGT 2025, including the members of our Programme Committee, our additional reviewers, and our invited speakers. We are grateful to Reiko Heckel, the Chair of the ICGT Steering Committee, for his valuable suggestions; the organising committee of STAF 2025, for hosting and supporting ICGT 2025; conf.researchr.org, for hosting our website; and EasyChair, for supporting the review process and the preparation of the proceedings.

April 2025

Matthias Tichy
Jörg Endrullis

Organization

Program Committee

Nicolas Behr	Université Paris Cité, CNRS, IRIF, France
Andrea Corradini	Università di Pisa, Italy
Joerg Endrullis	Vrije Universiteit Amsterdam, The Netherlands
Fabio Gadducci	Università di Pisa, Italy
Raffaela Groner	Chalmers — University of Gothenburg, Sweden
Reiko Heckel	University of Leicester, UK
Jens Kosiol	Philipps-Universität Marburg, Germany
Jean Krivine	CNRS, France
Harald König	FHDW Hannover, Germany
Leen Lambers	BTU Cottbus-Senftenberg, Germany
Detlef Plump	University of York, UK
Arend Rensink	University of Twente, The Netherlands
Andy Schürr	TU Darmstadt, Germany
Daniel Strüber	Chalmers — University of Gothenburg, Sweden
Gabriele Taentzer	Philipps-Universität Marburg, Germany
Matthias Tichy	Ulm University, Germany
Jens Weber	University of Victoria, Canada
Steffen Zschaler	King's College London, UK

Steering Committee

Paolo Bottoni	Sapienza University of Rome, Italy
Andrea Corradini	University of Pisa, Italy
Holger Giese	University of Potsdam, Germany
Reiko Heckeli	University of Leicester, UK
Barbara König	University of Duisburg-Essen, Germany
Hans-Jörg Kreowski	University of Bremen, Germany
Leen Lambers	BTU Cottbus Senftenberg, Germany
Fernando Orejas	Universitat Politècnica de Catalunya, Spain
Detlef Plump	University of York, UK
Arend Rensink	University of Twente, The Netherlands
Gabriele Taentzer	Philipps-Universität Marburg, Germany
Jens Weber	University of Victoria, Canada

Additional Reviewers

Ali, Syed Juned
Anwar, Muhammad Waseem
Borges, Heraldo
Bottoni, Paolo
Boutillier, Pierre
Cioffo, Cipriano
Conrardy, Aaron David
Drewes, Frank
Heithoff, Malte
Hofmann, Lukas Sebastian
Ismaili Alaoui, Ziad
Latifaj, Malvina
Lauer, Alexander
Marin, Adrian Costin
Muqeet, Asmar
Plösch, Reinhold
Sakizloglou, Lucas
Thillmann, Henrik
Will, Sebastian
Wu, Jiahui
Xavier, Felipe

Contents

Graph Transformation Theory: Semantics and Static Analysis

Termination of Injective DPO Graph Rewriting Systems Using Subgraph Counting

Qi Qiu[1,2(✉)]

[1] Universite Claude Bernard Lyon 1, CNRS, INSA Lyon, LIRIS, UMR5205, Villeurbanne 69622, France
[2] University of Grenoble Alpes, CNRS, Grenoble INP, LIG, Grenoble 38000, France
qi.qiu@univ-grenoble-alpes.fr

Abstract. We present a machine-checkable sufficient condition for relative termination of double-pushout graph rewriting systems with injective rules on edge-labeled multigraphs. Our method defines a graph's weight as the sum of weights of occurrences of a set of graphs within it. By ensuring (1) every rewriting step using rules in a set A strictly decreases the host graph's weight, and (2) every rewriting step using rules in a set B never increases it, we guarantee that rules in the set A can be applied only finitely many times in any rewriting sequence with rules in the union of A and B. Our method resolves termination cases that prior interpretation-based methods cannot. We also propose an implementation of our technique.

Keywords: Termination · Rewrite system · Automated reasoning · Program verification · Graph transformation

1 Introduction

Distributed systems play a crucial role in improving performance, scalability, and efficiency in modern computing. Ensuring their correctness is vital [15] but challenging due to the inherent complexity of these systems [18]. While interactive theorem proving [14] offers a promising approach to formally verifying properties of distributed systems [9,27,29], it requires proficiency in proof assistants, making the process both demanding and time-consuming. Automated methods that generate certificates verifiable by proof assistants provide a practical alternative [7,12].

In this paper, we present an automated method for relative termination [11] of double-pushout (DPO) graph rewriting systems with injective rules [8,13,17]. DPO graph rewriting provides an intuitive yet formal framework for modeling distributed systems, enabling rigorous analysis and automated verification of their properties. Relative termination ensures that, given two rule sets $\mathcal{A}$ and

This study was partially funded by the project SAPPORO, ANR, 2019-CE25-0005.

J. Endrullis and M. Tichy (Eds.): ICGT 2025, LNCS 15720, pp. 3–23, 2025.
https://doi.org/10.1007/978-3-031-94706-3_1

$\mathcal{B}$, rules in $\mathcal{A}$ can only be applied finitely many times in any sequence of rule applications from $\mathcal{A} \cup \mathcal{B}$.

Our method works as follows. Let $\mathbb{X}$ denote a set of graphs. We assign weights (natural numbers) to occurrences of graphs in $\mathbb{X}$, and the weight of a graph G is defined as the sum of the weights of all occurrences of graphs in $\mathbb{X}$ within G. Let $\mathcal{A}$ and $\mathcal{B}$ be sets of injective DPO rewriting rules. Our technique requires that, for every graph X in $\mathbb{X}$, and for every rewriting step $G \Rightarrow H$ using a rule from $\mathcal{A}$ or $\mathcal{B}$, there exists an injective mapping from the set of X-occurrences in the result graph H (those neither fully included in the right-hand side graph nor fully included in the context) to the set of X-occurrences in the host graph G (those neither fully included in the left-hand side graph nor fully included in the context).

Suppose the following conditions hold: (i) For every rule in $\mathcal{A}$, the left-hand side graph's weight is strictly greater than the right-hand side graph's weight; (ii) For every rule in $\mathcal{B}$, the left-hand side graph's weight is greater than or equal to the right-hand side graph's weight. Under these conditions, rewriting steps using rules in $\mathcal{A}$ strictly decrease the weights of the host graphs, while rewriting steps using rules in $\mathcal{B}$ do not increase them. Consequently, the rewriting rules in $\mathcal{A}$ can only be applied finitely many times: since the weight of a finite graph is a natural number and strictly decreases with each rewriting step using a rule in $\mathcal{A}$, the process must terminate after finitely many iterations.

Our method proves termination for systems such as the one in Example 43, which prior interpretation-based approaches [5,6,10,23,30] cannot handle. We also provide an implementation of our technique[1].

The remainder of this paper is organized as follows. § 2 recalls some essential preliminary definitions. § 3 presents a sufficient condition for termination of DPO graph rewriting systems. § 4 illustrates the method with some examples. § 5 compares our approach with some existing methods. Finally, § 6 concludes with remarks and future research directions. The missing proofs are available in the preprint version [28].

2 Preliminaries

We recall some definitions related to the double-pushout approach to edge-labeled directed multigraph rewriting and relative termination. For more detailed information, we refer the reader to [8,17] for the double-pushout approach to edge-labeled multigraphs rewriting, [2,24] for category theory concepts, and [11] for relative termination.

2.1 Graphs

Definition 1 ([2]). *An* ***unlabeled graph*** *G consists of a collection of* ***nodes*** *(also called* ***objects****) and a collection of* ***edges****, each equipped with a* ***source*** *(or*

[1] The source code is available at https://github.com/Qi-tchi/LyonParallel/tree/icgt2025.

***domain**) node and a **target** (or **codomain**) node. For an unlabeled graph G, we denote by G_0 its collection of nodes, G_1 its collection of edges,* $\mathrm{dom} : G_1 \to G_0$ *the domain function, and* $\mathrm{cod} : G_1 \to G_0$ *the codomain function. An unlabeled graph is **finite** if G_0 and G_1 are both finite sets. We write $a : s \to t$ to indicate that a is a directed edge from s to t.*

Definition 2. *Let G and H be unlabeled graphs. A **homomorphism of unlabeled graphs** $h : G \to H$ is a pair of functions $h_0 : G_0 \to H_0$ and $h_1 : G_1 \to H_1$ such that for every edge $a : s \to t$ in G, we have $h_1(a) : h_0(s) \to h_0(t)$ in H.*

Definition 3. *Let Σ be a finite set of labels. A **labeled graph** is an ordered pair (G, λ) where G is an unlabeled graph and $\lambda : G_1 \to \Sigma$ is an edge-labeling function. It is called **finite** if its underlying unlabeled graph is finite.*

By $a : s \xrightarrow{l} t$, we denote the arrow a labeled by l from s to t. Unless otherwise specified, the term "graph" refers to the graphs finite. Note that unlabeled graphs can be regarded as labeled graphs. A homomorphism of labeled graphs is a homomorphism of unlabeled graphs that preserves the labels assigned to the edges.

Definition 4. *Let (G, λ) and (H, λ') be labeled graphs. A **homomorphism of labeled graphs** $h : (G, \lambda) \to (H, \lambda')$ is a homomorphism of unlabeled graphs such that for each edge a in G, we have $\lambda(a) = \lambda'(h_1(a))$.*

2.2 Pushout and Pullback

Definition 5 (Category [2,24]). *A **category** is an unlabeled graph C together with a total function $u : C_0 \to C_1$ and a partial function $\star : C_1 \times C_1 \to C_1$ such that (i) for all edges $f : X \to Y$ and $g : Y \to Z$, the edge $f \star g : X \to Z$ is defined; (ii) for every node X, $u(X)$ is an edge from X to X; (iii) for every $f : X \to Y$, we have $u(X) \star f = f = f \star u(Y)$; (iv) for all edges f, g and h, we have $(f \star g) \star h = f \star (g \star h)$ whenever either side is defined. Edges are called **morphisms**. The function $\star$ is called **composition**. For all $X \in C_0$, the edge $u(X)$ is denoted* id_X *and is called the **identity** of the object X.*

Notation 6. *The composition of morphisms $f : X \to Y$ and $g : Y \to Z$ is written in diagrammatic order as $f \star g$, rather than in functional order $g \circ f$.*

Definition 7. (Monomorphism [2,24]). *A morphism $f : X \to Y$ is **monic** (or a **monomorphism**) if for all morphisms g and h, if $g \star f = h \star f$, then $g = h$. A monomorphism is denoted by $f : X \rightarrowtail Y$. The set of all monomorphisms from graph X to graph Y is denoted by* $\mathrm{Mono}(X, Y)$.

Example 8. Finite edge-labeled directed multigraphs and their homomorphisms form a category, hereafter denoted **Graph**. Its objects are labeled graphs, its morphisms are graph homomorphisms, and the monomorphisms are injective homomorphisms.

In category theory, a ordered pair ($\alpha : A \to B$, $\beta : A \to C$) of morphisms with a common domain is called a **span** [20], denoted by $B \overset{\alpha}{\leftarrow} A \overset{\beta}{\rightarrow} C$. Likewise, an ordered pair ($\beta' : B \to D$, $\alpha' : C \to D$) of morphisms with a common codomain is called a **cospan**, denoted by $B \overset{\beta'}{\rightarrow} D \overset{\alpha'}{\leftarrow} C$.

Definition 9 (Diagram [2]). *Let G be an unlabeled graph. A **diagram** (in $\mathcal{C}$ of shape G) is a homomorphism of unlabeled graphs $h : G \to C$ where C is the underlying unlabeled graph of the category $\mathcal{C}$. A diagram is **commutative** if, for all nodes u, v, and any two paths from u to v in the unlabeled graph G:*

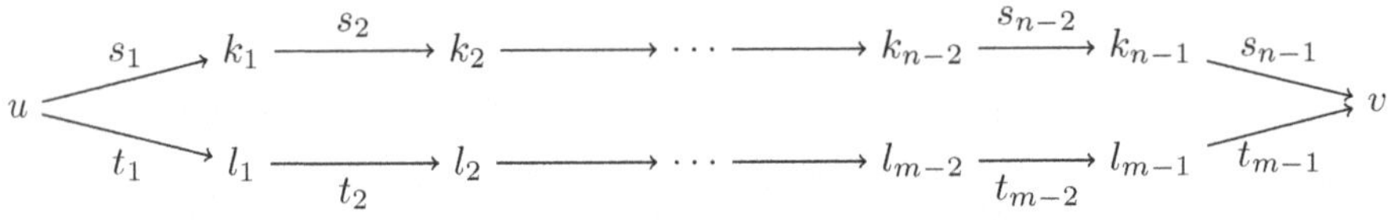

the equality $h(s_1) \star h(s_2) \star \cdots \star h(s_{n-1}) = h(t_1) \star h(t_2) \star \cdots \star h(t_{m-1})$ holds.

Definition 10 (Pushout [2]).
*A **pushout** of a span $B \overset{\alpha}{\leftarrow} A \overset{\beta}{\rightarrow} C$, as shown on the right, is defined as a cospan $B \overset{\beta'}{\rightarrow} D \overset{\alpha'}{\leftarrow} C$ such that $\alpha \star \beta' = \beta \star \alpha'$, and for every cospan $B \overset{\gamma'}{\rightarrow} E \overset{\gamma}{\leftarrow} C$, if $\alpha \star \gamma' = \beta \star \gamma$ holds, then there exists a unique morphism $\delta : D \to E$ such that $\gamma' = \beta' \star \delta$ and $\gamma = \alpha' \star \delta$.*

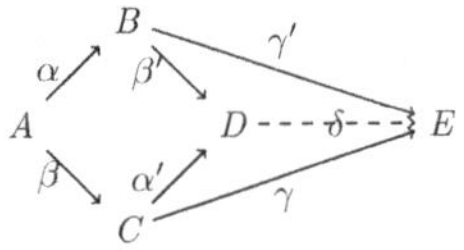

*The diagram involving $(\alpha, \beta, \alpha', \beta')$ is called the **pushout square**, with D as the **pushout object**. The existence of a unique morphism is known as the universal mapping property of the pushout.*

Definition 11 (Pullback [24]).
*A **pullback** of a cospan $B \overset{\beta'}{\rightarrow} D \overset{\alpha'}{\leftarrow} C$, as shown on the right, is defined as a span $B \overset{\alpha}{\leftarrow} A \overset{\beta}{\rightarrow} C$ such that $\alpha \star \beta' = \beta \star \alpha'$, and for every span $B \overset{\gamma'}{\leftarrow} E \overset{\gamma}{\rightarrow} C$ if $\gamma' \star \beta' = \gamma \star \alpha'$ holds, then there exists a unique morphism $\delta : E \to A$ such that $\gamma' = \delta \star \alpha$ and $\gamma = \delta \star \beta$.*

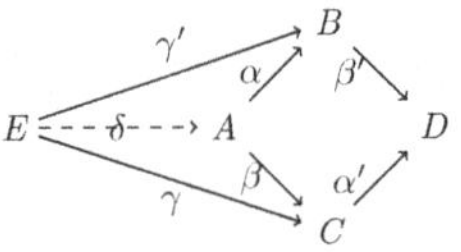

*The diagram involving $(\alpha, \beta, \alpha', \beta')$ is called the **pullback square**, with A as the **pullback object**. The existence of a unique morphism is known as the universal mapping property of the pullback.*

Notation 12 *When the context makes it clear, a morphism $h : A \to B$ will be denoted by h_{AB}, and diagrams will be referred to by their nodes, as is standard in geometry. For example, the diagram involving morphisms $(\alpha, \beta, \alpha', \beta')$ in Definition 10 will be denoted by $ACDB$ or $ABDC$.*

Remark 13. We work within the category **Graph**. The terms homomorphism and morphism will be used interchangeably; the same applies to monic and injective, and to injective homomorphism and monomorphism.

2.3 DPO Rewriting

Definition 14 (Rewriting rule and match [8]). *A* ***DPO rewriting rule*** *ρ is a span $L \xleftarrow{l} K \xrightarrow{r} R$, where K is the* ***interface****, L is the* ***left-hand side graph****, denoted* lhs(ρ)*, and R is the* ***right-hand side graph****, denoted* rhs(ρ)*. The rule is* ***left-monic*** *if the morphism l is monic,* ***right-monic*** *if the morphism r is monic,* ***monic*** *if it is left- and right-monic. A match of the rule in a graph G is a morphism $m : L \to G$.*

Definition 15 (Rewriting step [10]).

A DPO diagram δ is a diagram as shown on the right. This diagram δ is a witness for the ***rewriting step*** *from G to H using the rule ρ and* ***match*** *m, denoted $G \Rightarrow^m_\rho H$ or $G \Rightarrow^\delta_\rho H$. We denote by* left($\delta$) *and* right($\delta$) *the pushout squares $KLGC$ and $KRHC$, respectively.*

$$\begin{array}{ccccc} L & \xleftarrow{l} & K & \xrightarrow{r} & R \\ {\scriptstyle m}\downarrow & \text{PO} & \downarrow & \text{PO} & \downarrow{\scriptstyle m'} \\ G & \xleftarrow[l']{} & C & \xrightarrow[r']{} & H \end{array}$$

Definition 16 (Rewriting framework [10]). *A* ***DPO rewriting framework*** *$\mathfrak{F}$ is a mapping of DPO rewriting rules to classes of DPO diagrams such that, for every rule ρ, $\mathfrak{F}(\rho)$ is a class of DPO diagrams with top-span ρ. The* ***DPO rewriting relation*** *$\Rightarrow_{\rho,\mathfrak{F}}$* ***induced by a DPO rewriting rule*** *ρ* ***in*** *$\mathfrak{F}$ is defined as follows: $G \Rightarrow_{\rho,\mathfrak{F}} H$ iff $G \Rightarrow^\delta_\rho H$ for some $\delta \in \mathfrak{F}(\rho)$. The* ***DPO rewriting relation*** *$\Rightarrow_{\mathcal{R},\mathfrak{F}}$* ***induced by a set*** *$\mathcal{R}$* ***of DPO rewriting rules in*** *$\mathfrak{F}$ is given by: $G \Rightarrow_{\mathcal{R},\mathfrak{F}} H$ iff $G \Rightarrow_{\rho,\mathfrak{F}} H$ for some $\rho \in \mathcal{R}$. When $\mathfrak{F}$ is clear from the context, we suppress $\mathfrak{F}$ and write $\Rightarrow_\rho$ and $\Rightarrow_{\mathcal{R}}$.*

Let $\mathfrak{F}$ denote the DPO rewriting framework that associates any rule with the class of all DPO diagrams with the rule as the top span, and let $\mathfrak{M}$ denote the DPO rewriting framework that associates any rule with the class of all DPO diagrams having monic matches and the rule as the top span.

2.4 Relative Termination

Definition 17 (Rewriting sequence). *Let $\mathcal{R}$ be a set of DPO rewriting rules and $\mathfrak{F}$ a DPO rewriting framework in* ***Graph****. A* ***$(\mathcal{R}, \mathfrak{F})$-rewriting sequence*** *is a finite sequence $s_0, s_1, \ldots, s_m$ of objects such that for each $0 \leq n \leq m - 1$, we have $s_n \Rightarrow_{\mathcal{R},\mathfrak{F}} s_{n+1}$, or an infinite sequence $s_0, s_1, \ldots$ of objects such that $s_n \Rightarrow_{\mathcal{R},\mathfrak{F}} s_{n+1}$ for each $n \in \mathbb{N}$.*

We adapt the concept of relative termination presented in [11,16].

Definition 18 (Relative termination). *Let $\mathcal{A}$ and $\mathcal{B}$ be sets of rewriting rules, and $\mathfrak{F}$ a DPO rewriting framework in* ***Graph****. We say that $\Rightarrow_{\mathcal{A},\mathfrak{F}}$ is* ***terminating relative to*** *$\Rightarrow_{\mathcal{B},\mathfrak{F}}$ if any infinite $(\mathcal{A} \cup \mathcal{B}, \mathfrak{F})$-rewriting sequence has only a finite number of rewriting steps with rules in $\mathcal{A}$.*

3 A Termination Criterion

We present a method for termination analysis of DPO graph rewriting systems with injective rules. Building on the interpretation method framework [1], we first define the weight of a graph in § 3.1. Using this definition, we outline the general approach and identify the key challenge in § 3.2, and introduce the concept of non-increasing rules in § 3.3. A solution to the challenge is then proposed in § 3.4, culminating in our termination criterion in § 3.5.

3.1 Measuring Graphs by Counting Subgraphs

Let $\mathcal{R} = \mathcal{A} \cup \mathcal{B}$ be a set of injective rules. Define $\mathrm{Sub}(\mathrm{left}(\mathcal{R}))$ as the set of all subgraphs of left-hand sides graphs of rules in $\mathcal{R}$. We call elements of $\mathrm{Sub}(\mathrm{left}(\mathcal{R}))$ **ruler-graphs**. Analogous to measuring physical objects with rulers, we measure graphs by counting occurrences of these ruler-graphs.

Definition 19. *For a ruler-graph X and graph G, the* ***measurement*** *$m_X(G)$ is: $m_X(G) = |\,\mathrm{Mono}(X, G)|$.*

The occurrence count of ruler-graphs before and after a rewriting step provides a termination criterion: If every rewriting step using a rule in $\mathcal{A}$ strictly decreases the number of X's occurrences, and every rewriting step using a rule in $\mathcal{B}$ never increases it, then the rewriting relation induced by $\mathcal{A}$ is terminating relative to the rewriting relation induced by $\mathcal{B}$.

For some ruler-graphs, occurrence changes during a rewriting step using an injective rule are precisely computable by simply counting their occurrences in the left- and right-hand side graphs. Examples include: the ruler-graph $\bullet$ consisting of a single node; the ruler-graph $\bullet \circlearrowright x$ consisting of a single node and a single loop labeled by $x \in \Sigma$; the ruler-graph $\bullet \xrightarrow{x} \bullet$ consisting of two nodes and a single labeled edge between them.

However, systems like Example 28 require more complex ruler-graphs such as $\bullet \xrightarrow{a} \bullet \xrightarrow{a} \bullet$ as shown in Example 42. For such ruler-graph, occurrence changes during rewriting is unpredictable in general, as their occurrence creation and destruction depend on the context graph—a challenge we address in later sections.

To combine measurements from multiple ruler-graphs (analogous to combining measurements of physical objects from multiple rulers), we introduce:

Definition 20 (Weight function). *A* ***weight function*** *for a set of ruler-graphs $\mathbb{X}$ is a map $s_{\mathbb{X}} \colon \mathbb{X} \to \mathbb{N}$ assigning a weight in $\mathbb{N}$ to each ruler-graph.*

Example 44 demonstrates the usefulness of a weight function that assigns different weight to measurements from different ruler-graphs.

Definition 21 (Graph weight). *Let $\mathbb{X}$ be a set of ruler-graphs, $s_{\mathbb{X}}$ a weight function, and G a graph. The* ***weight of graph*** *G relative to $s_{\mathbb{X}}$, denoted $w_{s_{\mathbb{X}}}(G)$, is defined as:*

$$w_{s_{\mathbb{X}}}(G) = \sum_{X \in \mathbb{X}} s_{\mathbb{X}}(X) \cdot m_X(G).$$

3.2 General Idea and Key Challenge

For the remainder of this section, we assume: a set $\mathcal{R} = \mathcal{A} \cup \mathcal{B}$ of injective DPO rewriting rules, a set $\mathbb{X}$ of ruler-graphs, and a weight function $s_{\mathbb{X}}$. Let $\rho = (L \overset{l}{\leftarrowtail} K \overset{r}{\rightarrowtail} R)$ denote a rule in $\mathcal{R}$ and $X \in \mathbb{X}$.

The rewriting relation $\Rightarrow_{\mathcal{A},\mathfrak{M}}$ is terminating relative to $\Rightarrow_{\mathcal{B},\mathfrak{M}}$ if the following hold: (i) for all $G \Rightarrow_{\mathcal{A},\mathfrak{M}} H$, $w_{s_{\mathbb{X}}}(G) > w_{s_{\mathbb{X}}}(H)$, and (ii) for all $G \Rightarrow_{\mathcal{B},\mathfrak{M}} H$, $w_{s_{\mathbb{X}}}(G) \geq w_{s_{\mathbb{X}}}(H)$. However, directly verifying all rewriting steps is infeasible due to their potential infiniteness. Thus, we aim to provide a rule-based sufficient condition for termination.

Notation 22.

The disjoint union of two sets S and S' will be denoted by $S \uplus S'$. Let X, A, B, G be graphs, and let $\alpha\colon A \to G$ and $\beta\colon B \to G$ be morphisms as illustrated on the right. We define the following sets of monomorphisms from X in G based on their interaction with α and β:

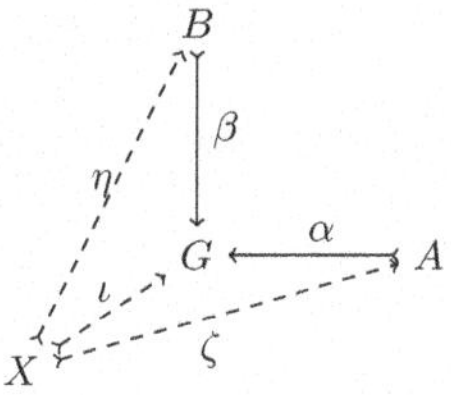

$$\mathrm{Mono}(X, G, \alpha) = \{\iota\colon X \rightarrowtail G \mid \exists \zeta\colon X \rightarrowtail A.\, \iota = \zeta \star \alpha\}$$
$$\mathrm{Mono}(X, G, \neg\alpha) = \{\iota\colon X \rightarrowtail G \mid \nexists \zeta\colon X \rightarrowtail A.\, \iota = \zeta \star \alpha\}$$
$$\mathrm{Mono}(X, G, \neg\alpha, \beta) = \left\{ \iota\colon X \rightarrowtail G \;\middle|\; \begin{array}{l} (\nexists \zeta\colon X \rightarrowtail A.\, \iota = \zeta \star \alpha) \\ \wedge (\exists \eta\colon X \rightarrowtail B.\, \iota = \eta \star \beta) \end{array} \right\}$$
$$\mathrm{Mono}(X, G, \neg\alpha, \neg\beta) = \left\{ \iota\colon X \rightarrowtail G \;\middle|\; \begin{array}{l} (\nexists \zeta\colon X \rightarrowtail A.\, \iota = \zeta \star \alpha) \\ \wedge (\nexists \eta\colon X \rightarrowtail B.\, \iota = \eta \star \beta) \end{array} \right\}$$

Consider a rewriting step $G \Rightarrow_{\rho,\mathfrak{M}} H$ defined by the DPO diagram illustrated on the right. The sets of monomorphisms $\mathrm{Mono}(X, G)$ and $\mathrm{Mono}(X, H)$ can be decomposed into the following disjoint unions:

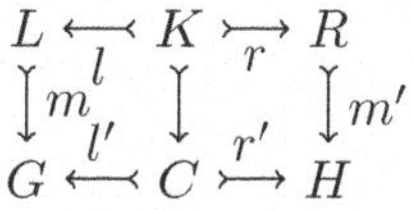

$$\mathrm{Mono}(X, G) = \mathrm{Mono}(X, G, l') \uplus \mathrm{Mono}(X, G, \neg l', m) \uplus \mathrm{Mono}(X, G, \neg l', \neg m)$$
$$\mathrm{Mono}(X, H) = \mathrm{Mono}(X, H, r') \uplus \mathrm{Mono}(X, H, \neg r', m') \uplus \mathrm{Mono}(X, H, \neg r', \neg m')$$

Thus, we have:

$$\begin{aligned} |\,\mathrm{Mono}(X, G)| &= |\,\mathrm{Mono}(X, G, l')| + |\,\mathrm{Mono}(X, G, \neg l', m)| \\ &\quad + |\,\mathrm{Mono}(X, G, \neg l', \neg m)| \\ |\,\mathrm{Mono}(X, H)| &= |\,\mathrm{Mono}(X, H, r')| + |\,\mathrm{Mono}(X, H, \neg r', m')| \\ &\quad + |\,\mathrm{Mono}(X, H, \neg r', \neg m')| \end{aligned}$$

Using the following Lemma 23 (see [28, Lemma 27] for a proof), we derive:

$$|\,\mathrm{Mono}(X, G)| = |\,\mathrm{Mono}(X, C)| + |\,\mathrm{Mono}(X, L, \neg l)| + |\,\mathrm{Mono}(X, G, \neg l', \neg m)|$$
$$|\,\mathrm{Mono}(X, H)| = |\,\mathrm{Mono}(X, C)| + |\,\mathrm{Mono}(X, R, \neg r)| + |\,\mathrm{Mono}(X, H, \neg r', \neg m')|$$

Lemma 23.
Let X be a ruler-graph. For a pushout square as shown on the right, we have

$$\begin{array}{ccc} C & \overset{\beta}{\leftarrowtail} & A \\ {\scriptstyle \alpha'}\downarrow & & \downarrow{\scriptstyle \alpha} \\ D & \underset{\beta'}{\leftarrowtail} & B \end{array}$$

$$\begin{aligned} |\,\mathrm{Mono}(X, B)| &= |\,\mathrm{Mono}(X, D, \beta')| \\ |\,\mathrm{Mono}(X, C, \neg\beta)| &= |\,\mathrm{Mono}(X, D, \neg\beta', \alpha')| \end{aligned}$$

Therefore, we obtain:

$$\begin{aligned} &|\,\mathrm{Mono}(X, G)| - |\,\mathrm{Mono}(X, H)| \\ =&|\,\mathrm{Mono}(X, L, \neg l)| - |\,\mathrm{Mono}(X, R, \neg r)| + \\ &|\,\mathrm{Mono}(X, G, \neg l', \neg m)| - |\,\mathrm{Mono}(X, H, \neg r', \neg m')| \end{aligned}$$

By Lemma 24 (see [28, Lemma 28] for a proof), it can be simplified to:

$$\begin{aligned} &|\,\mathrm{Mono}(X, G)| - |\,\mathrm{Mono}(X, H)| \\ =&|\,\mathrm{Mono}(X, L)| - |\,\mathrm{Mono}(X, R)| + \\ &|\,\mathrm{Mono}(X, G, \neg l', \neg m)| - |\,\mathrm{Mono}(X, H, \neg r', \neg m')| \end{aligned}$$

Lemma 24. *Let X be a graph. Let $L \overset{l}{\leftarrowtail} K \overset{r}{\rightarrowtail} R$ be an injective DPO graph rewriting rule. We have*

$$|\,\mathrm{Mono}(X, L, \neg l)| - |\,\mathrm{Mono}(X, R, \neg r)| = |\,\mathrm{Mono}(X, L)| - |\,\mathrm{Mono}(X, R)|$$

Consequently, the weight difference satisfies:

$$\begin{aligned} & w_{s_{\mathbb{X}}}(G) - w_{s_{\mathbb{X}}}(H) \\ \overset{\text{def}}{=} & \sum_{X \in \mathbb{X}} s_{\mathbb{X}}(X) * m_X(G) - \sum_{X \in \mathbb{X}} s_{\mathbb{X}}(X) * m_X(H) \\ \overset{\text{def}}{=} & \sum_{X \in \mathbb{X}} s_{\mathbb{X}}(X) * |\,\mathrm{Mono}(X, G)| - \sum_{X \in \mathbb{X}} s_{\mathbb{X}}(X) * |\,\mathrm{Mono}(X, H)| \\ = & \sum_{X \in \mathbb{X}} s_{\mathbb{X}}(X) * (|\,\mathrm{Mono}(X, G)| - |\,\mathrm{Mono}(X, H)|) \\ = & \sum_{X \in \mathbb{X}} s_{\mathbb{X}}(X) * (|\,\mathrm{Mono}(X, L)| - |\,\mathrm{Mono}(X, R)|) + \\ & \sum_{X \in \mathbb{X}} s_{\mathbb{X}}(X) * (|\,\mathrm{Mono}(X, G, \neg l', \neg m)| - |\,\mathrm{Mono}(X, H, \neg r', \neg m')|) \end{aligned}$$

Since $\sum_{X \in \mathbb{X}} s(X) * (|\,\mathrm{Mono}(X, L)| - |\,\mathrm{Mono}(X, R)|)$ can be precisely computed, the key challenge is to establish a lower bound for

$$\sum_{X \in \mathbb{X}} s_{\mathbb{X}}(X) * (|\,\mathrm{Mono}(X, G, \neg l', \neg m)| - |\,\mathrm{Mono}(X, H, \neg r', \neg m')|)$$

3.3 Non-increasing Rules

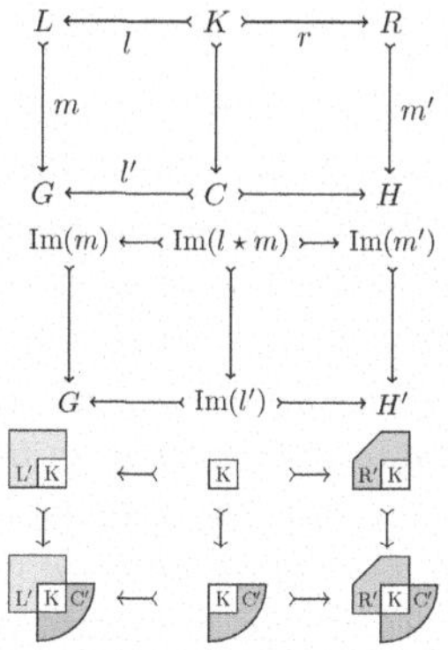

Let $\rho = (L \overset{l}{\hookleftarrow} K \overset{r}{\rightarrowtail} R)$ be an injective DPO rewriting rule. A rewriting step using ρ with injective match m is defined by a DPO diagram as shown on the upper right. It follows that both squares in the diagram depicted on the middle right, where all morphisms are inclusion functions and H' is a graph isomorphic to H, are pushouts. By viewing a graph as a set of nodes and edges and graph monomorphisms as set injections, the DPO diagram on the middle right can be considered as a DPO diagram in the category of sets, illustrated on the lower right. In the lower right diagram, L', C', R', K are mutually disjoint sets such that K, $K \uplus L'$, $K \uplus R'$, $K \uplus C'$, $L' \uplus K \uplus C'$ and $R' \uplus K \uplus C'$ are corresponding sets of $\mathrm{Im}(l \star m)$, $\mathrm{Im}(m)$, $\mathrm{Im}(m')$, $\mathrm{Im}(l')$, G and H' respectively, and all morphisms as inclusion functions. Thus, any subgraph of G (resp. H) can be considered as a union of subgraphs of C and L (resp. R), up to isomorphism.

With the above DPO diagrams in mind, we define:

Definition 25. *An X-**occurrence** in G is a subgraph of G isomorphic to X. An X-**occurrence** x **implicitly created by the rewriting step** is an X-occurrence in H which is included neither in R nor in C. The occurrence x can be decomposed as the union of graphs $x = R_x \cup C_x$ where $R_x = x \cap R$ and $C_x = x \cap C$. For every graph monomorphism $h : R_x \rightarrowtail L$ that preserves interface elements (i.e., $h(k) = k$ for all $k \in K$), we define the **corresponding X-occurrence of x relative to** h as the X-occurrence in G given by $h(R_x) \cup C_x$. An X-**occurrence implicitly destroyed by the rewriting step** is an X-occurrence in G which is included neither in L nor in C.*

We define the set $D(R, X)$ of all subgraphs of R which, when glued along some common interface elements with a subgraph C' of a context graph C, can form an X-occurrence implicitly created by the rewriting step.

Definition 26. *Let X be a graph and $L \overset{l}{\hookleftarrow} K \overset{r}{\rightarrowtail} R$ an injective DPO rewriting rule. The set $D(R, X)$ consists of all subgraphs $R' \subseteq R$ satisfying the following conditions:*

- *a diagram as shown on the right where $h_{K'K}\colon K' \rightarrowtail K$ and $h_{R'R}\colon R' \rightarrowtail R$ are inclusion functions can be constructed,*
- *R' is not a subgraph of $r(K)$,*
- *R' is not isomorphic to X.*

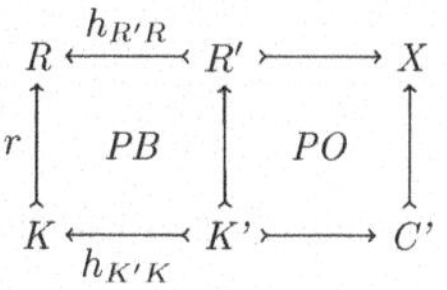

The first condition of the definition ensures that any $R' \in D(R, X)$ can be glued along some common interface elements with a subgraph C' of a context graph C to form an X-occurrence. The second and third conditions ensure that such X-occurrences are included neither in C nor in R, thereby guaranteeing that they are implicitly created as defined in Definition 25.

Notation 27. *We use the notation from [21, Notation 1] to visualize edge-labeled graph homomorphisms. Labeled graphs are enclosed in boxes with their names displayed in the top-left corner. Nodes and edges are assigned subsets of* $\mathbb{N}$ *as identifiers, and these identifiers are chosen such that: (i) Each node or edge* y *in the codomain graph is assigned the union of the identifiers of all nodes or edges in the domain graph that are mapped to* y*; (ii) The graph homomorphism is uniquely determined by this assignment.*
To further improve readability, we represent sets by listing their elements. Additionally, we omit identifiers when doing so does not cause confusion. This is illustrated in the following representation of a homomorphism $h : G \to H$*.*

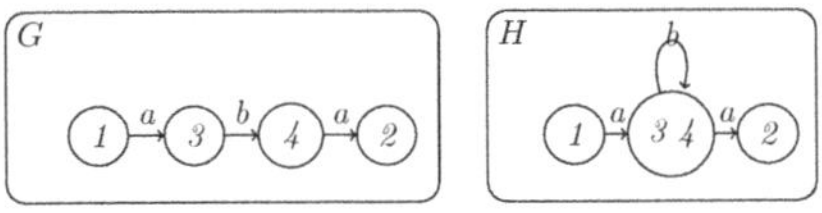

In this example, the sets $\{1\}$, $\{2\}$, $\{3\}$, $\{4\}$, *and* $\{3, 4\}$ *are represented as* 1, 2, 3, 4, *and* 3 4, *respectively. Edge identifiers are omitted.*

Example 28. Consider the injective DPO rule from [6, Example 6]:

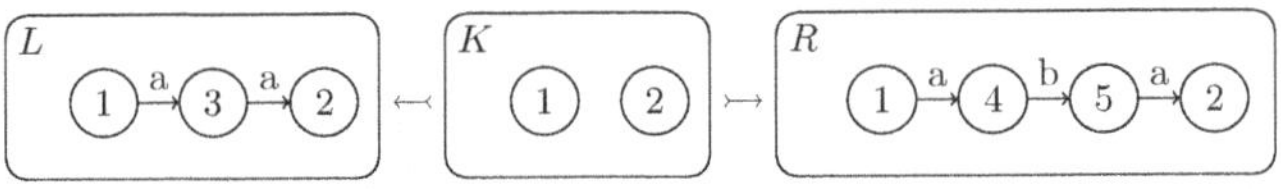

Let X be $\bullet \xrightarrow{a} \bullet \xrightarrow{a} \bullet$ and $R' \in D(R, X)$. R' is neither the empty graph nor isomorphic to X, by Conditions 2 and 3 of Definition 26. Therefore, R' must contain an interface node (either node 1 or node 2), because if it contains no interface node, Condition 1 of Definition 26 would be violated due to X's connectivity. Additionally, R' must contain node 4 or node 5 to avoid being a subgraph of $r(K)$ (Condition 2). Therefore, R' must contain either $1 \xrightarrow{a} 4$ or $5 \xrightarrow{a} 2$, because if node 4 or node 5 is isolated, Condition 1 would be violated due to X's connectivity. Furthermore, R' must not contain $4 \xrightarrow{b} 5$ by Condition 1. The following graph is not in $D(R, X)$ because it violates Condition 1.

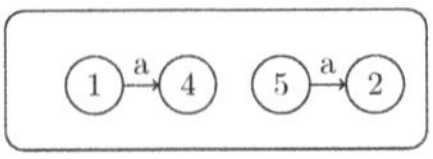

Thus, $D(R, X)$ consists of the four graphs R'_1, R'_2, R'_3 and R'_4 shown below. For each graph, the construction of the pullback square in Condition 1 is demonstrated in Example 31 and the construction of the pushout square in Condition 1 is straightforward.

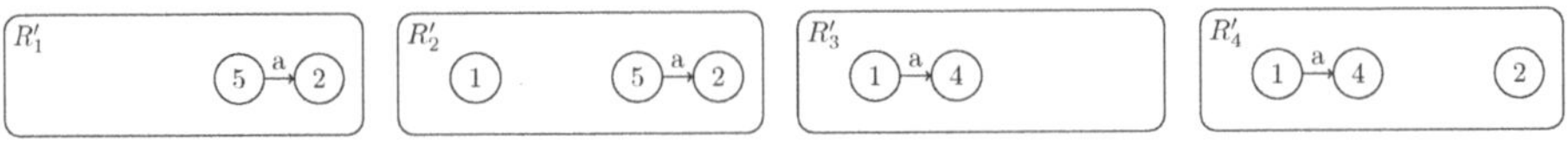

Example 29. Let X be the graph $\bullet \longrightarrow \bullet \longrightarrow \bullet$. Consider the following rule.

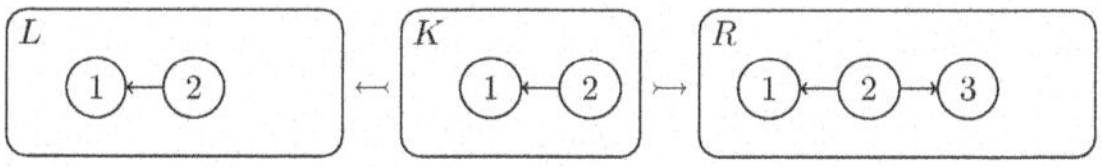

Let $R' \in D(R, X)$. R' must contain node 3, otherwise it would be a subgraph of $r(K)$, violating Condition 2 of Definition 26. Node 3 must be connected to an interface node, otherwise Condition 1 of Definition 26 would be violated. Therefore, R' must include ②→③. The graph R is not in $D(R, X)$ because it violates Condition 1. Thus, the set $D(R, X)$ consists of R'_1: ②→③ and R'_2: ① ②→③.

We define the notion of X-non-increasing rules. Intuitively, if a rule ρ is X-non-increasing, then for any rewriting step using ρ, there exists an injective mapping from the set of X-occurrences implicitly created by the step to the set of X-occurrences implicitly destroyed by the step. This will be proved in Lemma 36.

Definition 30 (Non-increasing rule). *Let* $\rho = (L \overset{l}{\hookleftarrow} K \overset{r}{\rightarrowtail} R)$ *be a rule and* X *a ruler-graph. Rule* ρ *is* X-non-increasing *if the following four conditions hold:*

1. *For all* $R' \in D(R, X)$, *there exists a morphism* $h_{R'L} : R' \rightarrowtail L$ *such that both squares in the diagram shown on the right are pullbacks, where all morphisms other than* $h_{R'L}$, l *and* r *are inclusion functions,*
2. *For all* $R' \in D(R, X)$, *for all node or edge* r' *in* R', *if* $r' \notin \mathrm{Im}(r)$ *then* $h_{R'L}(r') \notin \mathrm{Im}(l)$,
3. *For all* $R', R'' \in D(R, X)$, *for all edges* $x \in R'$ *and* $y \in R''$, *if* $x \neq y$, *then* $h_{R'L}(x) \neq h_{R''L}(y)$,
4. *If* X *has isolated nodes, then for all* $R', R'' \in D(R, X)$ *and nodes* $x \in R', y \in R''$, *if* $x \neq y$, *then* $h_{R'L}(x) \neq h_{R''L}(y)$.

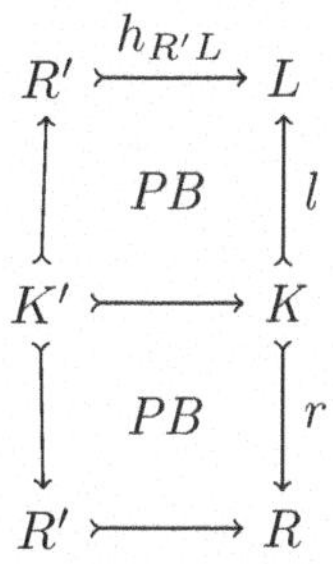

The first condition ensures that whenever $R' \in D(R, X)$ forms an X-occurrence in the result graph H, this occurrence has a corresponding X-occurrence relative to the morphism $h_{R'L}$ (as defined in Definition 25) in the host graph G. The second condition guarantees that any X-occurrence implicitly created by the rewriting step has its corresponding X-occurrence not included in the context (Consequently, its corresponding X-occurrence is implicitly destroyed by the rewriting step). Counterexamples are provided in Example 32 and Example 33. The third and fourth conditions ensure that distinct X-occurrences created implicitly have distinct corresponding X-occurrences. Counterexamples for these conditions are given in Example 34 and Example 35.

The following example shows an X-non-increasing rule.

Example 31. Let X be $\bullet \xrightarrow{a} \bullet \xrightarrow{a} \bullet$. Consider the rule presented in Example 28. By Example 28, the set $D(R, X)$ consists of exactly four graphs: R'_1, R'_2, R'_3 and R'_4. The rule is X-non-increasing because the following diagrams can be constructed, and $h_{R'_1L}, h_{R'_2L}, h_{R'_3L}$ and $h_{R'_4L}$ satisfy all conditions in Definition 30.

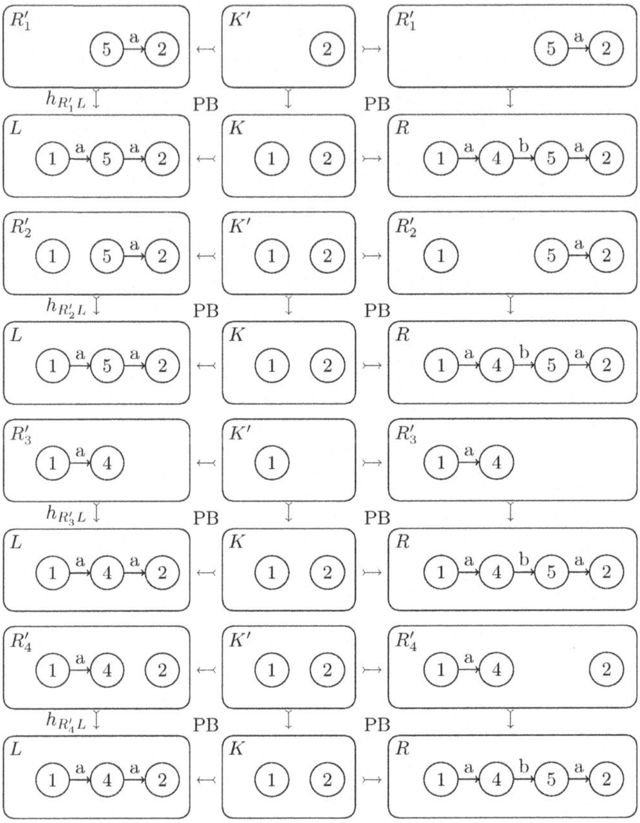

To illustrate the X-non-increasing property, consider the rewriting step defined by the DPO diagram below. Whenever R'_i (for $1 \leq i \leq 4$) forms an X-occurrence x that is included neither in R nor C with a subgraph $C' \subseteq C \cap x \subseteq C$, the graph $h_{R'_iL}(R'_i)$ forms a X-occurrence in G with C'. Furthermore, distinct X-occurrences created in H have distinct corresponding X-occurrences in G.

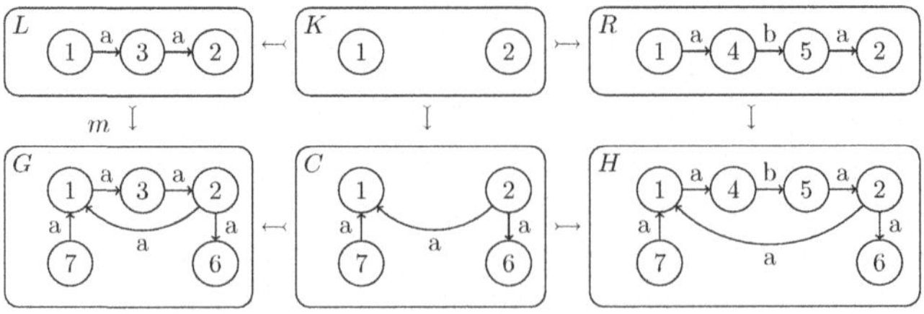

Example 32. Let X be the graph $\bullet \longrightarrow \bullet \longrightarrow \bullet$. The following rewriting rule is potentially X-increasing.

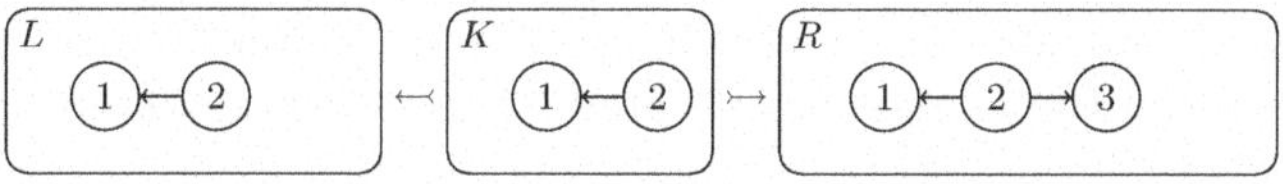

As shown in Example 29, $D(R, X)$ contains the graph R': $(2) \to (3)$. The unique monomorphism $h_{R'L} : R' \rightarrowtail L$ fails the second condition of Definition 30. For any rewriting step using this rule, the corresponding X-occurrence relative to $h_{R'L}$ of any implicitly created X-occurrence is included in the context.

Example 33. Let X be the graph $\bullet \longrightarrow \bullet \quad \bullet$. The following rewriting rule is potentially X-increasing.

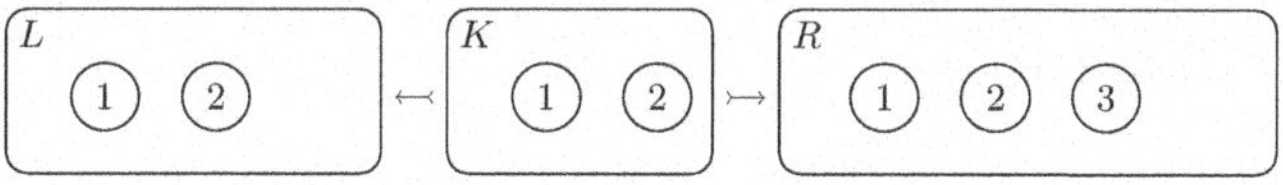

The set $D(R, X)$ contains R': $(1)\ (3)$. Despite the existence of a unique monomorphism $h_{R'L} : R' \rightarrowtail L$ preserving interface elements, this rule fails the second condition of Definition 30. For any rewriting step using this rule, the corresponding X-occurrence relative to $h_{R'L}$ of any implicitly created X-occurrence is included in the context.

Example 34. Let X be the graph $\bullet \longrightarrow \bullet \longrightarrow \bullet$. The following rewriting rule is potentially X-increasing.

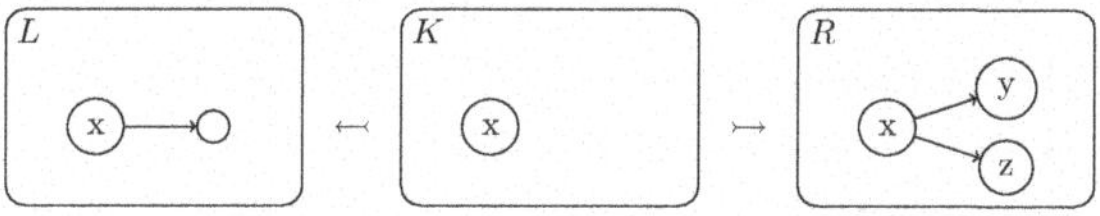

The set $D(R, X)$ contains exactly two elements R': $(x) \to (y)$ and R'': $(x) \to (z)$. Despite unique monomorphisms $h_{R'L} : R' \rightarrowtail L$ and $h_{R''L} : R'' \rightarrowtail L$ preserving interface elements, they fail the third condition of Definition 30. For any rewriting step using this rule, implicitly created X-occurrences with the same subgraph of the context have the same corresponding X-occurrence.

Example 35. Let X be the graph $\bullet \longrightarrow \bullet \quad \bullet$, which has an isolated node. The following rewriting rule is potentially X-increasing.

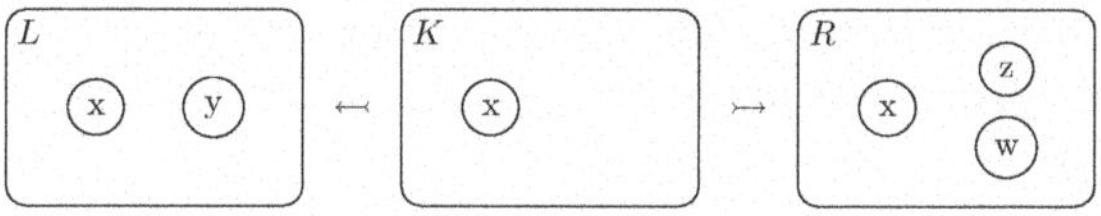

The set $D(R, X)$ contains exactly two elements R': $(x)\ (z)$ and R'': $(x)\ (w)$. There are unique monomorphisms $h_{R'L} : R' \rightarrowtail L$ and $h_{R''L} : R'' \rightarrowtail L$ preserving interface elements, but they fail the fourth condition of Definition 30. For any rewriting step using this rule, implicitly created X-occurrences with the same subgraph of the context have the same corresponding X-occurrence.

3.4 Solution to the Key Challenge

The following lemma (see [28, Lemma 40] for a proof) ensures that if ρ is X-non-increasing, then, for any rewriting step using ρ, more X-occurrences are implicitly destroyed than implicitly created by the rewriting step.

Lemma 36.
Let X be a ruler-graph and $\rho = (L \overset{l}{\leftarrowtail} K \overset{r}{\rightarrowtail} R)$ an injective DPO rewriting rule. Suppose that ρ is X-non-increasing. For every rewriting step induced by the DPO diagram (shown on the right), we have

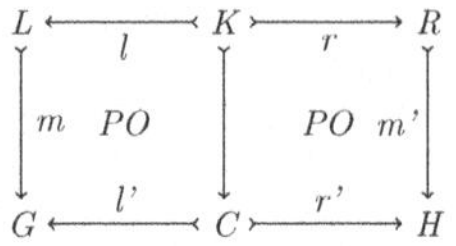

$$|\operatorname{Mono}(X, G, \neg m, \neg l')| \geq |\operatorname{Mono}(X, H, \neg m', \neg r')|$$

3.5 Termination Criterion

Finally, we present a lemma that allows us to overapproximate the change in weight of a graph when a rule is applied, followed by our termination criterion. See [28, Lemma 41, 42] for the proofs.

Lemma 37 (Decreasing step). *Let $\rho = (L \leftarrowtail K \rightarrowtail R)$ be an injective DPO rewriting rule, $\mathbb{X}$ a set of ruler-graphs, $s_{\mathbb{X}}$ a weight function, and $G \Rightarrow_{\rho,\mathfrak{M}} H$ a rewriting step. If ρ is X-non-increasing for every ruler-graph $X \in \mathbb{X}$, then $w_{s_{\mathbb{X}}}(G) - w_{s_{\mathbb{X}}}(H) \geq w_{s_{\mathbb{X}}}(L) - w_{s_{\mathbb{X}}}(R)$.*

Theorem 38 (Termination). *Let $\mathcal{A}$ and $\mathcal{B}$ be sets of injective DPO rewriting rules, $\mathbb{X}$ a set of ruler-graphs, and $s_{\mathbb{X}}$ a weight function. If the following hold:*

1. *for every $\rho \in \mathcal{A} \cup \mathcal{B}$ and for every $X \in \mathbb{X}$, the rule ρ is X-non-increasing,*
2. *for every $\rho \in \mathcal{A}$, we have $w_{s_{\mathbb{X}}}(\mathrm{lhs}(\rho)) > w_{s_{\mathbb{X}}}(\mathrm{rhs}(\rho))$,*
3. *for every $\rho \in \mathcal{B}$, we have $w_{s_{\mathbb{X}}}(\mathrm{lhs}(\rho)) \geq w_{s_{\mathbb{X}}}(\mathrm{rhs}(\rho))$.*

Then $\Rightarrow_{\mathcal{A},\mathcal{M}}$ is terminating relative to $\Rightarrow_{\mathcal{B},\mathcal{M}}$.

Our technique is presented for DPO rewriting with injective matches due to their expressiveness [13], but it also extends to non-injective matches.

Definition 39 (Quotient Rule [13]).
Given a rule $\rho = L \leftarrow K \rightarrow R$, a rule $\rho' = (L' \leftarrow K' \rightarrow R')$ is called a **quotient rule** *of ρ if there exists a DPO diagram (shown on the right) where all vertical graph morphisms are surjective. The set of quotient rules of ρ is denoted by $Q(\rho)$, and for a rule set $\mathcal{A}$, we write $Q(\mathcal{A}) = \bigcup_{\rho \in \mathcal{A}} Q(\rho)$.*

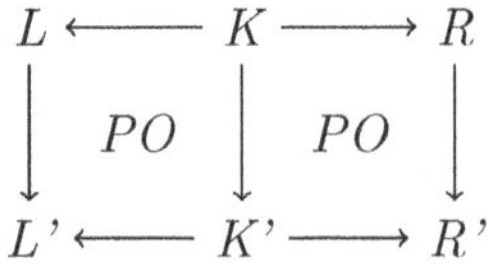

Lemma 40 (Quotient Lemma [13]). *Let ρ be a DPO rewriting rule. For any graphs G and H, $G \Rightarrow_{\rho,\mathfrak{F}} H$ if and only if $G \Rightarrow_{Q(\rho),\mathfrak{M}} H$.*

Corollary 41. *Let $\mathcal{A}$ and $\mathcal{B}$ be sets of injective DPO rewriting rules. The rewriting relation $\Rightarrow_{\mathcal{A},\mathfrak{F}}$ is terminating relative to $\Rightarrow_{\mathcal{B},\mathfrak{F}}$ if $\Rightarrow_{Q(\mathcal{A}),\mathfrak{M}}$ is terminating relative to $\Rightarrow_{Q(\mathcal{B}),\mathfrak{M}}$.*

4 Examples

While the type graph method [5,6,10] can prove termination for this example, it requires constructing an appropriate weighted type graph, a task that is difficult in general [5, §6]. In contrast, our criterion only requires checking simpler conditions.

Example 42. Consider the rewriting rule ρ from Example 28. Let X be the ruler-graph $\bullet \xrightarrow{a} \bullet \xrightarrow{a} \bullet$, $\mathbb{X} = \{X\}$ and $s_{\mathbb{X}}(X) = 1$. The rule is X-non-increasing as shown in Example 31. Since $w_{s_{\mathbb{X}}}(\text{lhs}(\rho)) = 1 > 0 = w_{s_{\mathbb{X}}}(\text{rhs}(\rho)$, it terminates by Theorem 38.

We consider an example for which the techniques from [5,6,10,23,26] fail.

Example 43. The rewriting rule below is a variant of a rule presented in [26, Example 6], obtained by removing all edges in the interface:

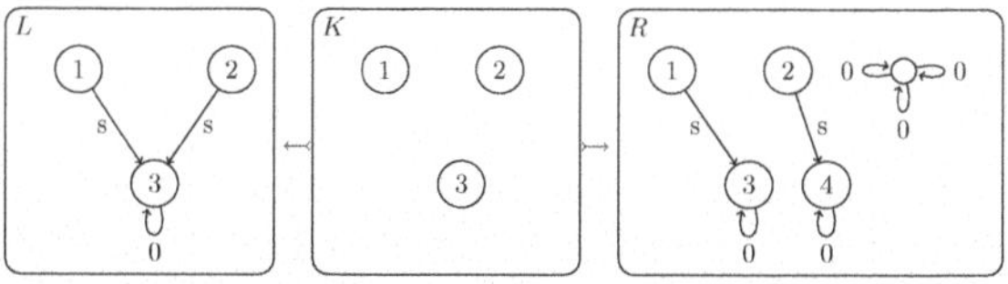

Let X be the ruler-graph $\bullet \xrightarrow{s} \bullet \xleftarrow{s} \bullet$, $\mathbb{X} = \{X\}$ and $s_{\mathbb{X}}(X) = 1$. The set $D(R, X)$ consists of two elements R'_1: $(1) \xrightarrow{s} (3)$ and R'_2: (2) $(1) \xrightarrow{s} (3)$. Each R'_i admits a unique monomorphism $h_{R'_i L} \colon R'_i \rightarrowtail L$ preserving interface elements. The rule is X-non-increasing with $h_{R'_1 L}$ and $h_{R'_2 L}$ because all conditions in Definition 30 are satisfied. Since $w_{s_{\mathbb{X}}}(L) = 1 > 0 = w_{s_{\mathbb{X}}}(R)$, it follows that the system terminates by Theorem 38.

The following example demonstrates the usefulness of a weight function that assigns distinct weight to measurements from different ruler-graphs.

Example 44. Consider the rewriting rules presented in [23, Example 5.6]:

$\rho =$ 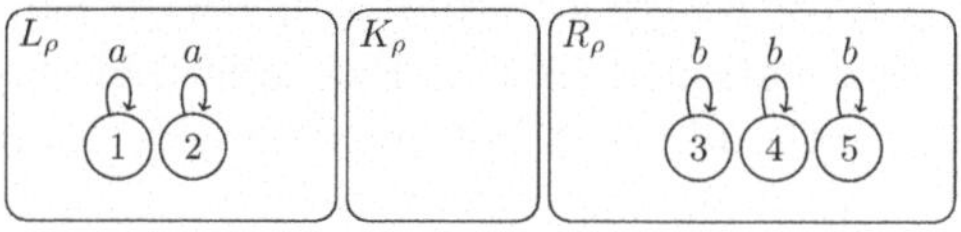

$\tau =$ 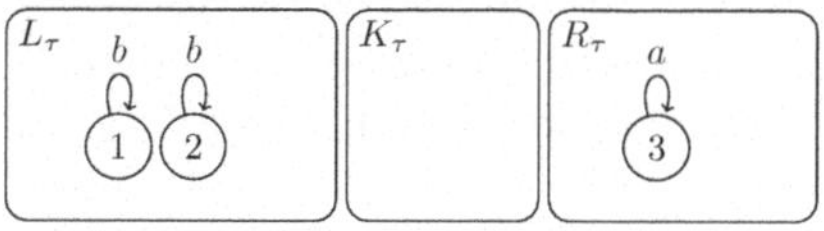

Let X be the ruler-graph $\bullet \circlearrowleft \text{a}$, Y the ruler-graph $\bullet \circlearrowleft \text{b}$ and $\mathbb{X} = \{X, Y\}$. The sets $D(R_\rho, X)$, $D(R_\rho, Y)$, $D(R_\tau, X)$ and $D(R_\tau, Y)$ are all empty. Therefore, both rules ρ and τ are X- and Y-non-increasing because all conditions in Definition 30 are trivially satisfied.

For $s_{\mathbb{X}}(X) = 5$ and $s_{\mathbb{X}}(Y) = 3$, we have $w_{s_{\mathbb{X}}}(L_\rho) = 10 > 9 = w_{s_{\mathbb{X}}}(R_\rho)$ and $w_{s_{\mathbb{X}}}(L_\tau) = 6 > 5 = w_{s_{\mathbb{X}}}(R_\tau)$. Thus, the system terminates by Theorem 38.

However, for $s'_{\mathbb{X}}(X) = 1$ and $s'_{\mathbb{X}}(Y) = 1$, we have $w_{s'_{\mathbb{X}}}(L_\rho) = 2 \not> 3 = w_{s'_{\mathbb{X}}}(R_\rho)$ which prevents us from applying Theorem 38 to prove termination of the rewriting system.

Example 45 (Limitation). Our method fails to prove termination for the rewriting rule in [10, Example D.3]. This failure occurs because, for any system $\mathcal{R}$, rewriting step $G \Rightarrow_{\mathcal{R},\mathfrak{M}} H$, set of ruler-graphs $\mathbb{X}$, and weight function $s_{\mathbb{X}}$, we have $w_{s_{\mathbb{X}}}(G) \geq w_{s_{\mathbb{X}}}(H)$ due to the existence of an injective graph homomorphism from G to H.

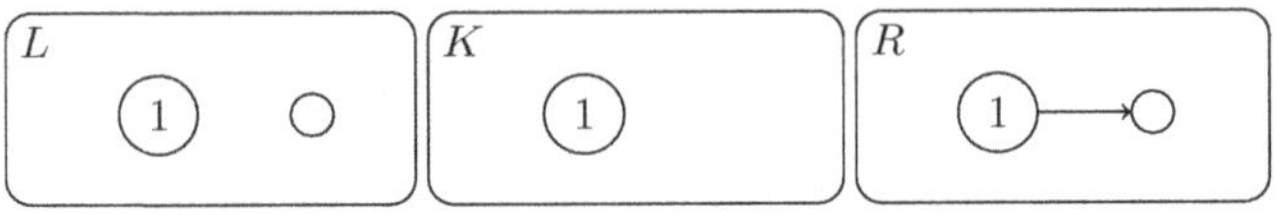

5 Related Work

The subgraph-counting method in [23] by Overbeek and Endrullis is designed for the PBPO+ [21,22]—a rewriting formalism capable of simulating left-injective DPO rewriting. It can prove termination for systems like [23, Examples 5.2, 5.7, 5.9] and [26, Example 6], which lie beyond the scope of our technique. Additionally, this method can be applied for many categories.

It is very closely related to our work in the setting where they are both defined. Both methods weigh objects by summing weighted morphisms targeting them, and, for both methods, the key challenge lies in estimating weights for morphisms whose images partially overlap with the rewriting context.

The two methods employ distinct strategies to overcome this challenge. The subgraph-counting method relies on type-morphisms (see [23, p. 9, remark 4.11, Lemma 4.23]), whereas our approach requires injective mappings from (i) subgraph occurrences partially overlapping the rewriting context in the result graph to (ii) those in the host graph.

Both approaches have limitations and strengths. Their method suffers from discrete interface, as mentioned in [23, Example 5.5]. For instance, it proves termination for [23, Example 5.5] but fails for [10, Example 6.3], where the rules differ only by a discrete interface. Similarly, it fails to prove termination for Example 43, but if an edge labeled by "s" from node 1 to node 3 is added, then it succeeds. Our approach, however, can handle all the aforementioned cases. Nevertheless, their method can address [25, Example 4.1], which remains beyond the scope of our technique. Finally, both approaches cannot handle [10, Example D.3] (see [23, Remark 6.2] and Example 45).

The type graph method, which weighs an object by summing the weights of morphisms from the object to a type graph, was initially introduced by Zantema, König and Bruggink [30] for cycle-rewriting systems. This method has since been

Table 1. Applicability of termination techniques to DPO rewriting examples. The symbol ✓ indicates termination can be proved by the technique, ✗ indicates it cannot be proved, and – denotes irrelevance or out-of-scope cases.

Examples \ Techniques		Forward closure [25]	Modular criterion [26]	Type graph [6]	Type graph [5]	Type graph [10]	Subgraph counting [23]	Our method
Current paper	Example 43	✓	✗	✗	✗	✗	✗	✓
[25]	Example 3.8	✓	–	–	–	–	–	✓
[26]	Example 3	–	✓	–	–	–	–	✓
	Example 5	–	✓	–	–	–	–	✓
[6]	Example 4 and 6	–	–	✓	–	–	–	✓
[5]	Example 2	–	–	–	✓	–	–	✓
	Example 4	–	–	–	✓	–	–	✓
[10]	Example 6.2	–	–	–	–	✓	–	✓
	Example 6.3	–	–	–	–	✓	✗	✓
	Example D.1	–	–	–	–	✓	–	✓
[23]	Example 5.3	–	–	–	–	–	✓	✓
	Example 5.5	–	–	–	–	–	✓	✓
	Example 5.6	–	–	–	–	–	✓	✓
	Example 5.8	–	–	–	–	–	✓	✓
[26]	Example 6	–	✓	–	–	–	–	–
[10]	Example 6.4	–	–	–	–	✓	–	–
	Example 6.5	–	–	–	–	✓	–	–
	Example D.4	–	–	–	–	✓	–	–
[23]	Example 5.2	–	–	–	–	–	✓	–
	Example 5.7	–	–	–	–	–	✓	–
	Example 5.9	–	–	–	–	–	✓	–
[25]	Example 4.1	✓	–	–	–	–	✓	✗
[26]	Example 4	–	✓	–	–	–	–	✗
[6]	Example 1	–	–	✓	–	–	–	✗
	Routing Protocol	–	–	✓	–	–	–	✗
	Example 5	–	–	✓	–	–	–	✗
[5]	Example 5	–	–	–	✓	–	–	✗
	Example 6	–	–	–	✓	–	–	✗
[10]	Example D.2	–	–	–	–	✓	–	✗
	Example D.3	–	–	–	–	✓	✗	✗

generalized to edge-labeled multigraphs by Bruggink et al. [6] for DPO rewriting with monic matches and injective rules, later extended to DPO rewriting in general by Bruggink et al. [5], and further adapted to more categories and different DPO variants by Endrullis et al. [10]. These methods are not directly comparable with our technique in general. On the one hand, the termination of Example 43 can be proved by our method but not by the type graph methods due to the exsitence of a surjection from the output graph to the input graph as explained in [10, Example D.4]. On the other hand, our method cannot prove the termination of [6, Example 1, 5 and Ad-hoc Routing Protocol], nor [5, Example 5, 6], nor [10, Examples D2 and D3].

Plump [25] introduced a necessary and sufficient termination condition for left-injective DPO hypergraph rewriting via forward closure, though verifying this condition is undecidable. While this method proves termination of Example 43, our approach succeeds on [25, Example 3.8] but fails in proving termination of [25, Example 4.1].

Plump [26] later proposed a modular critical pair-based strategy for left-injective DPO hypergraph rewriting with monic matches. Our method complements this: while modularity reduces global complexity, each subsystem requires individual termination proofs. For example, the measure based on the indegree proposed in [26] cannot prove the termination of Example 43 due to the additional loops. Specifically, if the context graph is identical to the interface graph, then we have $3^k < 2^k + 2^k + 3^k$ for all $k \in \mathbb{N}$. In contrast, our method succeeds.

Levendovszky et al. [19] propose a termination criterion for DPO rewriting (monic matches, injective rules, negative application condition), though automated verification is hard as explained in [19, §6].

Bottoni et al. [3] propose a termination criterion for DPO/SPO rewriting on high-level replacement units. Their method imposes a strongly constrained measuring function and the only concretes measuring function proposed are node-counting and edge-counting.

Bottoni et al. [4] presents a criterion for termination of DPO rewriting with monic matches, injective rules and negative application conditions, based on the construction of a labeled transition system.

A comparative analysis of termination techniques for DPO graph rewriting systems, drawn from prior work [5,6,10,23,25,26], is summarized in Table 1. Our approach successfully proves termination for 14 of these systems.

6 Conclusion and Future Work

We have presented a machine-checkable sufficient condition for relative termination of DPO systems with injective rules on edge-labeled multigraphs. This method resolves cases where prior interpretation-based techniques [5,6,10,23,30] fail (e.g., Example 43) and, unlike the subgraph-counting approach in [23], avoids limitations arising from discrete interfaces. The method has been implemented in an OCaml tool.

An interesting direction for future work is to extend our method to DPO systems with negative application conditions and to address the limitation highlighted in Example 45.

Acknowledgments. I thank my supervisor, Xavier Urbain, for his guidance; the anonymous reviewers for their valuable feedback; Rachid Echahed for proofreading; and my mother and wife for their unwavering support.

Disclosure of Interests. The author has no competing interests to declare that are relevant to the content of this article.

References

1. Baader, F., Nipkow, T.: Term rewriting and all that. Cambridge University Press (1998)
2. Barr, M., Wells, C.: Category theory for computing science (2. ed.). Prentice Hall international series in computer science, Prentice Hall (1995)
3. Bottoni, P., Koch, M., Parisi-Presicce, F., Taentzer, G.: Termination of high-level replacement units with application to model transformation. In: Minas, M. (ed.) Proceedings of the Workshop on Visual Languages and Formal Methods, VLFM 2004, Rome, Italy, September 30, 2004. Electronic Notes in Theoretical Computer Science, vol. 127, pp. 71–86. Elsevier (2004). https://doi.org/10.1016/J.ENTCS.2004.08.048
4. Bottoni, P., Parisi-Presicce, F.: A termination criterion for graph transformations with negative application conditions. Electron. Commun. Eur. Assoc. Softw. Sci. Technol. **30** (2010). https://doi.org/10.14279/TUJ.ECEASST.30.419
5. Bruggink, H.J.S., König, B., Nolte, D., Zantema, H.: Proving termination of graph transformation systems using weighted type graphs over semirings. CoRR **abs/1505.01695** (2015), http://arxiv.org/abs/1505.01695
6. Bruggink, H.J.S., König, B., Zantema, H.: Termination analysis for graph transformation systems. In: Díaz, J., Lanese, I., Sangiorgi, D. (eds.) Theoretical Computer Science - 8th IFIP TC 1/WG 2.2 International Conference, TCS 2014, Rome, Italy, September 1-3, 2014. Proceedings. Lecture Notes in Computer Science, vol. 8705, pp. 179–194. Springer (2014). https://doi.org/10.1007/978-3-662-44602-7_15
7. Contejean, E., Courtieu, P., Forest, J., Pons, O., Urbain, X.: Automated certified proofs with CiME3. In: Schmidt-Schauß, M. (ed.) Proceedings of the 22nd International Conference on Rewriting Techniques and Applications, RTA 2011, May 30 - June 1, 2011, Novi Sad, Serbia. LIPIcs, vol. 10, pp. 21–30. Schloss Dagstuhl - Leibniz-Zentrum für Informatik (2011). https://doi.org/10.4230/LIPICS.RTA.2011.21
8. Corradini, A., Montanari, U., Rossi, F., Ehrig, H., Heckel, R., Löwe, M.: Algebraic approaches to graph transformation - part I: basic concepts and double pushout approach. In: Rozenberg, G. (ed.) Handbook of Graph Grammars and Computing by Graph Transformations, Volume 1: Foundations, pp. 163–246. World Scientific (1997)
9. Courtieu, P., Rieg, L., Tixeuil, S., Urbain, X.: Certified universal gathering in $\mathbb{R}^2$ for oblivious mobile robots. In: Gavoille, C., Ilcinkas, D. (eds.) Distributed Computing - 30th International Symposium, DISC 2016, Paris, France, September 27-29, 2016. Proceedings. Lecture Notes in Computer Science, vol. 9888, pp. 187–200. Springer (2016). https://doi.org/10.1007/978-3-662-53426-7_14

10. Endrullis, J., Overbeek, R.: Generalized weighted type graphs for termination of graph transformation systems. In: Harmer, R., Kosiol, J. (eds.) Graph Transformation - 17th International Conference, ICGT 2024, Held as Part of STAF 2024, Enschede, The Netherlands, July 10-11, 2024, Proceedings. Lecture Notes in Computer Science, vol. 14774, pp. 39–58. Springer (2024). https://doi.org/10.1007/978-3-031-64285-2_3
11. Geser, A.: Relative Termination. Ph.D. thesis, University of Passau, Germany (1990). http://vts.uni-ulm.de/docs/2012/8146/vts_8146_11884.pdf
12. Giesl, J., et al.: Proving termination of programs automatically with AProVE. In: Demri, S., Kapur, D., Weidenbach, C. (eds.) Automated Reasoning - 7th International Joint Conference, IJCAR 2014, Held as Part of the Vienna Summer of Logic, VSL 2014, Vienna, Austria, July 19-22, 2014. Proceedings. Lecture Notes in Computer Science, vol. 8562, pp. 184–191. Springer (2014). https://doi.org/10.1007/978-3-319-08587-6_13
13. Habel, A., Müller, J., Plump, D.: Double-pushout graph transformation revisited. Math. Struct. Comput. Sci. **11**(5), 637–688 (2001). https://doi.org/10.1017/S0960129501003425
14. Harrison, J., Urban, J., Wiedijk, F.: History of interactive theorem proving. In: Siekmann, J.H. (ed.) Computational Logic, Handbook of the History of Logic, vol. 9, pp. 135–214. Elsevier (2014). https://doi.org/10.1016/B978-0-444-51624-4.50004-6
15. Heiser, G., Andronick, J., Elphinstone, K., Klein, G., Kuz, I., Ryzhyk, L.: The road to trustworthy systems. In: Xu, S., Asokan, N., Sadeghi, A. (eds.) Proceedings of the fifth ACM Workshop on Scalable Trusted Computing, STC@CCS 2010, Chicago, IL, USA, October 4, 2010, pp. 3–10. ACM (2010)
16. Klop, J.: Term rewriting systems: A tutorial. Bulletin of the European Association for Theoretical Computer Science (1987)
17. König, B., Nolte, D., Padberg, J., Rensink, A.: A tutorial on graph transformation. In: Heckel, R., Taentzer, G. (eds.) Graph Transformation, Specifications, and Nets - In Memory of Hartmut Ehrig. Lecture Notes in Computer Science, vol. 10800, pp. 83–104. Springer (2018). https://doi.org/10.1007/978-3-319-75396-6_5
18. Lamport, L., Shostak, R.E., Pease, M.C.: The byzantine generals problem. In: Malkhi, D. (ed.) Concurrency: the Works of Leslie Lamport, pp. 203–226. ACM (2019). https://doi.org/10.1145/3335772.3335936
19. Levendovszky, T., Prange, U., Ehrig, H.: Termination criteria for DPO transformations with injective matches. In: Rensink, A., Heckel, R., König, B. (eds.) Proceedings of the Workshop on Graph Transformation for Concurrency and Verification, GT-VC@CONCUR 2006, Bonn, Germany, August 31, 2006. Electronic Notes in Theoretical Computer Science, vol. 175, pp. 87–100. Elsevier (2006). https://doi.org/10.1016/J.ENTCS.2007.04.019
20. Löwe, M.: Graph rewriting in span-categories. In: Ehrig, H., Rensink, A., Rozenberg, G., Schürr, A. (eds.) Graph Transformations - 5th International Conference, ICGT 2010, Enschede, The Netherlands, September 27 - - October 2, 2010. Proceedings. Lecture Notes in Computer Science, vol. 6372, pp. 218–233. Springer (2010). https://doi.org/10.1007/978-3-642-15928-2_15
21. Overbeek, R., Endrullis, J.: A PBPO+ graph rewriting tutorial. In: Grabmayer, C. (ed.) Proceedings Twelfth International Workshop on Computing with Terms and Graphs, TERMGRAPH@FSCD 2022, Technion, Haifa, Israel, 1st August 2022. EPTCS, vol. 377, pp. 45–63 (2022). https://doi.org/10.4204/EPTCS.377.3

22. Overbeek, R., Endrullis, J., Rosset, A.: Graph rewriting and relabeling with PBPO$^+$: A unifying theory for quasitoposes. J. Log. Algebraic Methods Program. **133**, 100873 (2023). https://doi.org/10.1016/J.JLAMP.2023.100873
23. Overbeek, R., Endrullis, J.: Termination of graph transformation systems using weighted subgraph counting. Logical Methods in Computer Science **Volume 20, Issue 4**, 12 (Nov 2024). https://doi.org/10.46298/lmcs-20(4:12)2024
24. Pierce, B.C.: Basic Category Theory for Computer Scientists. MIT Press, Foundations of computing (1991)
25. Plump, D.: On termination of graph rewriting. In: Nagl, M. (ed.) Graph-Theoretic Concepts in Computer Science, 21st International Workshop, WG '95, Aachen, Germany, June 20-22, 1995, Proceedings. Lecture Notes in Computer Science, vol. 1017, pp. 88–100. Springer (1995). https://doi.org/10.1007/3-540-60618-1_68
26. Plump, D.: Modular termination of graph transformation. In: Heckel, R., Taentzer, G. (eds.) Graph Transformation, Specifications, and Nets - In Memory of Hartmut Ehrig. Lecture Notes in Computer Science, vol. 10800, pp. 231–244. Springer (2018). https://doi.org/10.1007/978-3-319-75396-6_13
27. Potop-Butucaru, M., Sznajder, N., Tixeuil, S., Urbain, X.: Formal methods for mobile robots. In: Flocchini, P., Prencipe, G., Santoro, N. (eds.) Distributed Computing by Mobile Entities, Current Research in Moving and Computing, Lecture Notes in Computer Science, vol. 11340, pp. 278–313. Springer (2019). https://doi.org/10.1007/978-3-030-11072-7_12
28. Qiu, Q.: Termination of Injective DPO Graph Rewriting Systems using Subgraph Counting. Tech. rep., Universite Claude Bernard Lyon 1 (Mar 2025). https://hal.science/hal-04955684
29. Söldner, R., Plump, D.: Formalising the double-pushout approach to graph transformation. Log. Methods Comput. Sci. **20**(4) (2024). https://doi.org/10.46298/LMCS-20(4:3)2024
30. Zantema, H., König, B., Bruggink, H.J.S.: Termination of cycle rewriting. In: Dowek, G. (ed.) Rewriting and Typed Lambda Calculi - Joint International Conference, RTA-TLCA 2014, Held as Part of the Vienna Summer of Logic, VSL 2014, Vienna, Austria, July 14-17, 2014. Proceedings. Lecture Notes in Computer Science, vol. 8560, pp. 476–490. Springer (2014). https://doi.org/10.1007/978-3-319-08918-8_33

Rewriting for Traced Monoidal Closed Categories

Alessandro Di Giorgio[1(✉)], Dan R. Ghica[2,3], and Fabio Zanasi[4]

[1] Tallinn University of Technology, Tallinn, Estonia
alessd@taltech.ee
[2] University of Birmingham, Birmingham, UK
d.r.ghica@bham.ac.uk
[3] Huawei Central Software Institute, Edinburgh, UK
[4] University College London, London, UK
f.zanasi@ucl.ac.uk

Abstract. Traced monoidal closed categories are a model for higher-order functional computation. We develop a formal language of string diagrams for these categories, and a faithful interpretation in terms of certain hypergraphs. We then use the interpretation to show that string diagram rewriting can be implemented as double-pushout rewriting in a sound and complete way. Finally, we showcase our approach on the λ-calculus with explicit recursion.

Keywords: String diagrams · Hypergraphs · Rewriting

1 Introduction

String diagrams are a graphical formal language for reasoning in symmetric monoidal categories, see e.g. [30,31] for a survey. In recent years, they have been increasingly adopted as a framework to reason about a variety of systems, including quantum processes [11], dynamical systems [4,9], digital circuits [17], relational databases [8], automata [29] and neural networks [13], amongst others. Compared to traditional syntax, string diagrams highlight resource-exchange between components, allowing to reason visually about subtle interactions such as quantum entanglement, concurrency, probabilistic dependency, and variable sharing. At the same time, string diagrams are a fully formal syntax, which may be reasoned about with the same tools as programming languages.

In a nutshell, string diagrams can be thought of as a circuit language, represented with boxes and wires, whose building blocks may be composed in sequence and in parallel to form more complex diagrams. In this setting, the laws of symmetric monoidal categories, rendered in Fig. 1, correspond to geometric manipulations, e.g. pulling of wires or sliding boxes.

Because of their compositional nature, string diagrams allow for a natural notion of rewriting consisting in replacing a sub-diagram with another. However, differently than traditional term rewriting, such replacement may not always be

J. Endrullis and M. Tichy (Eds.): ICGT 2025, LNCS 15720, pp. 24–43, 2025.
https://doi.org/10.1007/978-3-031-94706-3_2

immediately apparent. For example, consider the rule on the left and the first string diagram on the right.

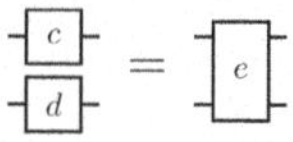

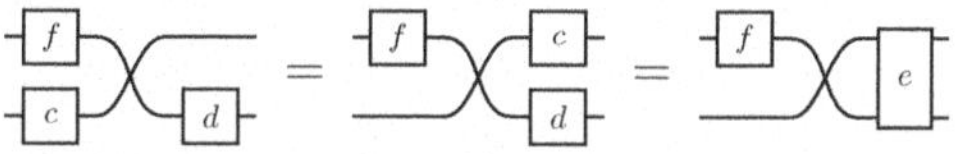

Before applying the substitution we need to manipulate the diagram, using the laws of symmetric monoidal categories, in order to get an exact match. Therefore, before reaching a match we might need to potentially inspect all those string diagrams equivalent to a given one. This makes string diagram rewriting computationally impractical. A recent line of works [5–7] proposes a solution to this problem by interpreting string diagrams as certain hypergraphs and their rewriting as double-pushout rewriting with interfaces (DPOI) [6], an extension of the classical double pushout rewriting (DPO) [12,14]. The advantage is that an equivalence class of string diagrams corresponds to exactly one hypergraph, making matching computationally more straightforward.

By now, this approach has been adapted to various kinds of categories and their string diagrams. Particularly relevant for our work are the hierarchical hypergraphs for monoidal closed categories [2] and the partial monogamous hypergraphs for traced monoidal categories [18].

The main contribution of our work is to develop a framework for string diagram rewriting in a more articulated setting: *traced monoidal closed categories*. This extends [2] with the handling of a trace structure, and [18] with the handling of a closed structure. The study of such extension is motivated by a computational understanding of the traced monoidal closed structure. First, traced monoidal categories, and notably traced cartesian categories, have been thoroughly studied as a model for recursion and iteration, starting with the work of Hasegawa [20]. Similarly, monoidal closed categories, and in particular cartesian closed categories, correspond to models of the simply typed λ-calculus [24]. Therefore, a combination of these structures provides a semantic framework to study computational processes featuring both recursion and higher-order functions. In particular, traced cartesian closed categories constitute models for the λ-calculus with explicit recursion [21]. The latter is essentially a version of the λ-calculus that features a `letrec` construct, typical of functional languages.

Our contribution will develop as follows. First, we will define a suitable notion of string diagram, accounting for both traces and the closed structure. Then we will introduce traced hierarchical hypergraphs, and show that they faithfully interpret string diagrams. Finally, we will be able to characterise string diagram rewriting in terms of double-pushout rewriting of these hypergraphs— with some subtleties due to the hierarchical nature of these structures. We will conclude by interpreting the λ-calculus with explicit recursion in our framework. This also allows us to discuss non-confluence of the calculus through the lenses of string diagram rewriting.

Synopsis: In §2 we introduce the string diagrams for traced monoidal closed categories. In §3 we recall the definition of hierarchical hypergraphs and their categorical structure. §4 contains our first contribution: we characterise exactly those hypergraphs that correspond to traced hierarchical string diagrams. As a second contribution, in §5 we develop a suitable notion of DPOI rewriting which adequately interprets rewriting for traced monoidal closed categories. Finally, in §6 we consider a case study: the string diagrammatic interpretation of the λ-calculus with explicit recursion.

2 String Diagrams for Traced Monoidal Closed Categories

In this section we introduce string diagrams for traced monoidal closed categories. These can be thought of as a combination of the usual graphical language for traced categories [22,31] and the hierarchical string diagrams for monoidal closed categories introduced in [18]. We first recall the notion of monoidal closed signature, adapted from [16].

Definition 1. *Given a set $\mathcal{S}$ of generating objects, the elements of $obj(\mathcal{S})$ are inductively generated as follows, where I is a designated object, $A \in \mathcal{S}$, and $\otimes$ and $\multimap$ are type forming operations:*

$$X, Y ::= I \mid A \mid X \otimes Y \mid X \multimap Y$$

A monoidal closed signature *is a pair $(\mathcal{S}, \Sigma)$ where Σ is a set of* generating morphisms *$s\colon X \to Y$ with $X, Y \in obj(\mathcal{S})$.*

A traced monoidal closed (tmc) category may be freely constructed starting from a monoidal closed signature, in the expected way. The morphisms of such category will be string diagrams accounting both for the trace and for the monoidal closed structure, which we now introduce.

Definition 2. *The traced monoidal closed $(\mathcal{S}, \Sigma)$-*terms *are generated by the following grammar, where $s\colon X \to Y$ is an element of Σ, and we use the typical boxes-and-wires notation of string diagrams [30] to represent terms pictorially:*

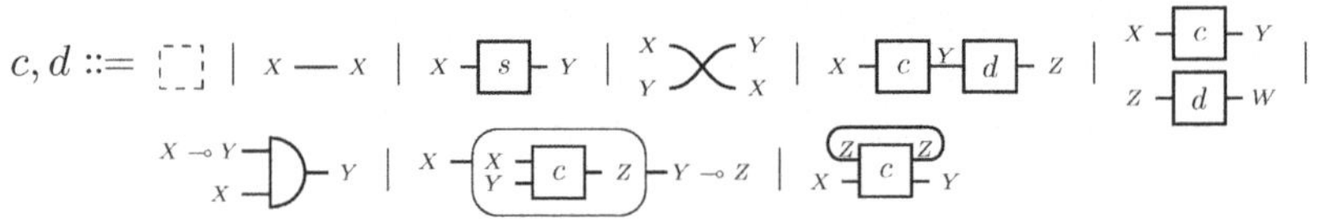

(1)

Definition 3. *Given a monoidal closed signature $(\mathcal{S}, \Sigma)$, the free traced monoidal closed category $\mathbf{T}_\Sigma$ on $(\mathcal{S}, \Sigma)$ is defined by letting objects be elements of $obj(\mathcal{S})$ and morphisms the tmc $(\mathcal{S}, \Sigma)$-terms quotiened by the axioms in Figs. 1, 2, 3.*

We write $\mathbf{C}_\Sigma$ for the monoidal closed wide subcategory of $\mathbf{T}_\Sigma$ with morphisms the $(\mathcal{S}, \Sigma)$-terms generated by (1) without the last production and quotiened only by the axioms in Figs. 1 and 3.

Remark 1. We are tacitly assuming that wires labelled solely with $X \otimes Y$ are always drawn as two separate wires stacked on top of each other. Thus, for example 

is 

, and 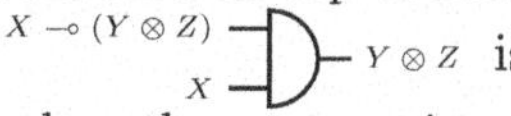

is 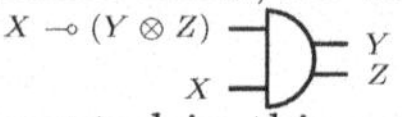

.

In other words, $\mathbf{T}_\Sigma$, as well as the other categories presented in this paper, is a coloured PROP [19] whose colours are either generating objects $A \in \mathcal{S}$ or elements of $obj(\mathcal{S})$ of the form $X \multimap Y$.

The reader may recognize the usual syntax of string diagrams for symmetric monoidal categories on the first line of (1). The second line introduces the additional structure needed for the categories of interest. In particular, the semicircle renders the morphism $ev_{X,Y}$, and the bubble around a diagram c is the operator $\Lambda_{X,Y,Z}(c)$, both of which are part of the definition of a monoidal closed category (see e.g. [16, Definition 3.5]). Similarly, a loop over c corresponds to the trace operator $Tr^{Z}_{X,Y}(c)$ of traced monoidal categories (see e.g. [22]). We refer to the arrows of $\mathbf{T}_\Sigma$ as *traced hierarchical string diagrams.* The fact that such syntax and equations suffice to make $\mathbf{T}_\Sigma$ and $\mathbf{C}_\Sigma$ monoidal closed is detailed in [16]. Intuitively, $\mathbf{C}_\Sigma$ is the category of hierarchical string diagrams without loops.

We now introduce the notion of string diagram rewriting that we intend to study. In the framework of [5], which we extend here, one is usually interest in

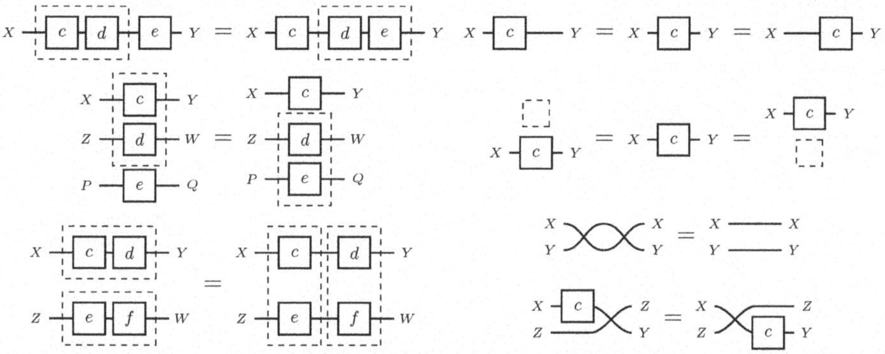

Fig. 1. Axioms of symmetric monoidal categories.

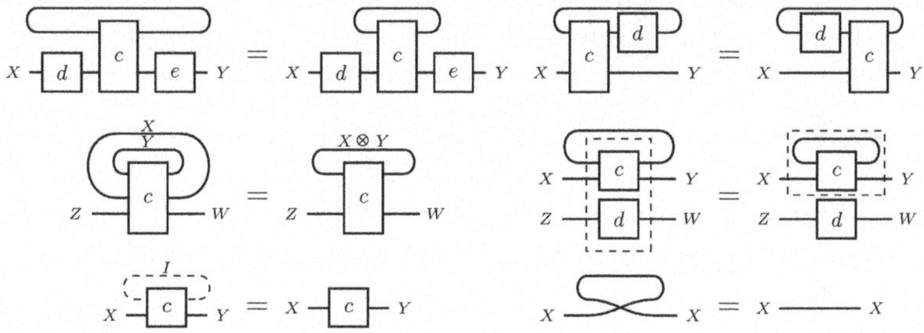

Fig. 2. Axioms of traced symmetric monoidal categories.

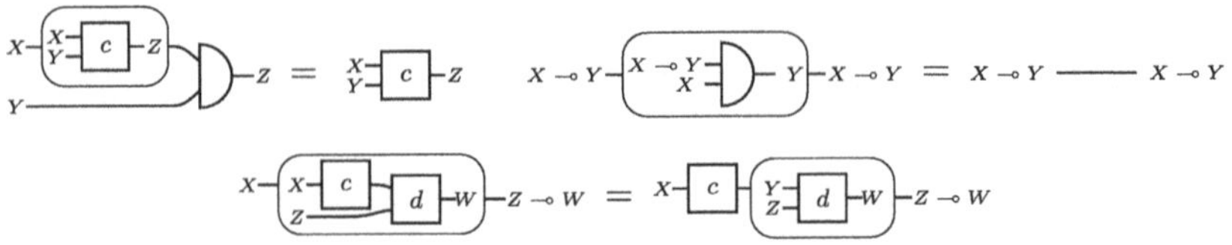

Fig. 3. Axioms of closed monoidal categories.

'domain-specific' rules, which describe some transformation of the computational processes represented by the string diagrams. Within this perspective, equations in Figs. 1, 2 are considered 'structural' for the purpose of rewriting—they yield alternative representation of the same system, rather than transformations to be accounted for as actual rewrite rules. Therefore, the notion of rewriting we seek works 'modulo traced monoidal structure'. We use TSMC to collectively refer to the axioms of traced symmetric monoidal categories, in Figs. 1, 2.

In order to soundly perform rewriting, we need to consider the categories of hierarchial string diagrams as TSMCs. Formally, we freely generate the categories $\mathbb{C}_\Sigma$ and $\mathbb{T}_\Sigma$ as done in Definition 3, but without quotienting by the axioms in Fig. 3. These axioms will be taken instead as rewrite rules.

Definition 4. *Let $f, g\colon X \to Y$ and $l, r\colon Z \to W$ be pairs of traced hierarchical string diagrams in $\mathbb{T}_\Sigma$. We say that f rewrites into g according to the rewrite rule $\langle l, r\rangle$, denoted as $f \Rightarrow_{\langle l,r\rangle} g$, if*

(2)

or

(3)

where a and b are traced hierarchical string diagrams, such that $a \Rightarrow_{\langle l,r\rangle} b$.

The above definition deserves some further remark. First, observe that, as anticipated, rewriting happens only modulo the axioms of TSMCs. Second, the definition is recursive: this is due to the fact that the string diagrams considered are hierarchical, and thus rewriting can happen at any level of the hierarchy.

For example, given $R = \langle$ 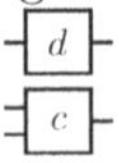 , 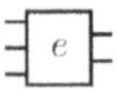 $\rangle$, we claim that 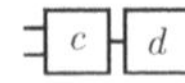 $\Rightarrow_R$ 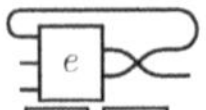 . At first, it might not be obvious why there is a match for R in 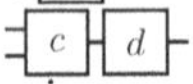 . The match appears clearly once we massage  using the axioms of TSMCs.

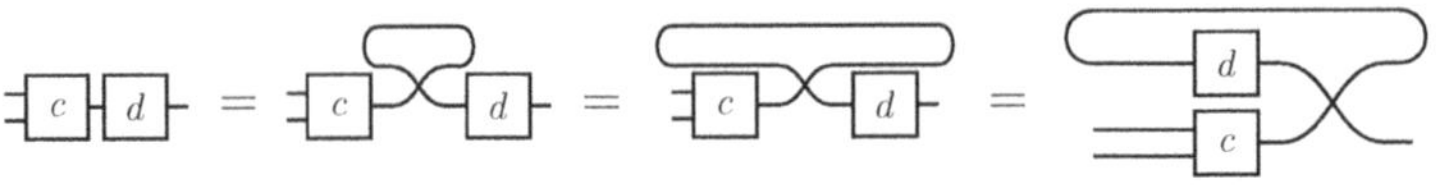

(4)

Now that the diagram is in the form of (2), we can directly substitute the occurence of 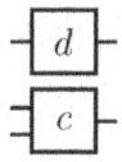 with 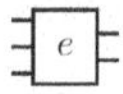 and obtain 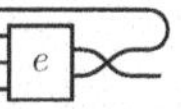.

Similarly, we can show that 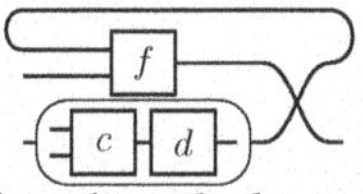 $\Rightarrow_R$ 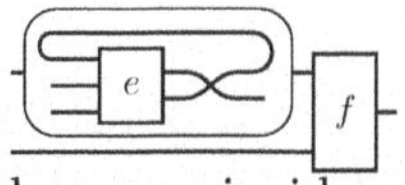, except that now the diagram is hierarchical and the rewirte happens inside a bubble. Therefore, this is the case of (3): we must bring the outer layer into the usual shape, and then proceed recursively for the diagram inside the bubble using (2).

Definition 4 can be easily extended to the case of multiple rewrite rules. In particular, given a set R of rewrite rules, we say that $f \Rightarrow_{\mathsf{R}} g$ if there is some $\langle l, r\rangle \in \mathsf{R}$ such that $f \Rightarrow_{\langle l,r\rangle} g$. As anticipated, to capture the monoidal closed axioms we need to consider them as rewrite rules. Thus, we build the set MC consisting of pairs $\langle l, r\rangle$, for each equation $l = r$ in Fig. 3.

Furthermore, if one is interested in *cartesian* closed categories it is necessary to add extra generators and equations to $\mathbf{T}_\Sigma$. Specifically, for each object $X \in obj(\mathcal{S})$, we add  $: X \to X \otimes X$ and  $: X \to I$ to Σ and quotient the morphisms by the axioms of natural cocommutative comonoids in Fig. 4. We call this category $\mathbf{T}_\Sigma^C$ and observe that, by Fox's theorem [15], it is a cartesian category, i.e. the monoidal product coincides with the categorical product. Again, these additional equations are not structural[1] and, for the purpose of rewriting in $\mathbb{T}_\Sigma$, they are considered as rewrite rules. We call $\mathbb{T}_\Sigma^C$ the category augmented with the generators for cocommutative comonoids and we add the equations in Fig. 4 to the set of rewrite rules MC, denoting the resulting set as CC.

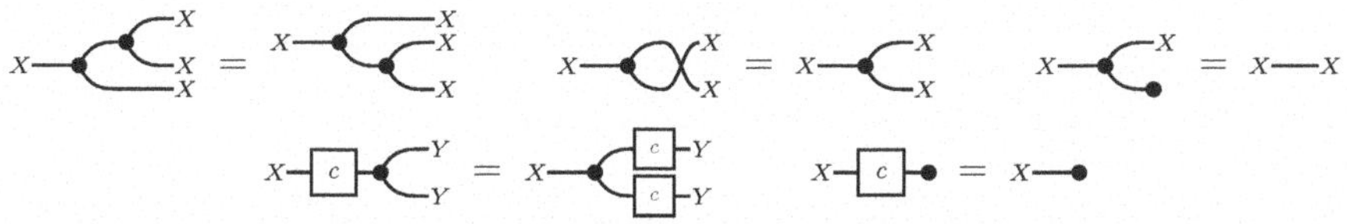

Fig. 4. Axioms of natural cocommutative comonoids.

3 Interpreting String Diagrams: Hierarchical Hypergraphs

String diagram rewriting as in Definition 4 is unpractical from a computational viewpoint: to find a matching we need to explore an equivalence class of objects—the string diagrams that are equivalent modulo the axioms of TSMCs to the

[1] More precisely, the axioms that are not structural are the naturality equations in the second row of Fig. 4. Those on the first row can be absorbed by a sensible variation [25] of the hypergraph interpretation introduced in Sect. 3.

given one. In the same style as [5], we introduce a notion of hypergraph with the purpose of implementing string diagram rewriting as hypergraph rewriting. The improvement is given by the fact that the string diagrams equivalent modulo the axioms of TSMCs are going to be represented by the same hypergraph, thus making matching computationally more straightforward. The notion we seek needs to accommodate the hierarchical structure. It is the same used in [2], with the difference that we do not require acyclicity, to reflect the presence of traces.

Definition 5. *Let $(\mathcal{S}, \Sigma)$ be a monoidal closed signature. A $(\mathcal{S}, \Sigma)$-labelled* hierarchical hypergraph *$\mathcal{H}$ is a tuple $(V, E, s, t, l_V, l_E, p_V, p_E)$ consisting of:*

- *finite sets of vertices V and edges E;*
- *source and target functions $s, t\colon E \to V^\star$, mapping an edge to a finite list of vertices, consistently with the type of the generators in Σ;*
- *a vertex labelling function $l_V : V \to obj(\mathcal{S})$;*
- *an edge labelling function $l_E\colon E \to \Sigma \cup \{\bot\}$, assigning to each edge a symbol in Σ or no symbol, i.e. $\bot$;*
- *parent functions $p_V : V \to E \cup \{\bot\}$ and $p_E\colon E \to E \cup \{\bot\}$, assigning to each vertex and edge a parent edge or no parent, i.e. $\bot$;*
- *the parent functions must satisfy some conditions: (1) an edge and any of its source and target vertices must have the same parent: $p_V(v) = p_E(e) = p_V(v')$ for all $v \in s(e)$ and $v' \in t(e)$, respectively; (2) the parent relation must be acyclic, in the sense that for all $e \in E$ there is some $k \geq 1$ such that $p^k_{E,\bot}(e) = \bot$ where $p_{E,\bot}\colon E \cup \{\bot\} \to E \cup \{\bot\}$ extends p_E with $p_{E,\bot}(\bot) = \bot$.*

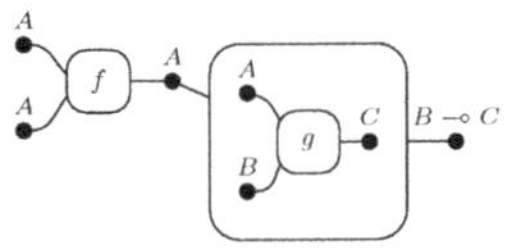

Example 1. Let $(\mathcal{S}, \Sigma)$ be the monoidal closed signature where $\mathcal{S} = \{A, B, C\}$ and $\Sigma = \{f\colon A \otimes A \to A,\ g\colon A \otimes B \to C\}$. The one on the right is a depiction of a $(\mathcal{S}, \Sigma)$-labelled hierarchical hypergraphs whose vertices are represented by the labelled black dots and the edges are the boxes.

Notice that there are three edges: those labelled with f and g, and one containing a sub-hypergraph, thus formally labelled with $\bot$.

In essence, a hierarchical hypergraph is a hypergraph with layers. The latter are determined by the parent function. The *outermost* layer 0 is formed by those vertices and edges that have no parent, i.e. their parent is $\bot$. Vertices and edges in layer $n + 1$ are those with parent edge sitting at layer n.

A hypergraph is *discrete* when its set of edges is empty. We use letters $v, w, \dots$ to denote discrete hypergraphs, with v, w standing for words over $obj(\mathcal{S})$.

It is important to also consider a notion of morphism between these structures. Note that a morphism may exist between hierarchical hypergraphs with a different number of layers: the outermost layer of the source hypergraph may be 'anchored' to any layer of the target one, as long as the hierarchy is respected.

Definition 6. *Given two $(\mathcal{S}, \Sigma)$-labelled hierarchical hypergraphs $\mathcal{H} = (V, E, s, t, l_V, l_E, p_V, p_E)$ and $\mathcal{G} = (V', E', s', t', l'_V, l'_E, p'_V, p'_E)$. A morphism of* *hierarchical hypergraphs* $\phi\colon \mathcal{H} \to \mathcal{G}$ *is a pair of functions* $\phi_V\colon V \to V'$ *and* $\phi_E\colon E \to E'$, *respecting the labelling of vertices and edges, the typing of the edges and the hierarchial structure, in the sense that:*

1. $l'_V(\phi_V(v)) = l_V(v)$ *and* $l'_E(\phi_E(e)) = l_E(e)$;
2. $s'(\phi_E(e)) = \hat{\phi}_V(s(e))$ *and* $t'(\phi_E(e)) = \hat{\phi}_V(t(e))$, *where* $\hat{\phi}_V\colon V^\star \to (V')^\star$ *is the obvious lifting of* $\phi_V\colon V \to V'$;
3. $\phi_V(p_V(v)) = p'_V(\phi_V(v))$ *if* $p_V(v) \neq \bot$;
4. $\phi_E(p_E(e)) = p'_E(\phi_E(e))$ *if* $p_E(e) \neq \bot$

A morphism $\phi\colon \mathcal{H} \to \mathcal{G}$ *is a monomorphism if* ϕ_V *and* ϕ_E *are both injective functions.* $(\mathcal{S}, \Sigma)$*-labelled hierarchical hypergraphs and morphisms between them form a category* $\mathbf{HHyp}_\Sigma$, *which is in fact symmetric monoidal.*

In order to interpret string diagrams, we need to regard hierarchical hypergraphs as morphisms of a monoidal category, and thus define a notion of sequential and parallel composition for them. As in the case of 'simple' hypergraphs [5], this is achieved by adding structures identifying a left and a right interface of the hierarchical hypergraph. Formally, this amounts to consider *discrete cospans*, i.e. cospans of the kind $v \to \mathcal{H} \leftarrow w$, where v and w are discrete hypergraphs. The idea is that such a hypergraph will interpret a string diagram of type $V \to W$, where V and W correspond to some words v and w over $obj(\mathcal{S})$.

Given two discrete cospans of hypergraphs $v \xrightarrow{f} \mathcal{H} \xleftarrow{g} u$ and $w \xrightarrow{h} \mathcal{G} \xleftarrow{i} z$, their parallel composition $v \xrightarrow{f} \mathcal{H} \xleftarrow{g} u \otimes w \xrightarrow{h} \mathcal{G} \xleftarrow{i} z$ is $v + w \xrightarrow{f+h} \mathcal{H} + \mathcal{G} \xleftarrow{g+i} u + z$ where $+$ is the coproduct in $\mathbf{HHyp}_\Sigma$.

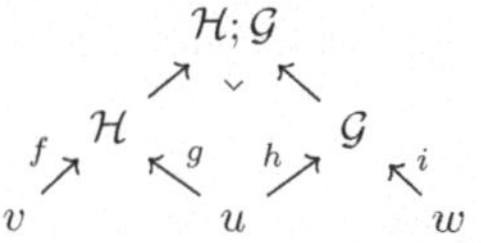

Similarly, given $v \xrightarrow{f} \mathcal{H} \xleftarrow{g} u$ and $u \xrightarrow{h} \mathcal{G} \xleftarrow{i} w$, their sequential composition $v \xrightarrow{f} \mathcal{H} \xleftarrow{g} u \,;\, u \xrightarrow{h} \mathcal{G} \xleftarrow{i} w$ is obtained via the pushout above.

Definition 7. *A hierarchical hypergraph* with interfaces *is a cospan in* $\mathbf{HHyp}_\Sigma$, *written* $v \to \mathcal{H} \leftarrow w$ *with* v *and* w *discrete. These can be composed in sequence via pushout, and in parallel via the coproduct in* $\mathbf{HHyp}_\Sigma$, *and, in fact, they form a SMC that we denote as* $\mathsf{Csp}_D(\mathbf{HHyp}_\Sigma)$.

Our graphical depiction of hierarchical hypergraphs with interfaces is as in Fig. 5. We depict the interfaces in a blue background and the rest of the structure in a grey background. When the hypergraphs has more than one layer, we distinguish between an *outer* interface and *inner* interfaces. For instance, consider the

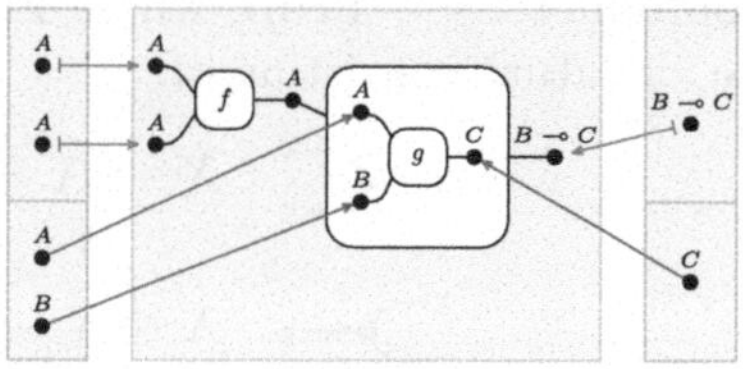

Fig. 5. A hierarchical hypergraph with interfaces.

one on the right. The horizontal dotted line between interfaces separates the outer interface with the inner one. This is also made explicit by using a red/blue colouring for the interface maps. Multiple sets of inner interfaces, e.g. when the hypergraphs have more than two layers, are separated by multiple dotted lines.

We now recall the hypergraph interpretation of hierarchical string diagrams, from [2], and we extend it to the case of traces.

Definition 8. *Let* $[\cdot] : \mathbb{C}_\Sigma \to \mathsf{Csp}_D(\mathbf{HHyp}_\Sigma)$ *be the symmetric monoidal functor defined as follows:*

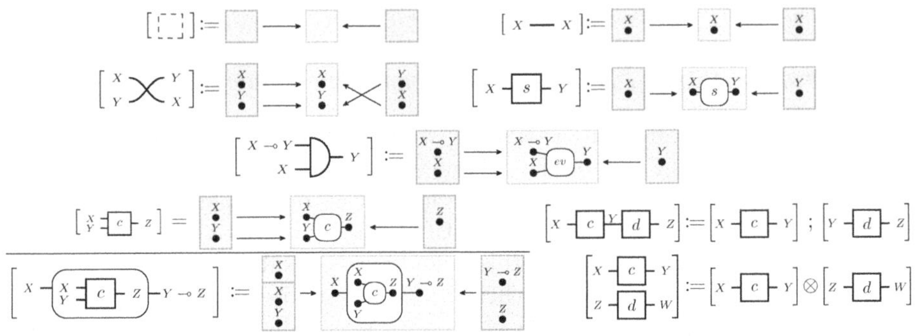

Proposition 1 ([2]). $[\cdot] : \mathbb{C}_\Sigma \to \mathsf{Csp}_D(\mathbf{HHyp}_\Sigma)$ *is a faithful functor of SMCs.*

The category of hypergraphs has a richer structure than $\mathbb{C}_\Sigma$. In particular, it is traced, in a way that is compatible with the functor $[\cdot]$. This yields an extension of the above interpretation to traced hierarchical string diagrams.

Lemma 1. *There is a faithful functor of TSMCs* $[\![\cdot]\!] : \mathbb{T}_\Sigma \to \mathsf{Csp}_D(\mathbf{HHyp}_\Sigma)$.

Proof. The category $\mathsf{Csp}_D(\mathbf{HHyp}_\Sigma)$ is a hypergraph category, that is every object is equipped with a Frobenius bimonoid [5], given by the following hypergraphs:

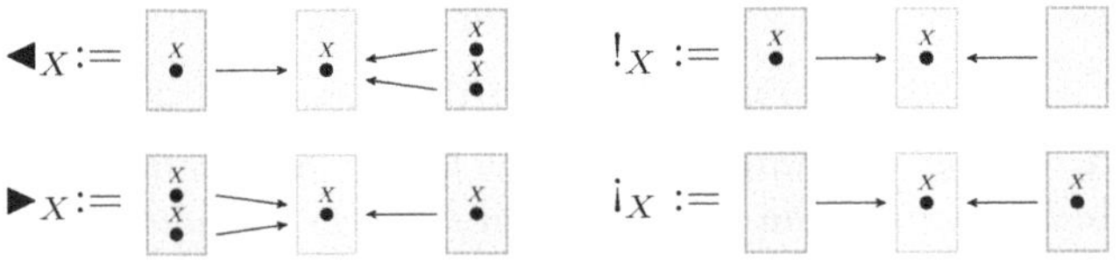

(5)

The category $\mathbb{C}_\Sigma$ can be augmented with a Frobenius structure, by taking the coproduct $\mathbb{C}_\Sigma + \mathbf{Frob}$, where **Frob** is the free traced hypergraph category on the signature consisting of:

 $: X \to X \otimes X$  $: X \to I$

 $: X \otimes X \to X$  $: I \to X$ (6)

Both $\mathsf{Csp}_D(\mathbf{HHyp}_\Sigma)$ and $\mathbb{C}_\Sigma + \mathbf{Frob}$ are TSMCs, with the trace being the canonical one arising from the Frobenius structure. For example, in $\mathbb{C}_\Sigma + \mathbf{Frob}$, the trace of 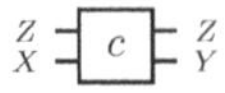 is 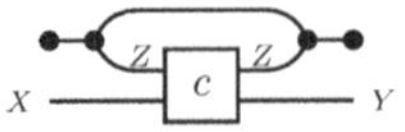.

Now observe that $\mathbb{T}_\Sigma$ is a subcategory of $\mathbb{C}_\Sigma + \mathbf{Frob}$ and that there is a faithful functor $F\colon \mathbf{Frob} \to \mathsf{Csp}_D(\mathbf{HHyp}_\Sigma)$, mapping the structure in (6) to the one in (5). The faithful functor $[\![\cdot]\!]\colon \mathbb{T}_\Sigma \to \mathsf{Csp}_D(\mathbf{HHyp}_\Sigma)$ is thus obtained as the composite $\mathbb{T}_\Sigma \xhookrightarrow{\iota} \mathbb{C}_\Sigma + \mathbf{Frob} \xrightarrow{[\cdot]+F} \mathsf{Csp}_D(\mathbf{HHyp}_\Sigma)$.

It is convenient to visualise the action of $[\![\cdot]\!]$ on string diagrams. If 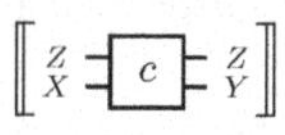 = 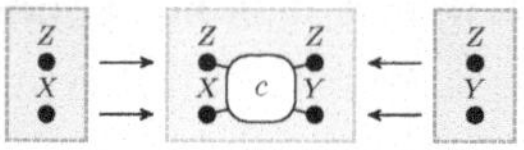, then 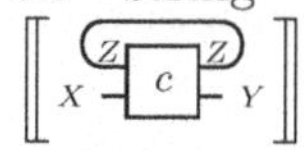 = 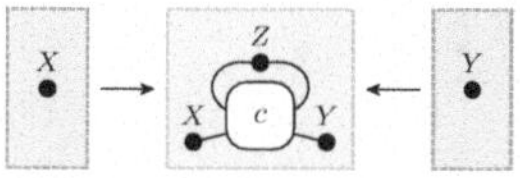.

4 The Image of the Interpretation: Hypernets

As noticed in previous work, starting with [6], double-pushout rewriting of hypergraphs representing string diagrams does not mirror exactly string diagram rewriting: a more restricted notion of DPOI rewriting is required to achieve soundness. The same issue needs to be tackled in our context. For this purpose, it is important to first understand which class of hierarchical hypergraphs are in the image of the functor $[\![\cdot]\!]$ we just introduced. The resulting notion will be the one of *hypernet*, generalising the acyclic hypernets defined in [2]. This generalisation is made possible by moving from *monogamous acyclic* to *partial monogamous* cospans of hypergraphs.

Definition 9 (Acyclicity). *For a hypergraph $\mathcal{H}$ and hyperedges e, e', a* path *from e to e' is a finite sequence of hyperdges $e_1, \ldots, e_n$ such that $e_1 = e$, $e_n = e'$ and for $i < n$ there exists a u that is in the target of e_i and in the source of e_{i+1}.*
A hypergraph is acyclic *if there exists no path from a vertex to itself.*

Acyclicity extends to hypergraphs with interfaces $v \to \mathcal{H} \leftarrow w$, by simply requiring $\mathcal{H}$ to be acyclic. However, cospan composition does not preserve acyclicity [6]. This issue is avoided by considering monogamous hypergraphs.

Definition 10 (Degree of a vertex). *The* in-degree *of a vertex u in a hypergraph $\mathcal{H}$ is the number of pairs (e, i) where e is a hyperedge with u as its i-th target. Similarly, the* out-degree *of u is the number of pairs (e, j) where e is a hyperedge with u as its j-th source.*

Definition 11 (Monogamy). *Let $v \xrightarrow{f} \mathcal{H} \xleftarrow{g} w$ be a hypergraph with interfaces, $in(\mathcal{H})$ the image of f and $out(\mathcal{H})$ the image of g. We say that the cospan is* monogamous *if f and g are mono and for all vertices u of $\mathcal{H}$:*

- *the in-degree of u is 0 if $u \in in(\mathcal{H})$, and 1 otherwise;*
- *the out-degree of u is 0 if $u \in out(\mathcal{H})$, and 1 otherwise.*

Example 2. The cospan in Fig. 5 is monogamous since the in- and out-degree of each vertex is at most 1 and, moreover, the in- (resp. out-) degree of the vertices connected to a left (resp. right) interface is 0. In contrast, the two cospans below are *not* monogamous. For the cospan on the left, the vertex labelled A has in-degree 0 but out-degree 2, as it is connected to an interface on the left but it appears twice as the source of f. For the one on the right, the vertex labelled C has both in- and out-degree 0, but it is not connected to any interface.

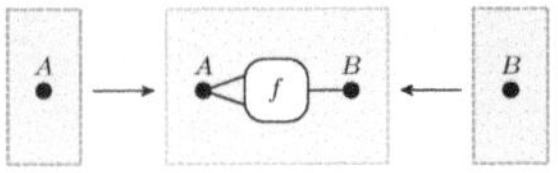

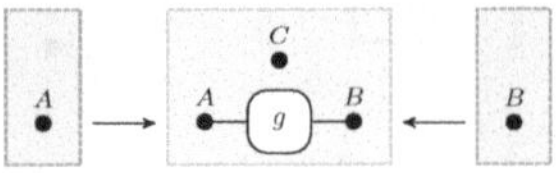

Monogamous acyclic cospans of hierarchical hypergraphs have been shown to be a suitable combinatorial characterization of hierarchical string diagrams. However, for the traced hierarchical setting, it is necessary to drop the acyclicity condition and require a weaker notion of monogamy.

Definition 12 (Partial monogamy). *A hypergraph with interfaces* $v \xrightarrow{f} \mathcal{H} \xleftarrow{g} w$, *is* partial monogamous *if* f *and* g *are mono and for all vertices* u *of* $\mathcal{H}$:

- *the in and out-degrees of* u *are both* 0 *if* $u \in in(\mathcal{H})$ *and* $u \in out(\mathcal{H})$;
- *the in-degree of* u *is* 0 *and the out-degree of* u *is* 1 *if* $u \in in(\mathcal{H})$;
- *the in-degree of* u *is* 1 *and the out-degree of* u *is* 0 *if* $u \in out(\mathcal{H})$;
- *the in and out-degrees of* u *are both* 1 *or* 0.

An example of partial monogamous cospan is the hypergraph with interfaces in Example 2 on the right. The intuition is that vertices like the one labelled C in the example represent traced identities, i.e. closed diagrams like  .

Definition 13. *A subgraph* $\mathcal{G}$ *of a hierarchical hypergraph* $\mathcal{F}$ *is* down-closed *if for all edges* e *in* $\mathcal{G}$, $\mathcal{F}_e \subseteq \mathcal{G}$. *Where* $\mathcal{F}_e$ *consists of vertices* $\hat{v}$ *and edges* $\hat{e}$, *such that* $p^k_{E,\perp}(\hat{e}) = e$ *and* $p^j_{E,\perp}(p_V(\hat{v})) = e$, *for some* $k \geq 1$ *and* $j \geq 0$. *In other words,* $\mathcal{F}_e$ *is the inner hypergraph sitting inside of* e.

Definition 14. *A hierarchical hypergraph with interfaces* $v \to \mathcal{H} \leftarrow w$ *is a* hypernet *if*

1. *it is partial monogamous when the hierarchical structure is forgotten,*
2. *if* $l_E(e) \neq \perp$ *then* e *has no children (it is not parent of any vertex nor hyperedge),*
3. *if* $l_E(e) = \perp$, *then* e *is a well-typed abstraction, that is: if the input and output interfaces of* $\mathcal{H}_e$ *are labelled with* X *and* Y, *respectively, then there exists* Z *such that* $s(e) \otimes Z = X$ *and* $t(e) = Z \multimap Y$.

A hierarchical hypergraph that only satisfies the first two conditions is called a weak hypernet. *We denote as* $\mathbf{Hynet}_\Sigma$ *the subcategory of* $\mathsf{Csp}_D(\mathbf{HHyp}_\Sigma)$ *consisting of hypernets and morphisms between them.*

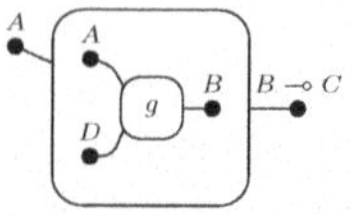

The well-typed abstraction condition in Definition 14 ensures that the interfaces of the hierarchical hyperedges are typed accordingly to the outer interfaces of the hypergraph sitting inside of it, and thus mimicking the abstraction operator of monoidal closed categories. For example, the hierarchical hyperedge in Fig. 5 respects this condition, whereas the one on the right does not, since the vertex labelled $B \multimap C$ should instead be $D \multimap B$.

Definition 15. *An* acyclic hypernet *is an hypernet whose underlying hypergraph structure is monogamous acyclic when the hierarchical structure is forgotten. We denote as* $\mathbf{AHynet}_\Sigma$ *the subcategory of* $\mathsf{Csp}_D(\mathbf{HHyp}_\Sigma)$ *consisting of acyclic hypernets and morphisms between them.*

Proposition 2. $\mathbf{Hynet}_\Sigma$ *is a traced symmetric monoidal category.*

Proof (sketch). Unlike $\mathsf{Csp}_D(\mathbf{HHyp}_\Sigma)$, $\mathbf{Hynet}_\Sigma$ is not a hypergraph category. In fact, the hypergraphs in (5) do not satisfy the partial monogamy requirement, and thus they are not hypernets. However, the Frobenius structure is only used to construct the canonical trace, which always yields a partial monogamous cospan.

The remaining of this section aims at showing the correspondence between traced hierarchical string digrams and hypernets.

Proposition 3. *Any hypernet* $v \to \mathcal{A} \leftarrow w$ *can be factored as*

$$Tr^i_{v,w}(i+v \to \mathcal{C}_1 \leftarrow u+v' \ ; \ (u \to u \leftarrow u \otimes v' \to \mathcal{G} \leftarrow w') \ ; \ u+w' \to \mathcal{C}_2 \leftarrow i+w)$$

where $i+v \to \mathcal{C}_1 \leftarrow u+v'$ *and* $u+w' \to \mathcal{C}_2 \leftarrow i+w$ *are acyclic hypernets and* $\mathcal{G}$ *is an outermost-layer subnet of* $\mathcal{A}$.

Proof. Since $\mathbf{Hynet}_\Sigma$ is a TSMC, $\mathcal{A}$ can be factored as $Tr^i_{v,w}(i+v \to \mathcal{B} \leftarrow i+w)$, where $\mathcal{B}$ is a hypernet whose outermost layer is trace-free. At this point the outermost layer of $\mathcal{B}$ is an acyclic hypernet that can be factored as $i+v \to \mathcal{C}_1 \leftarrow u+v' \ ; \ (u \to u \leftarrow u \otimes v' \to \mathcal{G} \leftarrow w') \ ; \ u+w' \to \mathcal{C}_2 \leftarrow i+w$ [3, Lemma 4.11].

The following result establishes the correspondence between hypernets and traced hierarchical string diagrams.

Lemma 2. *Every hierarchical hypergraph with interfaces is in the image of* $[\![\cdot]\!]$ *if and only if it is a hypernet.*

Proof. For the left to right direction it suffices to check, by induction on the syntax of $\mathbb{T}_\Sigma$, that the requirements of Definition 14 are met by $[\![c]\!]$.

For the other direction observe that, by Proposition 3, the factorisation of a hypernet directly yields a traced hierarchical string diagram in the same shape.

5 Rewriting for Traced Monoidal Closed Categories

In the previous section we established a correspondence between traced hierarchial string diagrams and hypernets. In this section we provide a notion of rewriting for hypernets as a suitable version of DPOI rewriting. Finally, we show a correspondence between rewriting of string diagrams (Definition 4) and its hypernet counterpart (Definition 16).

Definition 16 (Hypernet Rewriting). *Let $\mathcal{L}, \mathcal{R}$ be parallel hypernets, i.e. with same outermost interfaces v', w'. We say that the hypernet $\mathcal{F}$ with interfaces $\mathbf{v}, \mathbf{w}$ and outermost interfaces v, w, rewrites into the parallel hypernet $\mathcal{G}$ according to the rewrite rule $\langle \mathcal{L}, \mathcal{R} \rangle$, denoted as $\mathcal{F} \Rightarrow_{\langle \mathcal{L}, \mathcal{R} \rangle} \mathcal{G}$, if there is*

- *a morphism $m \colon \mathcal{L} \to \mathcal{F}$ whose image is a down-closed subgraph of $\mathcal{F}$, and*
- *a weak hypernet $\mathbf{v} + w' \xrightarrow{[\mathbf{c_i}, d_o]} \mathcal{C} \xleftarrow{[\mathbf{c_o}, d_i]} \mathbf{w} + v'$, such that*
- *the diagram on the right commutes and the two marked squares are pushouts.*

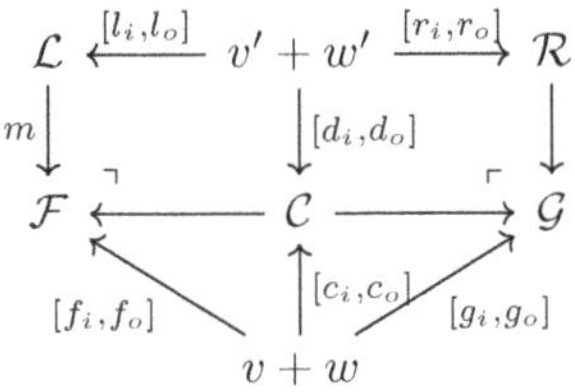

Notice that, unlike [2], we do not require the matching m to be mono, as this would cut off potential rewrites involving the trace structure, as explained in [18, Definition 58]. Similarly, we do not require that the image of m must be convex, i.e. any path between vertices must be captured, since we want to perform rewriting modulo the axioms of traces.

For example, if m was mono we could not match 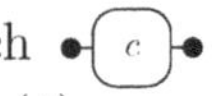 in 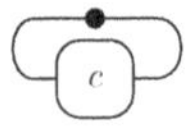. While requiring convexity does not permit rewrites as in (4).

It is important to note that not all pushouts exist when considering hierarchical hypergraphs, as witnessed by the span in [2, Figure 9c]. However, those needed for hypernet rewriting in Definition 16 are guaranteed to exist. In particular, the existence of the right pushout in the diagram above follows by [3, Lemma 4.7].

Proposition 4. *Let $\langle \mathcal{L}, \mathcal{R} \rangle$ be a hypernet rewrite rule, $\mathcal{F}$ a hypernet and $m \colon \mathcal{L} \to \mathcal{F}$ a morphism of hypernets. Then the object $\mathcal{C}$, called boundary complement, of Definition 16 exists and $\mathcal{G}$ is a hypernet.*

Proof (sketch). First, we obtain $\mathcal{C}$ in the usual way of DPOI rewriting, by erasing the subgraph $\mathcal{L}$ in $\mathcal{F}$. The erasure procedure does not affect partial monogamy, which is thus satisfied by $\mathcal{C}$. Note that, if $\mathcal{L}$ occurs inside an hierarchical hyperdge, then the well-typedness of the abstraction might not be preserved. This is harmless since we only require $\mathcal{C}$ to be a *weak* hypernet.

Now, the pushout with $\mathcal{R}$ restores the well-typedness of the abstraction, since $\mathcal{L}$ and $\mathcal{R}$ have the same outermost interfaces, and produces the hypernet $\mathcal{G}$ obtained by gluing $\mathcal{R}$ in $\mathcal{C}$ in place of the hole left by $\mathcal{L}$.

Example 3. Consider the following pairs of parallel hypernets:

$v' \to \mathcal{L} \leftarrow w' =$ 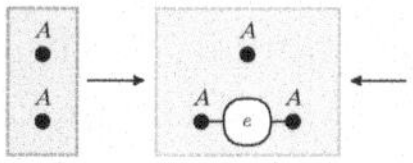 $v' \to \mathcal{R} \leftarrow w' =$ 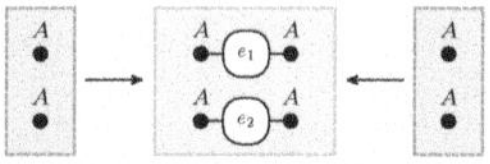

and

$v \to \mathcal{F} \leftarrow w =$ 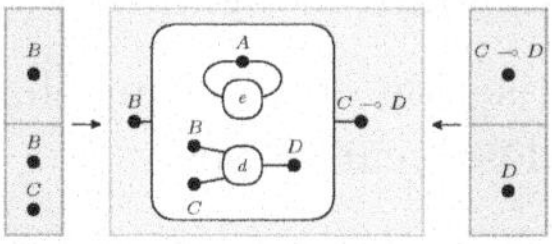 $v \to \mathcal{G} \leftarrow w =$ 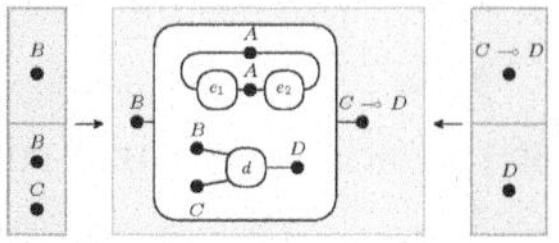 .

Now $\mathcal{F} \Rightarrow_{\langle \mathcal{L},\mathcal{R}\rangle} \mathcal{G}$ by the valid DPOI diagram below, where we name the A-labelled vertices to clarify how they are mapped by the morphisms.

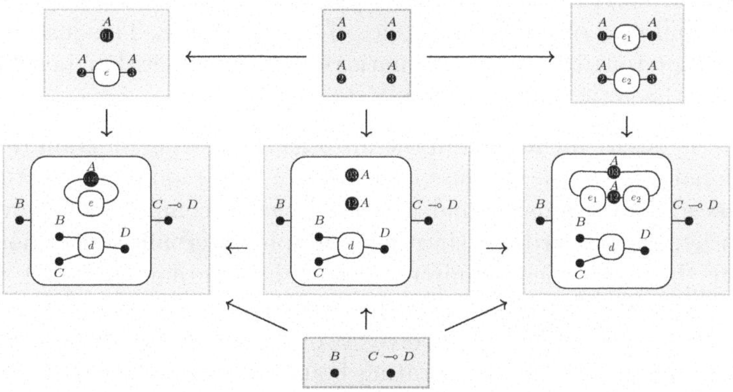

Notice that the boundary complement is only a weak hypernet, since the hypergraph in the middle does not respect the well-typed abstraction condition, i.e. the outermost interfaces are not typed accordingly to the hypergraph sitting inside the hierarchical hyperedge.

The hypernet rewriting above corresponds to a rewriting of string diagrams. Let l, r, f and g be the following diagrams in $\mathbb{T}_\Sigma$

$l =$ 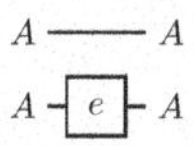 $r =$ 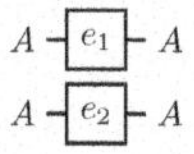 $f =$ 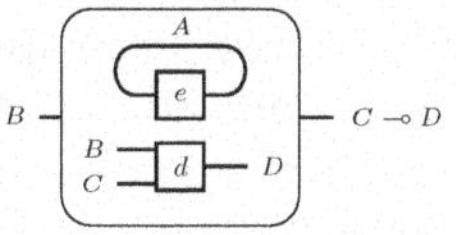 $g =$ 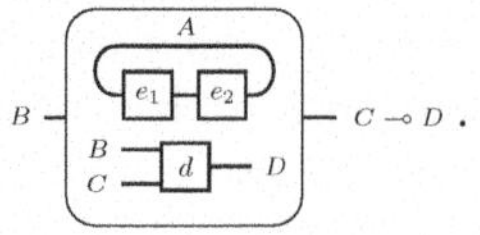 .

Then one can show that $f \Rightarrow_{\langle l,r\rangle} g$. The full derivation is similar to the one in (4), execept that it happens inside the hierarchical structure.

The correspondence hinted above between hypernet rewriting and string diagrams rewriting is formalised by the following completeness theorem.

Theorem 1. *Let $f, g\colon X \to Y$ and $l, r\colon Z \to W$ be pairs of traced hierarchical string diagrams in $\mathbb{T}_\Sigma$. Then, $f \Rightarrow_{\langle l,r\rangle} g$ if and only if $[\![f]\!] \Rightarrow_{\langle [\![l]\!],[\![r]\!]\rangle} [\![g]\!]$.*

Proof (sketch). For one direction assume $f \Rightarrow_{\langle l,r \rangle} g$, then f and g can be put in the form of Definition 4. Then we compute cospans $[\![l]\!], [\![r]\!], [\![f]\!], [\![g]\!]$. The weak hypernet $\mathcal{C}$ is computed starting from f. In particular, we erase the occurence of l in f (observe that this may happen at any level of the hierarchy, if we are in the case of Definition 4.(2)) and we rearrange the wires in the expected way. Call such diagram C, then $\mathcal{C} := [\![C]\!]$. Composition of cospans is by pushout and thus, by aptly combining this data, the DPOI diagram in Definition 16 is recovered.

As a corollary, we obtain the characterisation of rewriting that accounts also for the non-structural rules of monoidal closed categories.

Corollary 1. *Let $f, g \colon X \to Y$ and $l, r \colon Z \to W$ be pairs of traced hierarchical string diagrams in $\mathbb{T}_\Sigma$. Then, $f \Rightarrow_{\mathsf{MC}} g$ if and only if $[\![f]\!] \Rightarrow_{\{\langle [\![l]\!], [\![r]\!] \rangle \mid \langle l,r \rangle \in \mathsf{MC}\}} [\![g]\!]$.*

Note that equivalence in traced symmetric monoidal closed categories is not captured on the nose. However, observe that the rules in MC have an operational meaning, reminiscent of the η and β rules of the λ-calculus. This justifies making their application tangible in diagrammatic reasoning, rather than being absorbed by the graph-theoretic interpretation.

Remark 2. As briefly mentioned in the introduction, string diagram rewriting is computationally infeasible because each rewrite step requires inspecting equivalence classes of diagrams to find a redex. Such a computational overload is avoided in hypernet rewriting, since hypergraphs internalise the axioms corresponding to the geometrical manipulations of the diagrams in Figs. 1 and 2.

Let us give some more detail on how a match in DPOI rewriting may be effectively computed. It mainly involves two steps: (a) solving a subgraph isomorphism problem and (b) enumerating boundary complements. For (a) there exist efficient solutions to the problem, notably Ullmann's algorithm [32]. For (b), the situation is more articulated. In acyclic hypernet rewriting, the matching is required to be convex and the cospans are monogamous, making the boundary complements unique (up to iso). The presence of the trace in the hypernets studied in this paper forces the matching to be not necessarily convex, and the cospans under consideration to be partial monogamous rather than monogamous. In this setting the boundary complements are not necessarily unique, but can be enumerated. The number of boundary complements in hypergraph rewriting is tied to the notion of *adhesive* category [23]. Briefly, if the category of cospans under consideration is adhesive and certain additional conditions on rewriting are met, then the boundary complements are ensured to be unique. While $\mathbf{Hynet}_\Sigma$ is not adhesive, it might be related to the notion of $\mathcal{M},\mathcal{N}$-adhesivity studied in [10]. This could, in principle, allow us to define a bound on the number of boundary complements present in our setting. We leave a more detailed investigation of the implementation details to a follow up of this work.

6 λ-Calculus with Explicit Recursion

In the λ-calculus, recursion is usually represented by means of a fixed-point combinator. Although this method is theoretically sound, it is often impractical and does not reflect actual implementations. A more convenient way to represent recursive functions is provided by an explicit recursion operator, commonly found in functional languages as the `letrec` construct.

A version of the λ-calculus featuring explicit recursion, that we denote as $\lambda_{\texttt{letrec}}$, has been studied in [21]. In particular, its categorical interpretation extends the well-known correspondence between the simply typed λ-calculus and cartesian closed categories to $\lambda_{\texttt{letrec}}$ and traced cartesian closed categories.

In this section we develop a string diagrammatic interpretation of λ-terms featuring explicit recursion, crucially exploiting the graphical language introduced in this paper. To do so, we first recall the syntax of $\lambda_{\texttt{letrec}}$.

Fixed a set of variables X and a set of basic types $\mathcal{S}$, we build types as $T, T' := A \mid T \to T'$, where $A \in \mathcal{S}$. Terms are generated according to the following grammar, where $x \in X$:

$$u, v := x \mid \lambda x.u \mid v\ u \mid \texttt{letrec}\ x = u\ \texttt{in}\ v \tag{7}$$

Among terms generated by (7), we consider only those that can be typed according to type judgements of the form $\Gamma \vdash u \colon T$, where $\Gamma = x_1 \colon T_1, \dots, x_n \colon T_n$ is a variable type assignment. Then $\Gamma \vdash u \colon T$ is interpreted as u has type T if each x_i appearing in u has type T_i. Type judgements are defined according to the following rules:

$$\Gamma, x \colon T \vdash x \colon T \qquad \frac{\Gamma, x \colon T \vdash u \colon T'}{\Gamma \vdash \lambda x.u \colon T \to T'} \qquad \frac{\Gamma \vdash u \colon T \quad \Gamma \vdash v \colon T \to T'}{\Gamma \vdash v\ u \colon T'} \qquad \frac{\Gamma, x \colon T \vdash u \colon T \quad \Gamma, x \colon T \vdash v \colon T'}{\Gamma \vdash \texttt{letrec}\ x = u\ \texttt{in}\ v \colon T'}$$

Now, recall the traced cartesian closed category $\mathbf{T}^{C}_{\Sigma}$ and the traced symmetric monoidal category $\mathbb{T}^{C}_{\Sigma}$. As we want to perform rewriting on λ-terms, we interpret them as traced hierarchial string diagrams in the latter category. Thus, we define an encoding map $\epsilon \colon \lambda_{\texttt{letrec}} \to \mathbb{T}^{C}_{\Sigma}$ as follows.

Types are intepreted as objects of the category: in particular for every $A \in \mathcal{S}$, $\epsilon(A) = A$, $\epsilon(T \to T') = \epsilon(T) \multimap \epsilon(T')$ and $\epsilon(\Gamma) = \epsilon(x_1 \colon T_1, \dots, x_n \colon T_n) = \epsilon(T_1) \otimes \dots \otimes \epsilon(T_n)$. Typed terms are then interpreted inductively as follows:

$\epsilon(\Gamma, x \colon T \vdash x \colon T) =$ 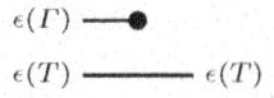

$\epsilon(\Gamma \vdash \lambda x.u \colon T \to T') =$ 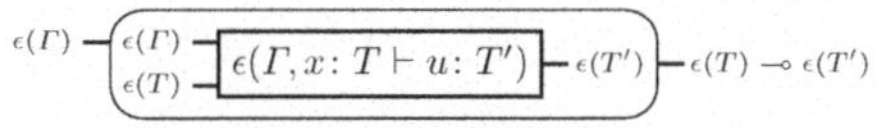

$\epsilon(\Gamma \vdash v\ u \colon T') =$ 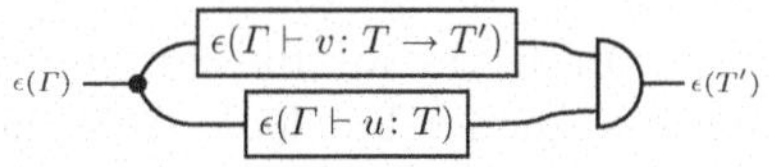

$\epsilon(\Gamma \vdash \texttt{letrec}\ x = u\ \texttt{in}\ v \colon T') =$ 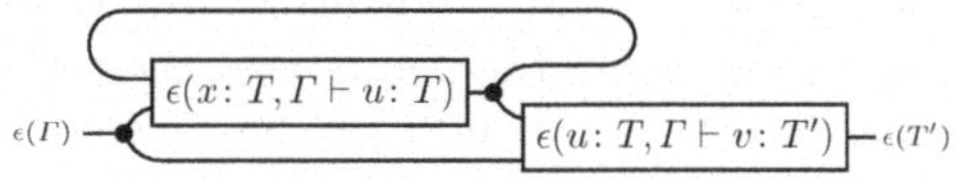

Notice from the diagrammatic encoding the various structures at play: the comonoids are necessary to deal with variable sharing; the hierachical structure is essential to express the λ-abstraction; and finally, the trace allows to model the explicit recursion operator.

By adding more generators and equations we can model other convenient features of a programming language such as arithmetic and logical operations or the `if-then-else` construct.

For example, the recursive program computing the factorial, below on the left, is represented as the diagram on the right, where `ite` corresponds to a function that selects a branch (second or third argument) according to the value of the guard (first argument).

$$\begin{aligned}&\texttt{letrec}\ f = \lambda x.\ \texttt{if}\ \ x = 0\\ &\qquad\qquad\quad \texttt{then}\ 1\\ &\qquad\qquad\quad \texttt{else}\ \ x \times f(x-1)\\ &\quad \texttt{in}\ f(n)\end{aligned}$$

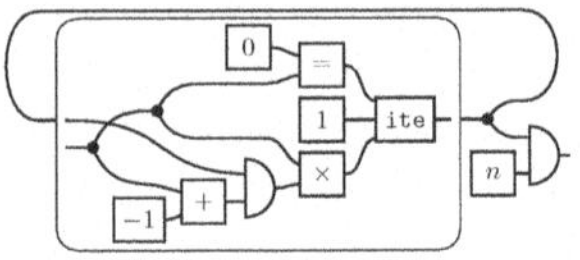

A direct consequence of the encoding of $\lambda_{\texttt{letrec}}$ is that we can reason diagrammatically to prove properties of functional programs. Even more, the results presented in §§3, 4 and 5, yield an interpretation of $\lambda_{\texttt{letrect}}$-terms as hierarchical hypergraphs whose rewriting can be studied in terms of hypernet rewriting.

6.1 On Confluence

A known result by Ariola and Klopp [1] shows that the λ-calculus with explict recursion is not confluent. In this section we discuss this result under the lens of string diagrams rewriting, leverging the algebraic nature of our axiomatisation.

Consider the following mutual-recursive program of $\lambda_{\texttt{letrec}}$ below on the left, together with its string diagrmmatic interpretation on the right.

$$\begin{aligned}\texttt{letrec}\ f &= \lambda x.\ g(\mathtt{S}x)\ ,\\ g &= \lambda y.\ f(\mathtt{S}x)\\ \texttt{in}\ f&\end{aligned}$$

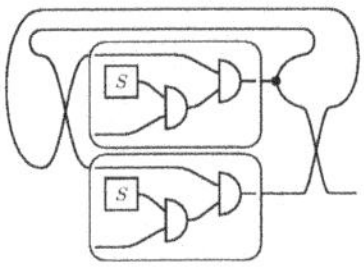

Using the axioms of traced symmetric monoidal categories, one can easily show that the diagram above is equivalent to the one in the middle below. Moreover their combinatorial representations coincide to the same hypernet.

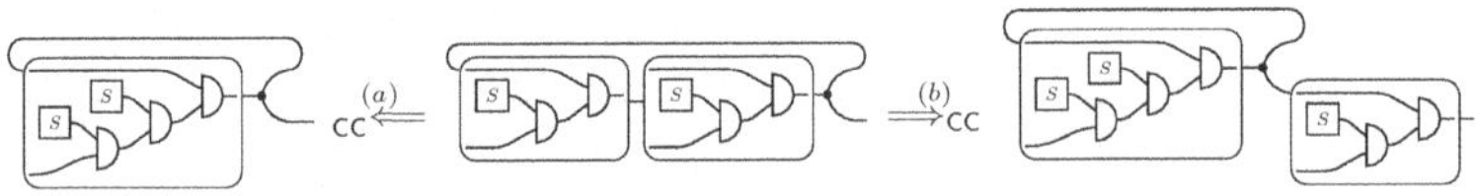

(8)

Note that the diagram in the middle can be rewritten into two non-joinable diagrams. The rewrite sequence labeled (a) applies the axioms of monoidal closed categories to perform a β-reduction, while the sequence labeled (b) first duplicates the bubble using the copier and then performs (a).

As noted in [1], non-confluence lies in the fact that the two diagrams are "out-of-sync". Specifically, any applicable rewrite will result in one diagram containing an odd number of `S`'s, while the other will have an even number.

Finally, note that in (8), we are working with string diagrams that include the additional generators and axioms from Fig. 4. Therefore (8) is not, in general, a counterexample to confluence of the rewriting system for hypernets.

7 Discussion and Future Work

Traced monoidal closed categories provide an expressive setting for reasoning about computational processes with both recursion and higher-order features. In this paper we provide a comprehensive account of the diagrammatic language for these categories. We represent their morphisms as a graphical syntax of hierarchical string diagrams with a trace forming operation. We then give a combinatorial interpretation of these structures, in terms of partial monogamous cospans of hierarchical hypergraphs. Importantly, string diagrams that are equivalent modulo the laws of traced symmetric monoidal categories are represented as the same cospan. This allows us to interpret string diagram rewriting as a certain variant of DPO hypergraph rewriting. Correctness of this characterisation is particularly subtle, as we need to handle both partial monogamy and the hierarchical structure of hypergraphs in the rewriting. As typical in this kind of result (see e.g. [6]), the key payoff is that a rewrite step modulo traced symmetric monoidal structure corresponds to a rewrite step 'on-the-nose' in the hypergraph interpretation. As a proof of concept, we show a diagrammatic representation of $\lambda_{\texttt{letrec}}$ and we discuss confluence of the resulting rewriting system.

As future work, we plan to investigate various properties of traced monoidal closed categories in our diagrammatic calculus. In particular, the relationship between the Int construction and the categories discussed in this paper. We are also interested in using our framework for program transformation, in the same spirit as string diagrams with additional structures are employed in [3] for automatic differentiation and in [16] for closure conversion. Finally, we intend to explore the link between our notion of rewriting and the graph rewriting techniques in [26–28] for compiler optimisation.

Acknowledgements. The authors acknowledge support from EPSRC EP/V002376/1. F. Zanasi is also supported by the ARIA Safeguarded AI TA1.1 programme. We thank the anonymous reviewers for their detailed comments and feedback.

Disclosure of Interests. The authors have no competing interests to declare that are relevant to the content of this article.

References

1. Cyclic lambda graph rewriting. In: Proceedings Ninth Annual IEEE Symposium on Logic in Computer Science, pp. 416–425. IEEE (1994)
2. Alvarez-Picallo, M., Ghica, D., Sprunger, D., Zanasi, F.: Rewriting for monoidal closed categories. In: 7th International Conference on Formal Structures for Computation and Deduction (FSCD 2022). Schloss Dagstuhl-Leibniz-Zentrum für Informatik (2022)
3. Alvarez-Picallo, M., Ghica, D.R., Sprunger, D., Zanasi, F.: Functorial string diagrams for reverse-mode automatic differentiation. arXiv preprint arXiv:2107.13433 (2021)
4. Baez, J., Erbele, J.: Categories in control. Theory Appl. Categories **30**, 836–881 (2015). http://www.tac.mta.ca/tac/volumes/30/24/30-24abs.html

5. Bonchi, F., Gadducci, F., Kissinger, A., Sobocinski, P., Zanasi, F.: String diagram rewrite theory I: rewriting with Frobenius structure. J. ACM **69**(2), 14:1–14:58 (2022). https://doi.org/10.1145/3502719
6. Bonchi, F., Gadducci, F., Kissinger, A., Sobocinski, P., Zanasi, F.: String diagram rewrite theory II: rewriting with symmetric monoidal structure. Math. Struct. Comput. Sci. **32**(4), 511–541 (2022)
7. Bonchi, F., Gadducci, F., Kissinger, A., Sobociński, P., Zanasi, F.: String diagram rewrite theory III: confluence with and without Frobenius. Math. Struct. Comput. Sci. **32**(7), 829–869 (2022)
8. Bonchi, F., Seeber, J., Sobocinski, P.: Graphical conjunctive queries. In: Ghica, D., Jung, A. (eds.) 27th EACSL Annual Conference on Computer Science Logic (CSL 2018). Leibniz International Proceedings in Informatics (LIPIcs), vol. 119, pp. 13:1–13:23. Schloss Dagstuhl–Leibniz-Zentrum fuer Informatik, Dagstuhl, Germany (2018). https://doi.org/10.4230/LIPIcs.CSL.2018.13, http://drops.dagstuhl.de/opus/volltexte/2018/9680
9. Bonchi, F., Sobocinski, P., Zanasi, F.: Full abstraction for signal flow graphs. In: Proceedings of the 42nd Annual ACM SIGPLAN-SIGACT Symposium on Principles of Programming Languages, pp. 515–526. POPL '15, Association for Computing Machinery, New York, NY, USA (2015). https://doi.org/10.1145/2676726.2676993
10. Castelnovo, D., Gadducci, F., Miculan, M.: A simple criterion for m, n-adhesivity. Theor. Comput. Sci. **982**, 114280 (2024). https://doi.org/10.1016/J.TCS.2023.114280
11. Coecke, B., Kissinger, A.: Picturing Quantum Processes: A First Course in Quantum Theory and Diagrammatic Reasoning. Cambridge University Press (2017). https://doi.org/10.1017/9781316219317
12. Corradini, A., Montanari, U., Rossi, F., Ehrig, H., Heckel, R., Löwe, M.: Algebraic approaches to graph transformation - part I: basic concepts and double pushout approach. In: Rozenberg, G. (ed.) Handbook of Graph Grammars and Computing by Graph Transformations, Volume 1: Foundations, pp. 163–246. World Scientific (1997)
13. Cruttwell, G.S., Gavranović, B., Ghani, N., Wilson, P., Zanasi, F.: Categorical foundations of gradient-based learning. In: European Symposium on Programming, pp. 1–28. Springer International Publishing Cham (2022)
14. Ehrig, H., Kreowski, H.J.: Parallelism of manipulations in information structures. In: MFCS 1976, pp. 284–293 (1976)
15. Fox, T.: Coalgebras and cartesian categories. Comm. Algebra **4**(7), 665–667 (1976). https://doi.org/10.1080/00927877608822127
16. Ghica, D., Zanasi, F.: String diagrams for λ-calculi and functional computation (2023)
17. Ghica, D.R., Jung, A.: Categorical semantics of digital circuits. In: 2016 Formal Methods in Computer-Aided Design (FMCAD), pp. 41–48 (2016). https://doi.org/10.1109/FMCAD.2016.7886659
18. Ghica, D.R., Kaye, G.: Rewriting modulo traced comonoid structure. In: 8th International Conference on Formal Structures for Computation and Deduction (FSCD 2023). Schloss Dagstuhl-Leibniz-Zentrum für Informatik (2023)
19. Hackney, P., Robertson, M.: On the category of props. Appl. Categ. Struct. **23**, 543–573 (2015)
20. Hasegawa, M.: Recursion from cyclic sharing: traced monoidal categories and models of cyclic lambda calculi. In: International Conference on Typed Lambda Calculi and Applications, pp. 196–213. Springer (1997)

21. Hasegawa, M.: Models of Sharing Graphs: A Categorical Semantics of let and letrec. Springer Science & Business Media (2012)
22. Joyal, A., Street, R., Verity, D.: Traced monoidal categories. Math Procs Cambridge Philos. Soc. **119**(3), 447–468 (1996)
23. Lack, S., Sobociński, P.: Adhesive and quasiadhesive categories. Theoret. Inf. Appl. **39**(3), 511–546 (2005)
24. Lambek, J.: Cartesian closed categories and typed λ-calculi. In: LITP Spring School on Theoretical Computer Science, pp. 136–175. Springer (1985)
25. Milosavljevic, A., Piedeleu, R., Zanasi, F.: Rewriting for symmetric monoidal categories with commutative (co)monoid structure. Logical Methods Comput. Sci. **21**(1), 12 (2025). https://doi.org/10.46298/lmcs-21(1:12)2025, https://lmcs.episciences.org/14937
26. Muroya, K.: Hypernet semantics of programming languages. Ph.D. thesis, University of Birmingham, UK (2020)
27. Muroya, K., Ghica, D.R.: The dynamic geometry of interaction machine: a call-by-need graph rewriter. arXiv preprint arXiv:1703.10027 (2017)
28. Muroya, K., Ghica, D.R.: The dynamic geometry of interaction machine: a token-guided graph rewriter. Logical Methods Comput. Sci. **15** (2019)
29. Piedeleu, R., Zanasi, F.: A finite axiomatisation of finite-state automata using string diagrams. Logical Methods Comput. Sci. **19**(1), 13 (2023). https://doi.org/10.46298/lmcs-19(1:13)2023, https://lmcs.episciences.org/10400
30. Piedeleu, R., Zanasi, F.: An introduction to string diagrams for computer scientists. CoRR abs/2305.08768 (2023)
31. Selinger, P.: A survey of graphical languages for monoidal categories. In: Coecke, B. (ed.) New Structures for Physics, Lecture Notes in Physics, vol. 813, pp. 289–355. Springer, Berlin, Heidelberg (2010). https://doi.org/10.1007/978-3-642-12821-9_4
32. Ullmann, J.R.: An algorithm for subgraph isomorphism. J. ACM (JACM) **23**(1), 31–42 (1976)

Parallel Rule Application with Doubling Avoidance

Hans-Jörg Kreowski[1] and Aaron Lye[2(✉)]

[1] Department of Computer Science, University of Bremen, P.O.Box 33 04 40, 28334 Bremen, Germany
kreo@uni-bremen.de

[2] German Aerospace Center (DLR), Institute for the Protection of Maritime Infrastructures, Fischkai 1, 27572 Bremerhaven, Germany
aaron.lye@dlr.de

Abstract. In this paper, the parallelization of parallel independent rule applications within the double-pushout approach over adhesive categories is generalized in such a way that doubling can be avoided. If two rule applications to the same object delete the same part of the commonly accessed part, then independence is restored by removing the deletion part of one of the rule applications. Similarly, one can avoid that certain insertions are doubled. It turns out that parallelization with doubling avoidance is closely related to synchronization by means of amalgamated rules.

1 Introduction

Parallelism and simultaneity are fundamental principles in nature and society. Most biological, chemical and physical processes on one hand and most processes in economy, administration, traffic, and every-day life run in parallel. Often coordination and regulation are needed to avoid collisions, conflicts or undesired effects. As one of the main concerns of Computer Science is the modeling of real-world processes, one encounters quite a spectrum of approaches to the modeling of parallel systems like, e.g., cellular automata, Lindenmayer systems, neural networks, reaction systems, swarm computing, and many more. The same applies, in particular, to the area of graph transformation. See, e.g., Taentzer's parallel and distributed graph transformation [1], Kniemeier's growth grammars [2], Metevier's and Sopena's graph relabeling systems [3], Boehm's, Fonio's and Habel's amalgamation of graph transformation [4], and various others. One of the earliest and simplest approaches to parallel graph transformation was introduced by Ehrig and Kreowski [5] and further studied in [6]. The basic concept is the coproduct of rules used as parallel rule. As this is a rule itself, parallel derivations are just derivations applying parallel rules. Moreover, some very helpful properties of the concept can be shown like sequentialization and parallelization. Given a parallel rule application, the component rules can be applied in arbitrary order yielding the same result as the given rule application.

J. Endrullis and M. Tichy (Eds.): ICGT 2025, LNCS 15720, pp. 44–62, 2025.
https://doi.org/10.1007/978-3-031-94706-3_3

Conversely, rule applications applied to the same graph that are pairwise independent can be applied in parallel. And this holds for two successive sequentially independent rule application, too. A comprehensive survey can be found in Corradini et al. [7], the generalization to adhesive categories in Ehrig et al. [8,9]. A complication is the fact that the search for a matching morphism is NP-complete if rules with arbitrarily large left-hand sides are allowed. Fortunately, a matching morphism of a parallel rule can be constructed from the matching morphisms of the component rules so that the obstacle can be circumvent. In this paper, we generalize this approach to rule-based parallel transformation in such a way that undesired doubling can be avoided. There are two types of doubling with respect to two rule applications to the same object. The first one concerns parallel dependent rule applications that - intuitively seen - delete exactly the same part of the intersection of the matchings. If one modifies one of the rule applications in such a way that the deletion part is removed while the other one is kept, it turns out that they become parallel independent and can be applied in parallel. Moreover, the modified rule application can be reconstructed so that no information gets lost in the process. The second case concerns rule applications that - again intuitively - insert isomorphic parts which may be merged. And again this can be achieved by removing the potentially common part from one of the rule applications while the other one is kept. Both types of doubling may occur together. In Sect. 3, the parallelization with doubling avoidance (dedoubling for short) of two rule applications is introduced and investigated while the concept is generalized to families of rule applications to the same object in Sect. 4. In Sect. 5 we relate our approach to amalgamation as studied in [4].

2 Preliminaries

In this section, the preliminaries are provided as far as needed for this paper. For the well-known categorical notions confer, e.g., Adamek et al. [10]. For the notion and properties of adhesive categories confer, e.g., Lack and Sobocinski [11]. For the notion of graph transformation confer, e.g., Ehrig et al. [8,9].

2.1 Categorical Prerequisites

We assume that the underlying category **C** is adhesive. In addition, we require an epi-mono factorization and a strict initial object.

Assumption.

1. **C** is adhesive, i.e., **C** has all pullbacks, **C** has pushouts along monomorphisms (meaning at least one of the two morphisms in the pushout span is a monomorphism), and pushouts along monomorphisms are Van Kampen squares.
2. **C** has a strict initial object $\emptyset$, i.e., an initial object $\emptyset$ with the property that every morphism in **C** with codomain $\emptyset$ is an isomorphism. This implies that the pushouts of spans of the form $G \leftarrow \emptyset \rightarrow H$ are the coproducts of G and H, denoted by $G + H$.

3. Every morphism in **C** has an epi-mono factorization, i.e., for every morphism f there is a factorization $f = m \circ e$ where e is an epimorphism and m is a monomorphism. The factorization is unique up to isomorphism in adhesive categories.

We use two adhesive categories explicitly: the category Σ-**Graphs** of directed edge-labeled graphs and the category **Rules** of graph transformation rules over an adhesive category **C**. Σ-**Graphs** is a special variant of a diagram category over the category **Sets** of sets. And **Rules** is a diagram category over **C**.

The category **Sets** of sets and mappings is the best-known adhesive category. Every diagram category over an adhesive category is adhesive. This remains true if some target objects are fixed. For instance, consider the diagram $\Sigma \leftarrow \cdot \rightrightarrows \cdot$ over **Sets** for a set Σ (of *labels*). The corresponding adhesive category is the category Σ-**Graphs** of directed edge-labeled graphs. An object is a system $G = (V, E, s, t, l)$ where V is a set of *vertices*, E is a set of *edges*, $s, t \colon E \to V$ and $l \colon E \to \Sigma$ are mappings assigning a *source*, a *target* and a *label* to every edge $e \in E$. An edge e with $s(e) = t(e)$ is called a *loop*. The empty graph is the initial object. Another adhesive category is given by the diagram $\cdot \leftarrow \cdot \rightarrow \cdot$ which remains adhesive if the arrows are restricted to monomorphisms. This provides the category **Rules** of rules over **C**.

2.2 The Double-Pushout Approach

The rewriting formalism which we use throughout this paper is the double-pushout (DPO) approach as introduced by Ehrig, Pfender and Schneider in [12]. It was originally introduced for graphs. However, it is well-defined in adhesive categories.

A *rule* is a span of monomorphisms $p = (L \underset{l}{\leftarrow} K \underset{r}{\rightarrow} R)$. L is called *left-hand side*, R is called *right-hand side*, and K is called *gluing object.*

A rule application to some object G is defined wrt a morphism $g \colon L \to G$ which is called *(left) matching morphism. G directly derives H* if the span $L \underset{l}{\leftarrow} K \underset{r}{\rightarrow} R$ and matching morphism g extend to the diagram

$$\begin{array}{ccccc}
L & \xleftarrow[l]{} & K & \xrightarrow[r]{} & R \\
{\scriptstyle g}\downarrow & (1) & {\scriptstyle z}\downarrow & (2) & {\scriptstyle h}\downarrow \\
G & \xleftarrow[m_{Z \to G}]{} & Z & \xrightarrow[m_{Z \to H}]{} & H
\end{array}$$

such that both squares are pushouts. Z is called *intermediate object* and h is called *right matching morphism.* Because rules and rule applications are symmetric, a rule $p = (L \underset{l}{\leftarrow} K \underset{r}{\rightarrow} R)$ induces an inverse rule $p^{-1} = (R \underset{r}{\leftarrow} K \underset{l}{\rightarrow} L)$ such that $G \underset{p}{\Longrightarrow} H$ implies $H \underset{p^{-1}}{\Longrightarrow} G$ and vice versa. The application of a rule p to G wrt g is called *direct derivation* and is denoted by $G \underset{p}{\Longrightarrow} H$ (where g is kept implicit). A *derivation* from G to H is a sequence of direct derivations

$G_0 \underset{p_1}{\Longrightarrow} G_1 \underset{p_2}{\Longrightarrow} \cdots \underset{p_n}{\Longrightarrow} G_n$ with $G_0 = G$, $G_n = H$ and $n \geq 0$. If $p_1, \cdots, p_n \in P$, then the derivation is also denoted by $G \overset{n}{\underset{P}{\Longrightarrow}} H$. If the length of the derivation does not matter, we write $G \overset{*}{\underset{P}{\Longrightarrow}} H$.

2.3 Local Church-Rosser Properties

Let $p = (L \underset{l}{\leftarrow} K \underset{r}{\rightarrow} R)$ and $\overline{p} = (\overline{L} \underset{\overline{l}}{\leftarrow} \overline{K} \underset{\overline{r}}{\rightarrow} \overline{R})$ be two rules.

1. Two direct derivations $G \underset{p}{\Longrightarrow} H$ and $G \underset{\overline{p}}{\Longrightarrow} \overline{H}$ with matching morphisms $g\colon L \rightarrow G$ and $\overline{g}\colon \overline{L} \rightarrow G$ are *parallel independent* if there exist two morphisms $f\colon L \rightarrow \overline{Z}$ and $\overline{f}\colon \overline{L} \rightarrow Z$ such that $g = m_{\overline{Z}\to G} \circ f$ and $\overline{g} = m_{Z\to G} \circ \overline{f}$. The situation is depicted in the following diagram.

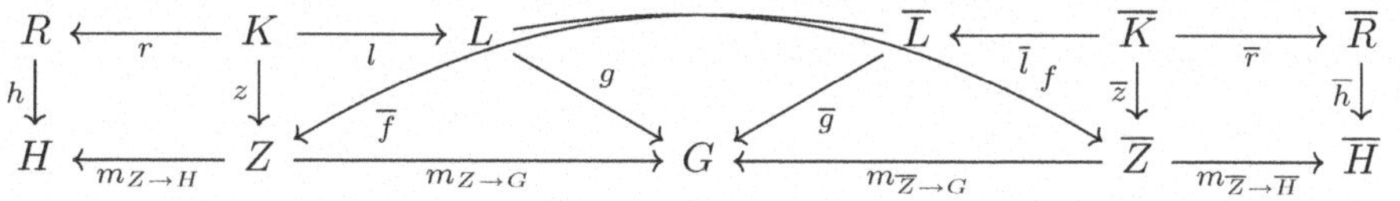

2. Successive direct derivations $G \underset{\overline{p}}{\Longrightarrow} \overline{H} \underset{p}{\Longrightarrow} X$ with the right matching morphism $\overline{h}\colon \overline{R} \rightarrow \overline{H}$ and the (left) matching morphism $g'\colon L \rightarrow \overline{H}$ are *sequentially independent* if $\overline{H} \underset{\overline{p}^{-1}}{\Longrightarrow} G$ and $\overline{H} \underset{p}{\Longrightarrow} X$ are parallel independent.

It is well-known that parallel independence induces the direct derivations $H \underset{\overline{p}}{\Longrightarrow} X$ and $\overline{H} \underset{p}{\Longrightarrow} X$ with matching morphisms $m_{Z\to H} \circ \overline{f}$ and $m_{\overline{Z}\to\overline{H}} \circ f$ and that sequential independence induces the derivation $G \underset{p}{\Longrightarrow} H \underset{\overline{p}}{\Longrightarrow} X$. The two constructions in the context of adhesive categories can be found in Lack and Sobocinski [11] (see also [8]).

Moreover, the coproduct of $p + \overline{p}$ of two rules p and $\overline{p}$ is called *parallel rule*. It is well-known that an application $G \underset{p+\overline{p}}{\Longrightarrow} X$ yields two sequentially independent sequentializations $G \underset{p}{\Longrightarrow} H \underset{\overline{p}}{\Longrightarrow} X$ and $G \underset{\overline{p}}{\Longrightarrow} \overline{H} \underset{p}{\Longrightarrow} X$ such that the two first steps are parallel independent. Conversely, parallel and sequentially independent rule applications induce corresponding parallel rule applications. The results concerning sequentialization and parallelization are based on the Butterfly Lemma as used in the proof of Theorem 3 in Sect. 4 (cf. [6–8] for more details). Moreover, in [6] is it shown that sequentialization is associative. The proof is done in the category **Graphs**. The same proof holds in adhesive categories.

3 Parallelization with Doubling Avoidance of Two Rule Applications

It is well-known that a set of pairwise parallel independent rule applications to the same object can be parallelized by applying the coproduct of the applied rules as parallel rule (see [6–8]). In this section, parallelization is generalized in such a way that so-called doubling can be avoided. Given two rule applications to the same object, a doubling is a common part of the applied rules on which the matching morphisms coincide. Intuitively, the two rule applications double the common part. Under certain conditions, the two rule applications can be parallelized while the doubling is avoided. This is formally achieved by covering and replacing each of the rule applications by the applications of suitable parallel rules such that their sequentializations and proper re-parallelizations yields the intended effect. This parallelization without doubling can be iterated for sets of rule applications with more than two elements (see Sect. 4). We start with an example that illustrates the idea.

Example 1. Our running example is inspired by pattern generation and tilings. We use two rules

$p_1 =$ 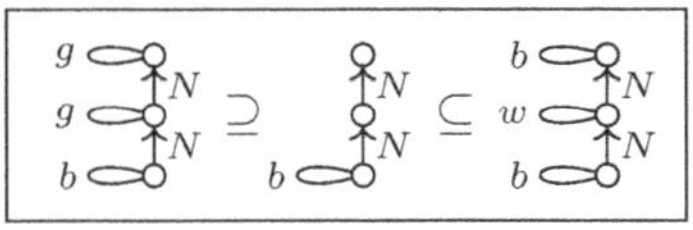 $p_2 =$ 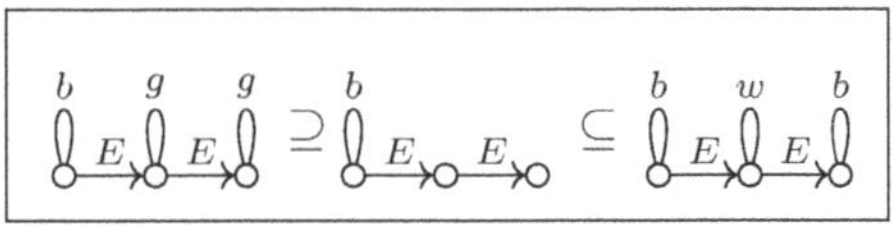

They are applied to finite subgraphs of an underlying infinite background graph defined by $B = (\mathbb{Z} \times \mathbb{Z}, \mathbb{Z} \times \mathbb{Z} \times \{b, g, w, E, N\}, s_B, t_B, l_B)$ where s_B is the projection to $\mathbb{Z} \times \mathbb{Z}$, l_B is the projection to $\{b, g, w, E, N\}$ and t_B is given by, for all $(x, y) \in \mathbb{Z} \times \mathbb{Z}, t_B((x, y), c) = (x, y)$ for $c \in \{b, g, w\}$, $t_B((x, y), E) = (x+1, y)$, $t_B((x, y), N) = (x, y + 1)$. This means that the vertices are the points in the plane with integer coordinates and each vertex is the source of five edges labeled with b, g, w, E and N respectively. The first three are loops colored by b(lack), g(ray), and w(hite). The fourth edge points to the right neighbor, the fifth to the upper neighbor. Each vertex $(x, y) \in \mathbb{Z} \times \mathbb{Z}$ represents a unit square $us(x, y)$ consisting of the four vertices $(x, y), (x + 1, y), (x, y + 1)$ and $(x + 1, y + 1)$ and the four edges $((x, y), E), ((x, y), N), ((x + 1, y), N)$ and $((x, y + 1), E)$. Each set T of vertices with one loop each represents a subgraph $sub(T)$ of B by closing T under the represented unit squares. It can be depicted as a grid of unit squares where each square is colored by the label of the loop at the corresponding vertex in T. The start graph is a rectangular triangle with legs of length 5. The color of the vertex at the bottom left is black; all other vertices have gray loops. A sample derivation applying the rules above to a sample grid is depicted in Fig. 1a. The only alternative result after five steps is depicted in Fig. 1b.

Both rules can be applied to the start pattern. As they are parallel independent, they can be applied in parallel as depicted in Fig. 1c. After the first parallel step, four rule applications are possible. But two of the matches overlap in the middle unit square of the hypotenuse. This is a loop that does not belong to the gluing graph so that the two rule application are parallel dependent. The other pairs are parallel independent so that two maximal parallel derivation steps are defined yielding the two result pattern in Fig. 1a and 1b.

A closer look at the dependence reveals that both rule applications do exactly the same on the overlap: They replace the g-labeled loop by a b-labeled one. If

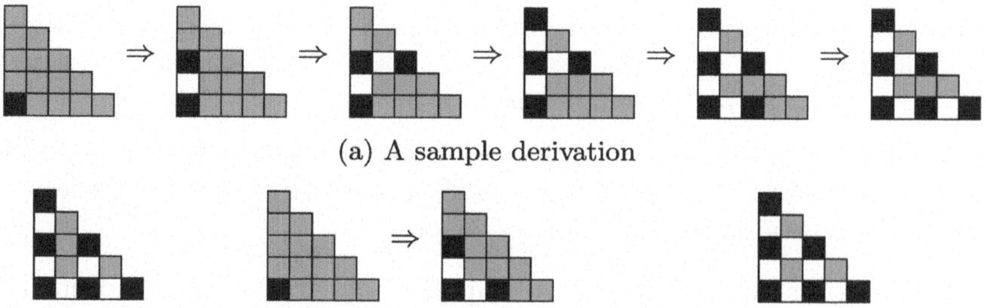

(a) A sample derivation

(b) Alternative result (c) Parallel rule application (d) Rule application with overlap

Fig. 1. Sample derivations

it would be allowed to apply this only once instead of twice, then all four rule applications could be parallelized yielding the grid depicted in Fig. 1d.

In order to formally describe this observation, we need the definition of a cover.

Definition 1. *Let* $p = (L \xleftarrow[l]{} K \xrightarrow[r]{} R), p' = (L' \xleftarrow[l']{} K' \xrightarrow[r']{} R')$ *and* $p'' = (L'' \xleftarrow[l'']{} K'' \xrightarrow[r'']{} R'')$ *be rules and* $\langle\lambda';\lambda''\rangle\colon L' + L'' \to L$, $\langle\kappa';\kappa''\rangle\colon K' + K'' \to K$ *and* $\langle\varrho';\varrho''\rangle\colon R' + R'' \to R$ *be epimorphisms such that the diagrams*

$$\begin{array}{ccccc}
L' + L'' & \xleftarrow{l'+l''} & K' + K'' & \xrightarrow{r'+r''} & R' + R'' \\
{\scriptstyle\langle\lambda';\lambda''\rangle}\downarrow & (3) & \downarrow{\scriptstyle\langle\kappa';\kappa''\rangle} \; (4) & & \downarrow{\scriptstyle\langle\varrho';\varrho''\rangle} \\
L & \xleftarrow[l]{} & K & \xrightarrow[r]{} & R
\end{array}$$

are pushouts. Then the pair (p', p'') *is a* cover *of* p. *The elements* p' *and* p'' *are called* cover components *of* p. *The rule* p'' *is called* complement rule *of* p' *and* p' *complement rule of* p''.

Theorem 1. *1. Let* (p', p'') *be a cover of* p *with the left epimorphism* $\langle\lambda';\lambda''\rangle\colon L'+L'' \to L$ *and* $G \underset{p}{\Longrightarrow} H$ *for some objects* G *and* H *with the matching morphism* $g\colon L \to G$ *be a rule application. Then* $p' + p''$ *can be applied to* G *with the matching morphism* $g \circ \langle\lambda';\lambda''\rangle\colon L' + L'' \to G$ *yielding* $G \underset{p'+p''}{\Longrightarrow} H$.

2. Let $G \underset{\overline{p}}{\Longrightarrow} \overline{H}$ *be another rule application that is parallel independent of* $G \underset{p}{\Longrightarrow} H$. *Then* $G \underset{\overline{p}}{\Longrightarrow} \overline{H}$ *is parallel independent of* $G \underset{p'+p''}{\Longrightarrow} H$, *too.*

Proof. 1. Let the diagrams (1) and (2) as given in Sect. 2.2 be the pushouts defining $G \underset{p}{\Longrightarrow} H$. Then the sequential compositions of (1) and (3) and (2) and (4) are pushouts defining $G \underset{p'+p''}{\Longrightarrow} H$.

2. Due to the assumption, there are morphisms $f\colon L \to \overline{Z}$ and $\overline{f}\colon \overline{L} \to Z$ such that $m_{\overline{Z}\to G} \circ f = g$ and $m_{Z\to G} \circ \overline{f} = \overline{g}$ where $\overline{Z}$ is the intermediate object of $G \underset{p}{\Longrightarrow} \overline{H}$ with the monomorphism $m_{\overline{Z}\to G}\colon \overline{Z} \to G$ and $\overline{g}\colon \overline{L} \to G$ is the matching morphism. Composing the first equation with $\langle\lambda';\lambda''\rangle$ yields $m_{\overline{Z}\to G} \circ f \circ \langle\lambda';\lambda''\rangle = g \circ \langle\lambda';\lambda''\rangle$. This proves (together with the second equation) the stated parallel independence. □

If Theorem 1 is applied to two rule applications to the same object in such a way that both covers share one of the component rules, then one gets the following observation by using the known results for sequentialization and parallelization.

Observation 1. Let $d_1 = (G \underset{p_1}{\Longrightarrow} H_1)$ and $d_2 = (G \underset{p_2}{\Longrightarrow} H_2)$ be two rule applications with matchings g_1 and g_2, respectively, and $p_1' + p_1''$ and $p_2' + p_2''$ covers of p_1 and p_2, respectively. Let $G \underset{p_1'+p_1''}{\Longrightarrow} H_1$ and $G \underset{p_2'+p_2''}{\Longrightarrow} H_2$ be the corresponding rule applications according to Theorem 1.1 and $g_1 \circ \lambda_1'' = g_2 \circ \lambda_2''$ where $\lambda_i''\colon L_i'' \to L_i$ for $i = 1, 2$ are the morphisms according to Definition 1. Then the following holds.

1. The sequentializations yield

$$G \underset{p_1'}{\Longrightarrow} H_1' \underset{p_1''}{\Longrightarrow} H_1 \quad G \underset{p_1''}{\Longrightarrow} X_1 \underset{p_1'}{\Longrightarrow} H_1 \quad G \underset{p_2'}{\Longrightarrow} H_2' \underset{p_2''}{\Longrightarrow} H_2 \quad G \underset{p_2''}{\Longrightarrow} X_2 \underset{p_2'}{\Longrightarrow} H_2$$

   for some objects H_1', H_2', X_1 and X_2.
2. If $p_0 = p_1'' = p_2''$ and $g_1 \circ \lambda_1'' = g_2 \circ \lambda_2''$ where $\lambda_i''\colon L_i'' \to G$ for $i = 1, 2$ are the morphisms mapping the left hand sides of p_i'' to the left-hand side of p_i according to the Definition 1. Then $X = X_1 = X_2$ and $(G \underset{p_0}{\Longrightarrow} X) = (G \underset{p_1''}{\Longrightarrow} X_1) = (G \underset{p_2''}{\Longrightarrow} X_2)$ such that the rule applications form the following diagram

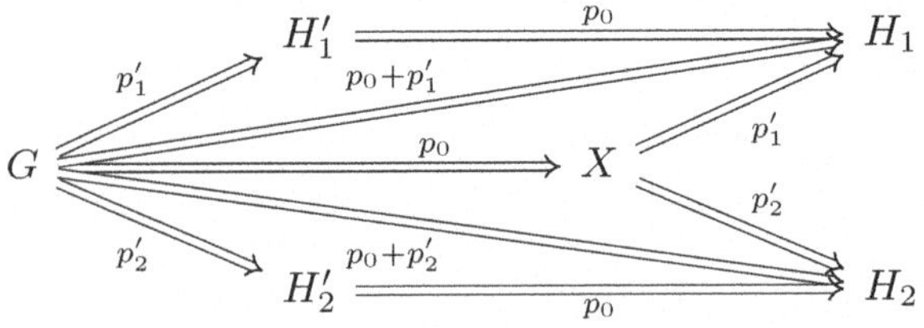

3. Moreover, $G \underset{p_0}{\Longrightarrow} X$ and $G \underset{p_1'}{\Longrightarrow} H_1'$ as well as $G \underset{p_0}{\Longrightarrow} X$ and $G \underset{p_2'}{\Longrightarrow} H_2'$ are parallel independent.
4. If, in addition, $G \underset{p_1'}{\Longrightarrow} H_1'$ and $G \underset{p_2'}{\Longrightarrow} H_2'$ are parallel independent, then the parallelization with $G \underset{p_0}{\Longrightarrow} X$ yields $G \underset{p_0+p_1'+p_2'}{\Longrightarrow} Y$ for some object Y and one obtains $G \underset{p_0+p_1'}{\Longrightarrow} H_1 \underset{p_2'}{\Longrightarrow} Y$ and $G \underset{p_0+p_2'}{\Longrightarrow} H_2 \underset{p_1'}{\Longrightarrow} Y$ by sequentialization.

The special role of p_0 is reflected in the following definition.

Definition 2. *Assume the situation of the previous observation. Moreover, let p_0 not be a span of isomorphisms.*

1. *Then* $G \underset{p_0}{\Longrightarrow} X$ *is called a* doubling *of* d_1 *and* d_2.
2. *If, in addition,* $G \underset{p'_1}{\Longrightarrow} H'_1$ *and* $G \underset{p'_2}{\Longrightarrow} H'_2$ *are parallel independent, then the doubling is* useful.

Remark 1. If a rule is a span of isomorphisms, then its application derives an object that is isomorphic to the host object meaning that is has no effect and – in particular – nothing is doubled.

Parallelization with dedoubling is captured in the following theorem.

Theorem 2. *Let* $G \underset{p_0}{\Longrightarrow} X$ *be a useful doubling of* $G \underset{p_1}{\Longrightarrow} H_1$ *and* $G \underset{p_2}{\Longrightarrow} H_2$ *with the corresponding cover components* $G \underset{p'_1}{\Longrightarrow} H'_1$ *and* $G \underset{p'_2}{\Longrightarrow} H'_2$, *respectively.*

1. *Then* $G \underset{p_1}{\Longrightarrow} H_1$ *and* $G \underset{p'_2}{\Longrightarrow} H'_2$ *as well as* $G \underset{p'_1}{\Longrightarrow} H'_1$ *and* $G \underset{p_2}{\Longrightarrow} H_2$ *are parallel independent.*
2. *Therefore, the parallelizations* $G \underset{p_1+p'_2}{\Longrightarrow} Y_{12}$ *for some object* Y_{12} *and* $G \underset{p_2+p'_1}{\Longrightarrow} Y_{21}$ *for some object* Y_{21} *are defined inducing the sequentialization*

$$G \underset{p_1}{\Longrightarrow} H_1 \underset{p'_2}{\Longrightarrow} Y_{12} \quad G \underset{p'_2}{\Longrightarrow} H'_2 \underset{p_1}{\Longrightarrow} Y_{12} \quad G \underset{p_2}{\Longrightarrow} H_2 \underset{p'_1}{\Longrightarrow} Y_{21} \quad G \underset{p'_1}{\Longrightarrow} H'_1 \underset{p_2}{\Longrightarrow} Y_{21}.$$

3. *Moreover,* $Y_{12} \cong Y_{21}$.

Proof. 1. According to the reasoning in Observation 1, the rule applications $G \underset{p_0+p'_2}{\Longrightarrow} H_2$ and $G \underset{p'_1}{\Longrightarrow} H'_1$ are parallel independent. Let

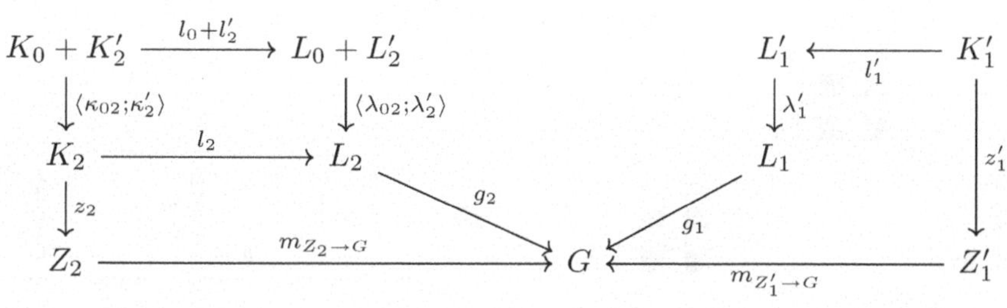

be the corresponding left pushouts. Then there are morphisms $f_{12}\colon L'_1 \to Z_2$ and $f_{21}\colon L_0 + L'_2 \to Z'_1$ such that

$$\text{(a) } m_{Z_2\to G} \circ f_{12} = g_1 \circ \lambda'_1$$
$$\text{(b) } m_{Z'_1\to G} \circ f_{21} = g_2 \circ \langle\lambda_0; \lambda'_2\rangle.$$

Let $g_2 = \overline{m} \circ \overline{e}$ for $\overline{e}\colon L_2 \to \overline{L}$ and $\overline{m}\colon \overline{L} \to G$ be the epi-mono factorization of g_2 and $f_{21} = \hat{m} \circ \hat{e}$ with $\hat{e}\colon L_0 + L'_2 \to \hat{L}$ and $\hat{m}\colon \hat{L} \to Z'_1$ be the epi-mono factorization of f_{21}. Then one gets the epi-mono factorization of the morphism of (b) in two variants by sequential composition with the monomorphism $m_{Z'_1 \to G}$ on one hand and with the epi $\langle \lambda_{02}; \lambda'_2 \rangle$ on the other hand, i.e., $m_{Z'_1 \to G} \circ \hat{m} \circ \hat{e} = \overline{m} \circ \overline{e} \circ \langle \lambda_{02}; \lambda'_2 \rangle$. The uniqueness of epi-mono factorization implies that, without loss of generality, $\overline{L} = \hat{L}$, $m_{Z'_1 \to G} \circ \hat{m} = \overline{m}$ and $\hat{e} = \overline{e} \circ \langle \lambda_{02}; \lambda'_2 \rangle$. Therefore, one get $m_{Z'_1 \to G} \circ \hat{m} \circ \overline{e} = \overline{m} \circ \overline{e} = g_2$. Combined with (a), this proves the parallel independence of $G \underset{p_2}{\Longrightarrow} H_2$ and $G \underset{p'_1}{\Longrightarrow} H'_1$.

The second case follows by symmetry.

2. The second statement follows from the first one.
3. Using Theorem 1, the first steps of $G \underset{p_1}{\Longrightarrow} H_1 \underset{p'_2}{\Longrightarrow} Y_{12}$ and $G \underset{p_2}{\Longrightarrow} H_2 \underset{p'_1}{\Longrightarrow} Y_{21}$ can be replaced by $G \underset{p_0+p'_1}{\Longrightarrow} H_1$ and $G \underset{p_0+p'_2}{\Longrightarrow} H_2$ respectively yielding two of the sequentializations of $G \underset{p_0+p'_1+p'_2}{\Longrightarrow} Y$ as shown in Observation 1. This means that $Y_{12} \cong Y \cong Y_{21}$. □

Example 2. We continue the discussion in Example 1. After the parallel step in Fig. 1c, the two parallel independent steps are depicted on the left in Fig. 2a. The common component rule p_0 is given at the top in Fig. 2b. It is applied to the middle of the hypotenuse. The complementary component rule p'_1 of the upper application of rule p_1 is given in the middle in Fig. 2b, and the complementary component rule p'_2 of the lower application of rule p_2 at the bottom in Fig. 2b. They are applied in such a way that all three rule applications overlap in the vertex corresponding to the middle hypotenuse so that they are pairwise parallel dependent. (p_0, p'_1) covers p_1, and (p_0, p'_2) covers p_2 so that the application of p_0 is a useful doubling, $p_0 + p'_i$ replaces p_i for $i = 1, 2$ and $p_0 + p'_1 + p'_2, p_1 + p'_2$ and $p_2 + p'_1$ can be applied to the start pattern in Fig. 2a all deriving the pattern on the right.

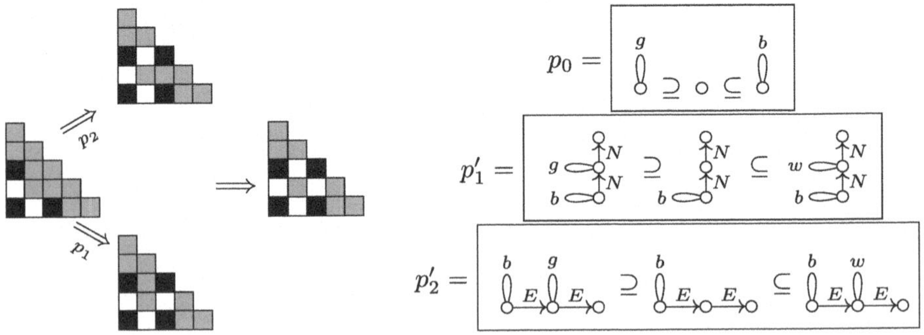

(a) Parallel rule application with dedoubling

(b) derived rules from dedoubling

Fig. 2. Sample derivations and rules of Example 2

Remark 2. Two rule applications d and d' to the same object are not parallel independent if at least one of the matching morphisms - say the matching morphism of d - cannot be restricted to the intermediate object of the other rule application d'. Intuitively, this means that d matches items that are deleted by d'. There are two cases: (1) d deletes the same items, or (2) d keeps some of them as part of the gluing object. In the second case, one faces an essential conflict of actions that are mutually exclusive. In the first case, deletion cannot be done twice. But if the deletion takes place at all, the purposes of both rule applications are served. With respect to deletion, this is the idea of parallelization with dedoubling. With respect to insertion, the situation is different as one has the choice between doubling or dedoubling.

4 Parallelization with Doubling Avoidance of Families of Rule Applications

In this section, the parallelization with dedoubling of two rule applications is generalized to families of rule applications. The parallelization with dedoubling as introduced in the previous section is used as an operator on families of rule applications so that it can be iterated. The operator chooses two rule applications of the family with a useful doubling non-deterministically and replaces one of the rule applications by its complement rule application. As shown in Theorem 2, the other rule application and the complement rule application are parallel independent. Moreover we show that the operator preserves parallel independence in the sense that a third rule application of the family which is parallel independent of the replaced rule application is also parallel independent of the replacing one. Therefore, the number of pairs of rule applications of the family cannot increase, and it decreases if the processed pair is not parallel independent. If one is interested in getting rid of the dependent pairs and applies the operator only to those, the iteration terminates either with a family of pairwise parallel independent rule applications or one gets stuck because there remain pairs without useful doublings. In the first case, the family can be parallelized. In the second case, parallelization fails.

Definition 3. *1. Let $\mathcal{F}$ be the class of all finite families of rule applications of the form $F = \{G \underset{p_i}{\Longrightarrow} H_i\}_{i \in I}$ for some common start object G and a finite index set I. Then the non-deterministic* dedoubling *operator $DD \in \mathcal{F} \times \mathcal{F}$ is defined as follows: Let $F = \{G \underset{p_i}{\Longrightarrow} H_i\}_{i \in I}, \overline{F} = \{G \underset{\overline{p}_i}{\Longrightarrow} \overline{H}_i\}_{i \in I} \in \mathcal{F}$. Then $(F, \overline{F}) \in DD$ if there are $j, k \in I$ and a useful doubling $G \underset{p_0}{\Longrightarrow} X$ of $G \underset{p_j}{\Longrightarrow} H_j$ and $G \underset{p_k}{\Longrightarrow} H_k$ with the complement component $G \underset{p'_j}{\Longrightarrow} H'_j$ of $G \underset{p_j}{\Longrightarrow} H_j$ such that $\overline{F}$ is obtained from F by replacing $G \underset{p_j}{\Longrightarrow} H_j$ by $G \underset{p'_j}{\Longrightarrow} H'_j$. $(F, \overline{F}) \in DD$ is denoted by $F \rightsquigarrow \overline{F}$.*

2. *As a binary relation of $\mathcal{F}$, DD can be iterated $F_0 \rightsquigarrow F_1 \rightsquigarrow \cdots \rightsquigarrow F_n$ for some $n \in \mathbb{N}$. This is denoted by $F \overset{n}{\rightsquigarrow} \overline{F}$ with $F = F_0$ and $\overline{F} = F_n$ or by $F \overset{*}{\rightsquigarrow} \overline{F}$ if the length of the dedoubling sequence is not needed explicitly.*

An interesting question concerning iterated processes is about termination. In many cases, a sufficient condition that guarantees termination is a natural variable which decreases in every process step. In the case of dedoubling, this applies if one wants to get rid of parallel dependence because the processed pair of rule applications results in a parallel independent pair (cf. Theorem 2). Moreover, it can be shown that dedoubling preserves parallel independence.

Theorem 3. *Let $G \underset{p_i}{\Longrightarrow} H_i$ for $i = 1, 2, 3$ be rule applications. Let $G \underset{p_0}{\Longrightarrow} X$ be a useful dedoubling of $G \underset{p_1}{\Longrightarrow} H_1$ and $G \underset{p_2}{\Longrightarrow} H_2$ with the corresponding cover component $G \underset{p_1'}{\Longrightarrow} H_1'$ of $G \underset{p_1}{\Longrightarrow} H_1$. Let $G_1 \underset{p_1}{\Longrightarrow} H_1$ and $G \underset{p_3}{\Longrightarrow} H_3$ be parallel independent. Then $G \underset{p_1'}{\Longrightarrow} H_1'$ and $G \underset{p_3}{\Longrightarrow} H_3$ are parallel independent, too.*

Proof. Consider the following diagram

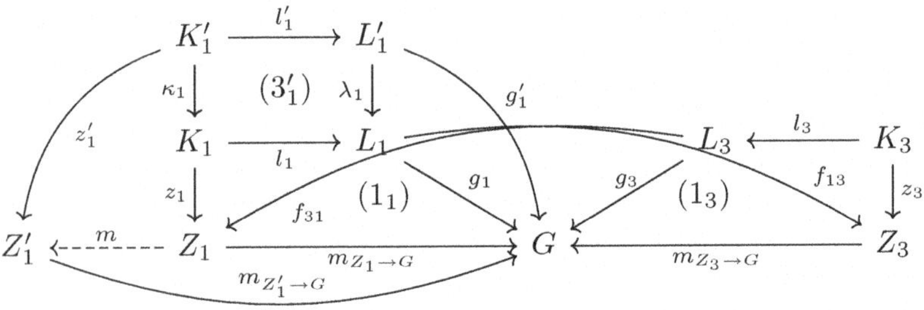

$G \underset{p_1}{\Longrightarrow} H_1$ and $G \underset{p_3}{\Longrightarrow} H_3$ provide the pushouts (1_1) and (1_3). As they are parallel independent, there are morphisms $f_{13}\colon L_1 \to Z_3$ and $f_{31}\colon L_3 \to Z_1$ such that (a) $m_{Z_3 \to G} \circ f_{13} = g_1$ and (b) $m_{Z_1 \to G} \circ f_{31} = g_3$. $G \underset{p_1'}{\Longrightarrow} H_1'$ provides the pushout depicted on the left outer square above. Choosing (c) $f_{13}' = f_{13} \circ \lambda_1$ yields

$$\text{(d) } m_{Z_3 \to G} \circ f_{13}' \underset{(c)}{=} m_{Z_3 \to G} \circ f_{13} \circ \lambda_1 \underset{(a)}{=} g_1 \circ \lambda_1 = g_1'.$$

If there is a morphism $m\colon Z_1 \to Z_1'$ with (e) $m_{Z_1' \to G} \circ m = m_{Z_1 \to G}$, then choosing (f) $f_{31}' = m \circ f_{31}$ yields

$$\text{(g) } m_{Z_1' \to G} \circ f_{31}' \underset{(f)}{=} m_{Z_1' \to G} \circ m \circ f_{31} \underset{(e)}{=} m_{Z_1 \to G} \circ f_{31} \underset{(b)}{=} g_3.$$

(d) and (g) show that $G \underset{p_1'}{\Longrightarrow} H_1'$ and $G \underset{p_3}{\Longrightarrow} H_3$ are parallel independent.

The morphism m is obtained by applying the Butterfly Lemma to the left pushout of $G \underset{p_1'+p_0}{\Longrightarrow} H_1$:

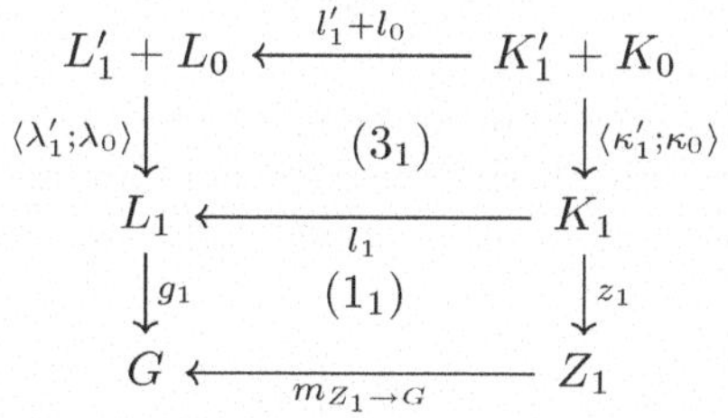

implies

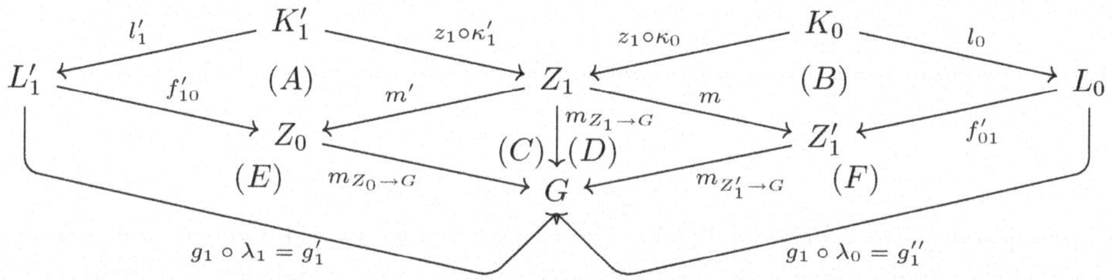

where the diagram $(C)+(D)$ is a pushout. As l_1' is a monomorphism, the pushout complement of $g_1' \circ l_1'$ is unique (up to isomorphism). Moreover, (D) is commutative so that $m\colon Z_1 \to Z_1'$ is the morphism needed above. This completes the proof. □

Definition 4. *Let* $F = \{G \underset{p_i}{\Longrightarrow} H_i\}_{i \in I} \in \mathcal{F}$. *Then* $di(F)$ *is the number of pairs of* F *that are parallel dependent, called* dependence index *of* F.

Corollary 1. *1. Let* $F \rightsquigarrow \overline{F}$ *for* $F, \overline{F} \in \mathcal{F}$ *and the processed pair of rule applications be parallel dependent. Then* $di(F) > di(\overline{F})$.

2. Let $F_0 \overset{n}{\rightsquigarrow} F_n$ *be a sequence of dedoublings where all processed pairs are parallel dependent. Then* $di(F_0) \geq n$.

Proof. 1. According to Theorem 3, $di(F)$ does not increase after dedoubling. According to Theorem 2, it decreases at least by 1 wrt the processed pair.

2. According to Point 1, the length of a dedoubling is bounded by $di(F_0)$ if only dependent pairs are processed. □

The situation is more difficult if the processed pairs of dedoublings are parallel independent because no general termination criterion is known. But whenever a family is reached by a sequence of dedoublings such that each pair of members is parallel independent, one may stop the process. The result provides a parallelization with doubling avoidance of the initial family.

Another interesting question about iterated dedoubling concerns the semantic relation between a given family of rule applications and one obtained by a sequence of dedoublings. A family member remains unchanged or is repeatedly replaced by a component rule application that is complementary to the removed component rule application. A component is only removed if it is a doubling so that it remains present in another member of the family and no information gets lost. This is reflected in the following theorem stating that each member of the start family is not changed or can be reconstructed as parallel rule application of the resulting member and all the removed component rule applications that are involved in the processing of all members with the same index as the initial member.

Theorem 4. *Let* $F = \{d_i = (G \underset{p_i}{\Longrightarrow} H_i)\}_{i \in I}, \overline{F} = \{\overline{d}_i = (G \underset{\overline{p}_i}{\Longrightarrow} \overline{H}_i)\}_{i \in I} \in \mathcal{F}$ *with* $dd = (F \overset{*}{\rightsquigarrow} \overline{F})$*. Then for each* $i \in I$*, either* $d_i = \overline{d}_i$ *or there is a parallelization* $G \underset{(\sum_{l=1}^{m(i)} p_{0l}) + \overline{p}_i}{\Longrightarrow} H_i$ *of* $G \underset{p_i}{\Longrightarrow} \overline{H}_i$ *and the dedoublings* $G \underset{p_{0l}}{\Longrightarrow} H_{0l}$ *for* $l = 1, \ldots, m(i)$ *which are used to diminish the ith component of dd.*

Proof. Induction on the length of dd.

Base for $F \overset{0}{\rightsquigarrow} \overline{F}$: Then $F = \overline{F}$ and $d_i = \overline{d}_i$ for all $i \in I$.

Step for $(F \overset{n+1}{\rightsquigarrow} \overline{F}) = (F \overset{n}{\rightsquigarrow} \hat{F} \overset{1}{\rightsquigarrow} \overline{F})$ for some $\hat{F}$: Then the last step is given by a dedoubling of $\hat{d}_j = (G \underset{\hat{p}_j}{\Longrightarrow} \hat{H}_j)$ and $\hat{d}_k = (G \underset{\hat{p}_k}{\Longrightarrow} \hat{H}_k)$ with a useful dedoubling $\hat{d}_0 = (G \underset{\hat{p}_0}{\Longrightarrow} \hat{H}_0)$ corresponding cover components $\hat{d}'_j = (G \underset{\hat{p}'_j}{\Longrightarrow} \hat{H}'_j)$ and $\hat{d}'_k = (G \underset{\hat{p}'_k}{\Longrightarrow} \hat{H}'_k)$ as well as $\overline{d}_j = \hat{d}'_j$ and $\overline{d}_k = \hat{d}'_k$. This means for $i \neq j$ that $\overline{d}_i = \hat{d}_i$ so hat the statement holds by induction hypothesis. For $j \in I$, one has (a) $d_j = \hat{d}_j$ and (b) $D_j = (G \underset{(\sum_{l=1}^{m(j)} p_{0l}) + \hat{p}_j}{\Longrightarrow} \hat{H}_j)$ being the parallelization of $\hat{d}_j = (G \underset{\hat{p}_j}{\Longrightarrow} \hat{H}_j)$ and the dedoubling $d_{0l} = (G \underset{p_{0l}}{\Longrightarrow} H_{0l})$ for $l = 1, \ldots, m(j)$ by induction hypothesis. By construction, one has $\hat{d}_j = (G \underset{\hat{p}_0 + \hat{p}'_j}{\Longrightarrow} \hat{H}_j)$. The first step of the sequentialization of $\hat{D}_j$ are $\hat{d}'_j = (G \underset{\hat{p}'_j}{\Longrightarrow} \hat{H}'_j)$ and $\hat{d}_i = (G \underset{\hat{p}}{\Longrightarrow} \hat{H}_j)$. In the case (a), this proves the statement. In the case (b), $\hat{d}_j$ and all d_{0l} are pairwise parallel independent as components of a parallelization. $\hat{D}_j$ covers $\hat{d}_j$ so that $\hat{D}_j$ is parallel independent of all d_{0l} according to Theorem 2. Therefore, the parallelization $(G \underset{(\sum_{l=1}^{m(j)} p_{0l}) + \hat{p}_0 + \hat{p}_j}{\Longrightarrow} H_j)$ is defined proving the statement for the case (b) and completing the proof. □

5 Relating Parallelization with Doubling Avoidance to Amalgamation

Boehm, Fonio and Habel [4] introduced and studied amalgamation of graph transformation as a generalization of parallelization of parallel independent rule applications. Their approach is generalized to adhesive categories in Ehrig et al. [9]. They consider the same basic situation as in our approach, i.e., two rules p_1 and p_2 sharing the same component rule p_0 together with the embedding morphisms $e_{01}\colon p_0 \to p_1$ and $e_{02}\colon p_0 \to p_2$ both accompanied with complementary component rules p_1' and p_2' resp., which are called remainders. Then the amalgamated rule p_{12} is defined as the pushout of the span given by e_{01} and e_{02} (provided that it exists). The Amalgamation Theorem states the following: Let $G \underset{p_1}{\Longrightarrow} H_1$ and $G \underset{p_2}{\Longrightarrow} H_2$ be two rule applications the matching morphisms of which coincide on the embeddings of the left-hand side of p_0. Moreover, let $G \underset{p_1'}{\Longrightarrow} H_1'$ and $G \underset{p_2'}{\Longrightarrow} H_2'$ be parallel independent rule applications defined by the matching morphisms that are composed of the given matching morphisms and the embeddings of left-hand sides of p_1' and p_2' into the left-hand sides of p_1 and p_2 resp. Then there are derivations $G \underset{p_{12}}{\Longrightarrow} Y, G \underset{p_1}{\Longrightarrow} H_1 \underset{p_2'}{\Longrightarrow} Y$ and $G \underset{p_2}{\Longrightarrow} H_2 \underset{p_1'}{\Longrightarrow} Y$ for some object Y. We show that the amalgamated rule p_{12} is covered by our parallel rule $p_0 + p_1' + p_2'$ (Theorem 5) such that the Amalgamation Theorem turns out to be a kind of corollary of Theorem 2. The proof is based on the relation among the component rules of a useful doubling as summarized in Observation 2.

Observation 2. The following diagram displays the relations among the rules that are involved in a useful doubling where the given epimorphisms are denoted by e_1 and e_2, the injections into coproducts are indicated by i with proper indices and the restrictions of e_1 and e_2 to the coproduct components by e with proper indices. Moreover, the pushout p_{12} of the span given by the common component rule p_0 of p_1 and p_2 with the embeddings $e_{0k}\colon p_0 \to p_k$ for $k = 1, 2$ is added. The pushout morphisms are denoted by $f_k\colon p_k \to p_{12}$ for $k = 1, 2$. And $f_0\colon p_0 \to p_{12}$ is defined as $f_1 \circ e_{01} = f_2 \circ e_{02}$. Altogether all triangles are commutative.

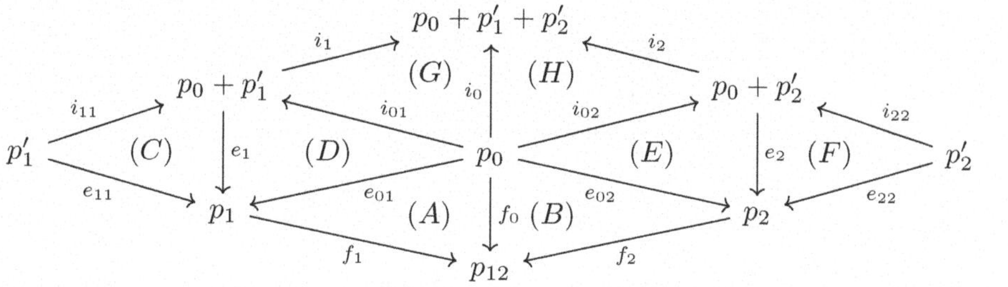

Using the coproduct property of $p_0 + p_1' + p_2'$ the morphisms $f_1 \circ e_{11}, f_0 = f_1 \circ e_{01} = f_2 \circ e_{02}$ and $f_2 \circ e_{22}$ induce $e\colon p_0 + p_1' + p_2' \to p_{12}$ with

$$\begin{aligned}
&\text{(I)} \quad e \circ i_k \circ i_{kk} = f_k \circ e_{kk}, \quad k = 1,2\\
&\text{(J)} \quad e \circ i_2 \circ i_{02} = e \circ i_1 \circ i_{01} = f_1 \circ e_{01} = f_2 \circ e_{02}\\
&\text{(K)} \quad e \circ i_0 = f_0.
\end{aligned}$$

Theorem 5. *Let $G \underset{p_0}{\Longrightarrow} X$ be a useful doubling of $G \underset{p_1}{\Longrightarrow} H_1$ and $G \underset{p_2}{\Longrightarrow} H_2$ for some rule p with the embeddings $e_{0i}\colon p_0 \to p_i$ and the corresponding complementary rule applications $G \underset{p_i'}{\Longrightarrow} H_i'$ for $k = 1,2$. Let p_{12} with the rule morphisms $f_i\colon p_i \to p_{12}$ for $k = 1,2$ be the pushout of the span $p_1 \underset{e_{01}}{\longleftarrow} p_0 \underset{e_{02}}{\longrightarrow} p_2$ and $e\colon p_0 + p_1' + p_2' \to p_{12}$ be the rule morphism constructed in the observation above. Then*

1. *e is an epimorphism and*
2. *the diagram*

$$\begin{array}{ccccc}
L_0 + L_1' + L_2' & \xleftarrow{l_0 + l_1' + l_2'} & K_0 + K_1' + K_2' & \xrightarrow{r_0 + r_1' + r_2'} & R_0 + R_1' + R_2' \\
{\scriptstyle e_L}\downarrow & & \downarrow{\scriptstyle e_K} & & \downarrow{\scriptstyle e_R} \\
L_{12} & \xleftarrow[l_{12}]{} & K_{12} & \xrightarrow[r_{12}]{} & R_{12}
\end{array}$$

is a double pushout.

Proof. 1. Let $f, f'\colon p_{12} \to p$ with (X) $f \circ e = f' \circ e$. Then using equations C, F, I and X and D, E, B, A, K and X respectively, one gets $f \circ f_k \circ e_k \circ i_{kk} = f' \circ f_k \circ e_k \circ i_{kk}$ and $f \circ f_k \circ e_k \circ i_{0k} = f' \circ f_k \circ e_k \circ i_{0k}$ for $k = 1,2$. Using the coproduct property of $p_0 + p_k'$ one gets $f \circ f_k \circ e_k = f' \circ f_k \circ e_k$. As e_k is an epimorphism this implies

$$(Y_k) \quad f \circ f_k = f' \circ f_k.$$

Summarizing, one gets

$$\begin{array}{ccccc}
f \circ f_1 \circ e_{01} & \underset{A}{=} & f \circ f_0 & \underset{B}{=} & f \circ f_2 \circ e_{02} \\
{\scriptstyle Y_1}\| & & & & {\scriptstyle Y_2}\| \\
f' \circ f_1 \circ e_{01} & \underset{A}{=} & f' \circ f_0 & \underset{B}{=} & f' \circ f_2 \circ e_{02}
\end{array} .$$

Because $(A) + (B)$ is a pushout, this implies $f = f'$ such that e is an epimorphism.

2. Consider the following diagram for $k = 1,2$:

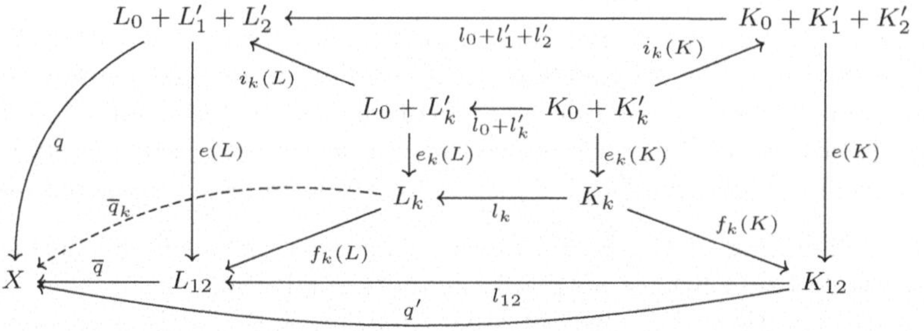

where the square in the middle is a pushout and the squares surrounding the pushout commute. Let q, q' with $q \circ (l_0 + l'_1 + l'_2) = q' \circ e(K)$ for some object X. Then this implies

$$q \circ i_k(L) \circ (l_0 + l'_k) = q \circ (l_0 + l'_1 + l'_2) \circ i_k(K) = q' \circ e(K) \circ i_k(K) = q' \circ f_k(K) \circ e_k(K).$$

The pushout property of L_k yields that there is a morphism $\overline{q}_k : L_k \to X$ with $q \circ i_k(L) = \overline{q}_k \circ e_k(L)$ and $q' \circ f_k(K) = \overline{q}_k \circ l_k$. This implies

$$\begin{aligned} & \overline{q}_1 \circ e_{01}(L) = \overline{q}_1 \circ e_1(L) \circ i_{01}(L) = q \circ i_1(L) \circ i_{01}(L) = q \circ i_0(L) \\ = \; & q \circ i_2(L) \circ i_{02}(L) = \overline{q}_2 \circ e_2(L) \circ i_{02}(L) = \overline{q}_2 \circ e_{02}(L). \end{aligned}$$

The pushout property of L_{12} (as component of p_{12}) yields that there is a morphism $\overline{q}\colon L_{12} \to X$ with $\overline{q} \circ f_k(L) = \overline{q}_k$. It remains to prove $\overline{q} \circ e(L) = q$ and $\overline{q} \circ l_{12} = q'$. Using the available commutativities, one gets, for $k = 1, 2$,

$$\overline{q} \circ e(L) \circ i_k(L) = \overline{q} \circ f_k(L) \circ e_k(L) = \overline{q}_k \circ e_k(L) = q \circ i_k(L)$$

and

$$\overline{q} \circ e(L) \circ i_k(L) = \overline{q} \circ f_0 = \overline{q} \circ f_k(L) \circ e_{0k}(L) = \overline{q}_k \circ e_{0k} = q \circ i_k(L).$$

As $L_0 + L'_1 + L'_2$ is a coproduct, this implies $\overline{q} \circ e(L) = q$. Moreover, one gets

$$\overline{q} \circ l_{12} \circ f_k(K) = \overline{q} \circ f_k(L) \circ l_k = \overline{q}_k \circ l_k = q' \circ f_k(K)$$

such that the pushout property of K_{12} implies $\overline{q} \circ l_{12} = q'$. Both together prove the stated equations. Because of the symmetry of double pushouts, the same reasoning works for the right-hand side completing the proof. □

Related Work 1.

1. As the coproduct is associative and commutative, the theorem shows that the pairs $(p_0, p'_1 + p'_2)$, $(p_0 + p'_1, p'_2)$ and $(p_0 + p'_2, p'_1)$ are covers of p_{12}. Therefore, Theorem 2 can be used showing that $p_0 + p'_1 + p'_2$ plays the same role as the amalgamated rule p_{12} in the Amalgamation Theorem (see [4,9]).

2. Let $F = \{G \underset{p_i}{\Longrightarrow} H_i\}_{i \in I} \in \mathcal{F}$ be a family of rule applications and $g_i \colon L_i \to G$ be the left matching morphisms for $i \in I$ with the particular property that each two members share a useful doubling, meaning that there is a common component rule p_0 and complement rules p'_i for $i \in I$ such that (p_0, p'_i) covers p_i. Then, obviously, $n - 1$ dedoublings are possible if I has n elements. Moreover $G \underset{p_0 + \sum_{i \in I} p'_i}{\Longrightarrow} Y$ for some object Y is a parallelization with dedoubling of the initial family. This very situation is considered and investigated by Ehrig et al. [9,13] as multi-amalgamation.
3. It turns out that the effect of amalgamation and multi-amalgamation can be obtained by means of particular sequentializations and parallelizations leading to parallelization with dedoubling. In both approaches, one needs to find special component rules (a common one and complements). One may argue that sequentialization and parallelization are slightly simpler and more basic constructions than amalgamation. Moreover, the general amalgamated rule does not always exist in adhesive categories as arbitrary pushouts do not exist in general. A further difference between both approaches concerns the scaling from two rule applications to an arbitrary number of them. Multi-amalgamation in [9,13] is only considered for the special case that all rule applications delete the same part of the host object.

6 Conclusion

In this paper, we have studied the parallelization of rule applications that are constructed by double-pushouts in adhesive categories. The well-known parallelization of parallel independent rule applications by means of coproducts of rules used as parallel rules is generalized in such a way that doubling can be avoided. If the dependence of two rule applications to the same object is caused by the fact that - intuitively - both delete that same part of the commonly accessed part, then independence can be restored by removing the deletion part of one of the rule applications provided that there are proper component rules complementing the deletion parts. Similarly, one can avoid that certain insertions are doubled. We have given sufficient conditions to guarantee that doubling avoidance - called dedoubling for short - works. The construction combines the sequentialization of parallel rule applications and certain re-parallelizations of the resulting rule applications. As a main result, we have shown that dedoubling within a family of rule applications preserves parallel independence. Consequently, a sequence of dedoubling steps yields a family of pairwise parallel independent rule applications that can be applied in parallel, or it gets stuck because none of the pairs of rule applications allow a further dedoubling.

To obtain a better insight into the parallelization with doubling avoidance, the following aspects may be considered:

1. A dedoubling of two rule applications requires to cover them by parallel rules with a common component rule. To find such covers is guesswork so far. Therefore, it would be helpful to come up with a systematic construction of

covers. According to the definition, the component rules are embedded into the given rules by arbitrary rule morphisms. If one assumes monomorphic embeddings and finitary categories, then there is only a finite number of component rules so that all covers can be enumerated.
2. If a family of rule applications has two members $G \underset{p_1}{\Longrightarrow} H_1$ and $G \underset{p_2}{\Longrightarrow} H_2$, then dedoubling may not be possible or dedoubling yields either $G \underset{p_1'}{\Longrightarrow} H_1'$ and $G \underset{p_2}{\Longrightarrow} H_2$ or $G \underset{p_1}{\Longrightarrow} H_1$ and $G \underset{p_2'}{\Longrightarrow} H_2'$. In Sect. 3, it is shown that $G \underset{p_1'+p_2}{\Longrightarrow} Y, G \underset{p_1+p_2'}{\Longrightarrow} Y$, and $G \underset{p_0+p_1'+p_2'}{\Longrightarrow} Y$ for some object Y. This means that both dedoublings are closely related from a semantic point of view. We expect that a result like this holds for arbitrary families and arbitrary sequences of dedoublings that does not require to store the doublings and is stronger than Theorem 4 in this way.
3. A dedoubling applied to two parallel dependent rule applications yields parallel independence so that all dedoublings have the same effect with respect to the deletion part. Dedoubling with respect to the insertion part behaves quite different as one has various options between a common component rule that does not insert anything and some insertion part that identifies the maximal mergable parts of the two involved right-hand sides. The question is whether there are more interesting standard cases between the two extremes.
4. It would be interesting to study applications of the proposed concepts. As our example in Sects. 3 and 4 indicates, the area of pattern generation is a potential candidate as many patterns, tilings, and fractals can be generated in a parallel mode of construction. Rule-based collaborative text editing (e.g. refactoring source code by multiple people) fits into this view. The authors of [13–15] have pointed out that amalgamation and multi-amalgamation are very helpful concepts for the specification of model transformations. Therefore, it may be interesting to see whether parallelization with dedoubling can play a similar role.

Acknowledgements. We are grateful to the anonymous reviewers for their valuable comments that led to various improvements.

Disclosure of Interests. The authors have no competing interests to declare that are relevant to the content of this article.

References

1. Taentzer, G.: Parallel and distributed graph transformation: formal description and application to communication-based systems. Ph.D. thesis, TU Berlin, Shaker Verlag (1996)
2. Kniemeyer, O.: Design and implementation of a graph grammar based language for functional-structural plant modelling. Ph.D. thesis, Brandenburg University of Technology, Cottbus - Senftenberg, Germany (2008)
3. Métivier, Y., Sopena, E.: Graph relabelling systems: a general overview. Comput. Artif. Intell. **16**(2), 167–185 (1997)

4. Boehm, P., Fonio, H.-R., Habel, A.: Amalgamation of graph transformations with applications to synchronization. In: Ehrig, H., Floyd, C., Nivat, M., Thatcher, J. (eds.) CAAP 1985. LNCS, vol. 185, pp. 267–283. Springer, Heidelberg (1985). https://doi.org/10.1007/3-540-15198-2_17
5. Ehrig, H., Kreowski, H.-J.: Parallelism of manipulations in multidimensional information structures. In: Mazurkiewicz, A. (ed.) MFCS 1976. LNCS, vol. 45, pp. 284–293. Springer, Heidelberg (1976). https://doi.org/10.1007/3-540-07854-1_188
6. Kreowski, H.-J.: Manipulationen von Graphmanipulationen. Ph.D. thesis, Technische Universität Berlin (1978)
7. Corradini, A., Montanari, U., Rossi, F., Ehrig, H., Heckel, R., Löwe, M.: Algebraic approaches to graph transformation part I: basic concepts and double pushout approach. In: Rozenberg, G. (ed.) Handbook of Graph Grammars and Computing by Graph Transformation, Vol. 1: Foundations, pp. 163–245. World Scientific, Singapore (1997)
8. Ehrig, H., Ehrig, K., Prange, U., Taentzer, G.: Fundamentals of Algebraic Graph Transformation. Monographs in Theoretical Computer Science. An EATCS Series. Springer, Heidelberg (2006). https://doi.org/10.1007/3-540-31188-2
9. Ehrig, H., Ermel, C., Golas, U., Hermann, F.: Graph and Model Transformation - General Framework and Applications. Monographs in Theoretical Computer Science. An EATCS Series. Springer, Heidelberg (2015). https://doi.org/10.1007/978-3-662-47980-3
10. Adámek, J., Herrlich, H., Strecker, G.E.: Abstract and Concrete Categories - The Joy of Cats. Dover Publications (2009)
11. Lack, S., Sobociński, P.: Adhesive and quasiadhesive categories. RAIRO Theor. Inform. Appl. **39**(3), 511–545 (2005)
12. Ehrig, H., Pfender, M., Schneider, H.J.: Graph-grammars: an algebraic approach. In: Proceedings of the 14th Annual Symposium on Switching and Automata Theory, Iowa City, Iowa, USA, 15–17 October 1973, pp. 167–180. IEEE Computer Society (1973)
13. Biermann, E., Ehrig, H., Ermel, C., Golas, U., Taentzer, G.: Parallel independence of amalgamated graph transformations applied to model transformation. In: Engels, G., Lewerentz, C., Schäfer, W., Schürr, A., Westfechtel, B. (eds.) Graph Transformations and Model-Driven Engineering. LNCS, vol. 5765, pp. 121–140. Springer, Heidelberg (2010). https://doi.org/10.1007/978-3-642-17322-6_7
14. Born, K., Taentzer, G.: An algorithm for the critical pair analysis of amalgamated graph transformations. In: Echahed, R., Minas, M. (eds.) ICGT 2016. LNCS, vol. 9761, pp. 118–134. Springer, Cham (2016). https://doi.org/10.1007/978-3-319-40530-8_8
15. Taentzer, G., Golas, U.: Towards local confluence analysis for amalgamated graph transformation. In: Parisi-Presicce, F., Westfechtel, B. (eds.) ICGT 2015. LNCS, vol. 9151, pp. 69–86. Springer, Cham (2015). https://doi.org/10.1007/978-3-319-21145-9_5

Granular Conflict Analysis for Transformation Rules with Application Conditions

Alexander Lauer(✉), Jens Kosiol, and Gabriele Taentzer

Philipps-Universität Marburg, Marburg, Germany
alexander.lauer@uni-marburg.de,
{kosiolje,taentzer}@mathematik.uni-marburg.de

Abstract. *Conflict and dependency analysis* is an important static analysis tool for gaining an overview of potential interactions of graph transformation rules. The analysis is based on *critical pairs*, which represent conflicting transformations in a minimal context. For many applications, however, the computation of complete critical pairs is unnecessarily complex. Instead, the crucial information about a conflicting pair of transformations is already contained in much smaller structures, called *conflict reasons* or *disabling essences* in existing research. So far, these structures have only been introduced for *plain rules*; in practical scenarios, however, transformation rules are regularly equipped with *(application) conditions*. In this paper, we fill this gap by lifting the computation of *disabling essences* to rules with conditions. It turns out that the situation is much more complex than for plain rules, where a pair of transformations is in conflict if and only if a non-trivial disabling essence embeds into the pair. However, we can show that, for every conflicting pair of transformations via rules with conditions, there is a disabling essence that embeds into this pair. For the special case of *negative application conditions* we can even show that a disabling essence is embeddable if and only if the transformation pair is in conflict. We work within the general context of adhesive HLR categories, which include several types of graph-like structures.

Keywords: Graph transformation · Critical pair analysis · Static analysis · Adhesive HLR categories

1 Introduction

In the algebraic approach to graph transformation [6,7], *(transformation) rules* declaratively describe changes of graphs. The approach has a broad range of interesting applications, especially in software engineering [13]. Put simply, a rule

This work was partially funded by the German Research Foundation (DFG), projects "Triple Graph Grammars (TGG) 3.0" and "Model-Driven Optimization in Software Engineering".

J. Endrullis and M. Tichy (Eds.): ICGT 2025, LNCS 15720, pp. 63–90, 2025.
https://doi.org/10.1007/978-3-031-94706-3_4

specifies (i) the local context that must exist for the rule to be applicable, (ii) the elements to be deleted from that local context upon application of the rule, and (iii) the elements to be created. In addition, a rule may have an *(application) condition* that expresses further conditions on the global context in which this rule is applicable. To apply a rule to a graph, an image of the local context required by the rule is searched for in the graph, it is validated whether or not the rule's condition is satisfied by the selected image, and, if so, the deletions and creations specified by the rule are performed.

Importantly, different applications of rules to a graph can interact with each other. The most basic interactions are called *conflicts* and *dependencies*. In a conflict, the application of one rule prevents another rule application, e.g., by creating an element that is forbidden to exist by the condition of the second rule. Analyzing rules for their potential interactions and, in particular, for potential conflicts has long been an important topic in graph transformation research (see Sect. 2). These analyses are part of an elaborate theory and also play a crucial role in several applications such as feature interaction detection, model versioning, test case generation, or graph parsing (see [23] for an overview of such applications).

Many analysis techniques aim to compute an overview of all potential conflicts for a given set of rules. A classical format for representing a conflict is a *critical pair*, which consists of two conflicting transformations in a minimal context [11]. However, a critical pair explicitly contains a large amount of information that is unnecessary in many applications and may overwhelm users [23]. Therefore, research has been conducted to extract only the core information about potential conflicts between rules [2,18,21,23]. Two approaches to computing only the core information of potential conflicts have recently been developed, using the terms *conflict reason* [18] and *conflict essence* [2], respectively, for this core information. Both approaches extract exactly the conflict-inducing elements (along with an embedding in the involved rules) – which is much lighter than a critical pair, without losing any important information. Both approaches have been developed in the very general setting of adhesive HLR categories. However, both only consider plain rules without conditions. Since practical applications usually involve rules with conditions, and such conditions can be involved in conflicts, this is a significant gap in the research. *In this paper, we fill this gap by extending the approach of Azzi et al.* [2] *to rules with conditions:*

1. We introduce *proto-essences (by deletion and insertion) for rules with conditions*, which contain core information about how one rule application can interfere with the validity of another rule's condition, indicating potential conflicts. Unlike the theory for plain rules, which only examines the context and deleted elements of the rules, we must also consider the created elements, since conditions can be violated either by inserting forbidden structure or by deleting structure required by the condition.
2. In order to detect whether a proto-essence by insertion occurs in a transformation pair (whether a forbidden structure has been inserted), the result object of one of the transformations must be inspected. To overcome this, we

introduce the notion of a *shifted proto-essence* and show that a proto-essence by insertion is embeddable in the result of one of the transformations if and only if its shifted proto-essence is embeddable before both transformations take place.

3. Finally, we define *disabling essences* by filtering proto-essences (excluding some cases that can never indicate a conflict) and show that, given a pair of conflicting transformations, there is a disabling essence that can be embedded in that pair. Conversely, however, in general the embeddability of a disabling essence does not imply that the transformations are in conflict. For the special case of *negative application conditions* (NACs), though, where the condition only specifies a forbidden structure, we show that a disabling essence is embeddable if and only if the corresponding transformations are in conflict.

The paper is structured as follows: In Sect. 2, we present related work on conflict analysis, and in Sect. 3 we introduce the example used throughout the paper. In Sect. 4, we present formal preliminaries, and in Sect. 5, we introduce *proto-essences by deletion (insertion)*, a precursor to *disabling essences*, in the setting of adhesive HLR categories. In Sect. 6, we introduce *shifted proto-essences* and *disabling essences* for rules with conditions, and show our results regarding the embedding of *disabling essences*. Section 7 concludes the paper and Appendix A contains the proofs of our propositions and theorems.

2 Related Work

We briefly review the research on *critical pairs* as this is the line of research that our work directly continues. Critical pairs, as known from term rewriting [15], have been adapted to different graph-like structures and to rules with conditions [10,11,25]. A critical pair is a pair of conflicting transformations in a minimal context. Given two rules, the set of their critical pairs is *complete* in the sense that, for every pair of conflicting transformations, there is a critical pair embedded in it [8, Theorem 6.1]. The Critical Pair Lemma states that strict confluence of critical pairs implies local confluence of a transformation system [8, Theorem 7.1].

Since critical pairs tend to be large objects, and even small transformation systems can induce large sets of critical pairs, two orthogonal research directions have been pursued: (i) reducing the number of critical pairs that have to be computed (in such a way that completeness and a corresponding Critical Pair Lemma still hold), and (ii) finding objects that contain the core information about conflicting transformations in a more compact way than critical pairs. With respect to (i), first *essential critical pairs* [20] and then *initial conflicts* [19] were introduced, finally leading to the result that initial conflicts are the smallest set of critical pairs that is still complete. This also holds for rules with conditions [22]. Orthogonally, Campbell and Plump introduced the notion of *confluence up to garbage*, where only a subset of critical pairs needs to be checked for strict confluence to guarantee local confluence of a transformation system, at least on a class of graphs of interest [4]. Regarding the core information of conflicts, [3] introduced the notions of *conflict atoms* and *(minimal) conflict reasons* as objects that consist of the structure that actually causes a conflict (or individual elements thereof in the case of atoms), together with an embedding into the underlying rules. This was followed by an alternative computation of the core information of critical pairs, so-called *disabling* and *conflict essences* [1], and by a more elaborate representation of conflict reasons, also relating them to initial conflicts [18]. A basic implementation and empirical study of this multi-granular conflict analysis was presented in [23] and algorithmic considerations for the computation of conflict reasons in [21]. Finally, an in-depth study of disabling and conflict essences was presented in [2]. For typed graphs, the notions of conflict reasons and conflict essences coincide [2, Corollary 6.9], but in general, conflict essences can be smaller and allow for an easier computation. Therefore, in this paper, we take up the construction by Azzi et al. [2] and extend it to the case of rules with conditions.

3 Running Example

To illustrate our work, we consider a snippet of a project administration model. Projects in different departments of a company are run by teams of project members. Project management is agile in the sense that new projects can be created and assigned to some department, members can be created and added to projects, and members can start and stop quality assurance (QA). These actions are specified in a graph transformation system, six rules of which are shown in Fig. 1.

Some of the displayed rules are equipped with *(application) conditions*. Often, such conditions serve to ensure that given constraints are not violated. For example, the two NACs of rule *addMember* (the blue structures annotated with forbid) serve to prevent that a Member is allocated to multiple Projects or multiple times to the same Project. A more complex condition for rule *stopQA* is shown in Fig. 2. This condition serves to ensure that in every Department there is a Project whose Members all perform QA. In its presentation, we omit morphisms and only present their codomain graphs, whenever the domain graph can be unambiguously inferred. The morphisms are indicated via the names of elements such as m1 and q1. We will use these rules and this condition to illustrate our approach.

4 Preliminaries

In this section, we present the preliminaries of our work. We focus on *adhesive HLR categories*, *transformation rules with (application) conditions* and on *parallel independence*. For space reasons, we omit some technical background that is needed in the proofs but not necessary to understand the main ideas of the paper. An introduction that covers all the details needed in this paper can be found in [2].

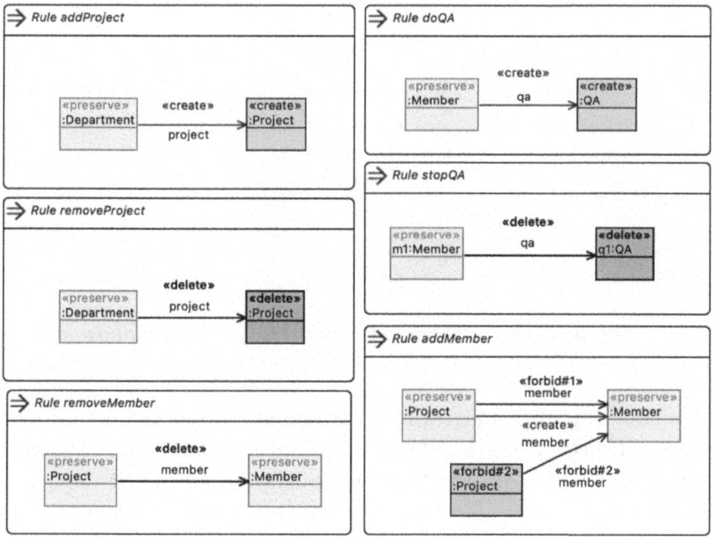

Fig. 1. Henshin rules of our running example

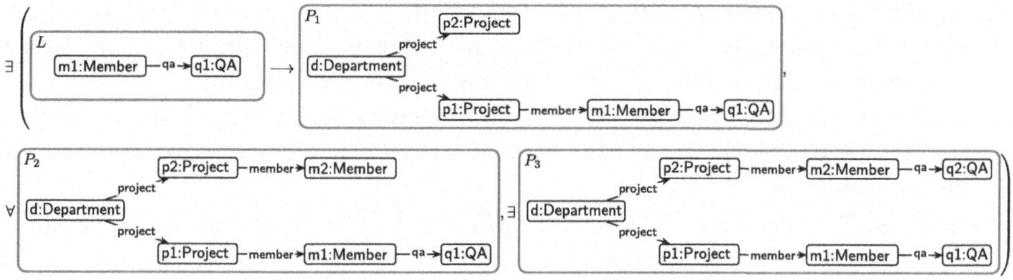

Fig. 2. Additional application condition for rule *stopQA*

Adhesive HLR Categories. Adhesiveness has been defined in several variants; in these kinds of categories, pullbacks and pushouts behave particularly well with respect to a *stable* class of monomorphisms $\mathcal{M}$ [2,7,17]. A class of monomorphisms $\mathcal{M}$ is *stable* (and we depict its elements via hooked arrows $m : A \hookrightarrow B$) if it is stable under pullbacks, contains all isomorphisms, and is closed under composition. Notions of adhesiveness are based on the concept of *van Kampen squares.*

Definition 1 (Van Kampen square. Adhesive HLR category). *A pushout is a* van Kampen square *if, for any commutative cube with this pushout as bottom square (compare Fig. 3) and where the back faces are pullbacks, the top square is a pushout if and only if both front faces are pullbacks.*

A category $\mathcal{C}$ with a stable class of monomorphisms $\mathcal{M}$ is adhesive HLR *if $\mathcal{C}$ has pushouts along $\mathcal{M}$-morphisms, $\mathcal{M}$ is stable under pushouts, and the pushouts along $\mathcal{M}$-morphisms are van Kampen squares.*

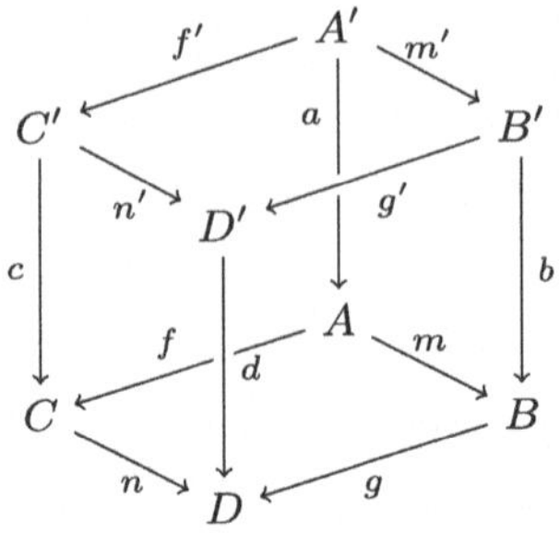

Fig. 3. Commutative cube over the pushout square

A lot of recent research in graph transformation has been performed in the even more general setting of $\mathcal{M}$-adhesive categories [7–9], where the van Kampen property only needs to hold for cubes over pushouts along $\mathcal{M}$-morphisms where the vertical morphisms are also from $\mathcal{M}$. In our proofs, however, we regularly need to invoke the van Kampen property for arbitrary cubes over pushouts along $\mathcal{M}$-morphisms. For this reason, we decided to work in the setting of adhesive HLR categories. This already encompasses a wide range of examples of practical interest (like different kinds of attributed graphs) and, moreover, is also the setting in which the work of Azzi et al. [2] was done, on which we build. (Note, however, that they call $\mathcal{M}$-adhesive what here and in other works is called adhesive HLR.) Alternatively, we could have done our work in the setting of $\mathcal{M}$-adhesive categories – at the cost of restricting ourselves to the case of $\mathcal{M}$-matching of rules.

Throughout, we assume an ambient adhesive HLR category with all pullbacks, initial pushouts over $\mathcal{M}$-morphisms, and an $\mathcal{M}$-initial object $\bot$, whose outgoing arrows we denote as $!_A \colon \bot \hookrightarrow A$ for any object A.

Conditions. In the context of graph transformation, *nested conditions and constraints* have been developed as a suitable (graphical) formalism for expressing properties of graphs and of graph homomorphisms. On graphs, they have been shown to be expressively equivalent to first-order logic [12,26], but they can be defined for arbitrary categories. Here, to simplify the presentation, we address the *linear fragment (without equality)* of this logic, i.e., we exclude branching of conditions via logical connectives like 'and' and require conditions to be built from $\mathcal{M}$-morphisms. Moreover, we assume that conditions are given in *alternating quantifier normal form* (ANF) [27].

Definition 2 (Condition). *A* nested condition in ANF *over an object P_0 is defined recursively as follows:* true *is a universal condition over P_0 and* false *is an existential one. If d is an existential (universal) condition over P_1 and $p \colon P_0 \hookrightarrow P_1$ is an $\mathcal{M}$-morphism that is not iso, then $\forall\,(p \colon P_0 \hookrightarrow P_1, d)$ (resp. $\exists\,(p \colon P_0 \hookrightarrow P_1, d)$) is a universal (existential) condition over P_0. The* nesting level *of a condition c is recursively defined as $\mathrm{nl}(c) := \mathrm{nl}(d)+1$ if $c = Q\,(p \colon P_0 \hookrightarrow P_1, d)$ (where $Q \in \{\forall, \exists\}$) and $\mathrm{nl}(\mathsf{true}) = \mathrm{nl}(\mathsf{false}) = 0$.*

A morphism $q\colon P_0 \to G$ *satisfies a condition* c*, denoted as* $q \models c$*, if one of the following is true: 1.)* $c = \mathsf{true}$*; 2.)* $c = \exists\,(p\colon P_0 \hookrightarrow P_1, d)$ *and there exists a morphism* $q'\colon P_1 \to G$ *with* $q' \circ p = q$ *such that* $q' \models d$*; or 3.)* $c = \forall\,(p\colon P_0 \hookrightarrow P_1, d)$ *and for every morphism* $q'\colon P_1 \to G$ *with* $q' \circ p = q$ *it holds that* $q' \models d$*.*

Rules and Transformations. A transformation rule declaratively specifies the modification of objects (graphs in our running example), and its application is formalized by computing a double pushout.

Definition 3 (Transformation rule and transformation). *A* (transformation) rule $\rho = (L \overset{l}{\hookleftarrow} K \overset{r}{\hookrightarrow} R, ac)$ *consists of a span of* $\mathcal{M}$*-morphisms, called* plain rule *(of* ρ*), and a condition* ac *over* L*;* L *and* R *are called the* left- *and the* right-hand side *(LHS and RHS) of* ρ*,* ac *its* application condition *or just* condition*, for short. An* application *of a rule* ρ *to an object* G *at a morphism* $m\colon L \to G$ *consists of two pushouts, as shown at the left bottom of Fig. 4. Such an application is possible if the pushout complement of* $m \circ l$ *exists and* $m \models ac$*. Then, it is also called a* transformation*, the morphism* m *is called a* match*, the morphism* $n\colon R \to H$ *is called the* comatch*, and the whole transformation* t *is denoted as* $G \Longrightarrow_{\rho,m} H$*. A transformation is* ac-disregarding *if it is a transformation of the plain rule of* ρ*, i.e., a transformation at a morphism* m*, for which* $m \not\models ac$ *might hold.*

The inverse rule ρ^{-1} *of the plain rule of* ρ *is the plain rule* $(R \overset{r}{\hookleftarrow} K \overset{l}{\hookrightarrow} L)$*. The* inverse transformation t^{-1} *of a transformation* t *is the application of the inverse rule to the comatch* n *of* t*.*

Parallel Independence and Disabling Essence. Two *parallel independent* transformations do not interfere with each other; this is formally captured by the Local Church–Rosser Theorem, which states that two such transformations can be applied in any order. We present two equivalent definitions of parallel independence (see Theorem 2 in [5]) and recapitulate the notion of a *disabling essence* for analyzing transformations that *conflict* with each other, i.e., that are not parallel independent.

Definition 4 (Parallel independence). *Given two rules* $\rho_j = (L_j \overset{l_j}{\hookleftarrow} K_j \overset{r_j}{\hookrightarrow} R_j, ac_j)$*, with* $j = 1, 2$*, two transformations* $t_1 : G \Longrightarrow_{\rho_1,m_1} H_1$ *and* $t_2 : G \Longrightarrow_{\rho_2,m_2} H_2$ *via the rules* ρ_1 *and* ρ_2 *are* parallel independent *if there exist two morphisms* $d_1 : L_2 \to D_1$ *and* $d_2 : L_1 \to D_2$ *(as shown in Fig. 4) such that* $m_1 = g_2 \circ d_2$ *and* $m_2 = g_1 \circ d_1$ *and* $h_1 \circ d_1 \models ac_2$ *and* $h_2 \circ d_2 \models ac_1$*.*

Equivalently, for plain rules, two transformations are parallel independent if the squares (1), (2) and (4) in Fig. 4 are pullbacks and the morphisms q_1 *and* q_2 *are isomorphisms.*

If the morphism d_j *does not exist or, equivalently, the morphism* q_j *is not an isomorphism, or* $h_j \circ d_j$ *does not satisfy the condition of the other rule, we say that the transformation* t_j causes a conflict *for the other transformation. We say that two transformations are* parallel independent ac-disregarding *if they are*

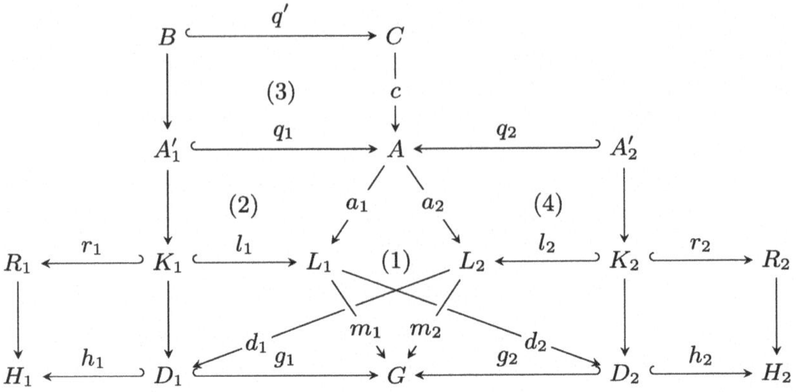

Fig. 4. Parallel independence and construction of proto-essence/disabling essence

parallel independent when the conditions of the rules are ignored (i.e., d_1 and d_2 exist in Fig. 4 but $h_1 \circ d_1$ and $h_2 \circ d_2$ may not satisfy the respective conditions).

A transformation via a plain rule causes a conflict for another such transformation if it deletes structure that the second transformation also needs. A *disabling essence* captures exactly this problematic structure and its occurrence in the respective rules of the conflicting transformations.

Definition 5 (Disabling essence [2]**).** *Given two plain rules $\rho_j = L_j \xhookleftarrow{l_j} K_j \xhookrightarrow{r_j} R_j$ and morphisms $m_j \colon L_j \to G$, with $j = 1, 2$, the* proto-essence for (m_1, m_2) under l_1, *denoted by $\mathrm{ess}_{l_1}(m_1, m_2)$, is defined as $(a_1 \circ c, a_2 \circ c)$, where the squares (1) and (2) in Fig. 4 are pullbacks and (3) is the initial pushout over q_1. C is called the* essence object. *A proto-essence $\mathrm{ess}_{l_1}(m_1, m_2)$ is a* disabling essence *if m_1 and m_2 are matches for transformations t_1 and t_2; in that case, it is denoted as $\mathrm{ess}_{\mathrm{dbl}}(t_1, t_2)$ and called the* disabling essence of (t_1, t_2).

A disabling essence $\mathrm{ess}_{\mathrm{dbl}}(t_1, t_2)$ computes the structure which is deleted by the transformation t_1 although it is needed by the transformation t_2, and shows its (not necessarily monomorphic) occurrences in the LHSs of both rules.

Importantly, a transformation t_1 causes a conflict for transformation t_2 if and only if $\mathrm{ess}_{\mathrm{dbl}}(t_1, t_2) \neq (!_{L_1}, !_{L_2})$ [2, Theorem 3.13].

5 Proto-Essences for Rules with Conditions

In this section, we present our construction of *proto-essences* for rules with conditions. When considering conditions, new types of conflicts can arise. Intuitively, a transformation t_1 via ρ_1 can now additionally cause a conflict for a transformation t_2 via ρ_2 if (i) t_1 deletes structure that ac_2 requires to exist (via an existentially quantified morphism) or (ii) t_1 creates structure whose existence is restricted by ac_2 (via a universally quantified morphism). Therefore, we have to consider the interference of a condition with both the right and left morphism of a rule, which can lead to proto-essences *by deletion* or *insertion*, respectively.

The intuition described above suggests using the known computation of proto-essences also in this more general situation – where one of the inputs is a morphism starting from the object of a condition instead of the LHS of a rule. In this way, instead of computing structure that is deleted by one rule and needed by another, we compute structure that is *deleted or created* by one rule and might affect the validity of another rule's condition. Our computation proceeds recursively along the morphisms of conditions.

Intuitively, we start with an overlap L_1P_n of the LHS L_1 of the first rule and the largest object P_n of the condition of the second rule (see Fig. 5). For proto-essences by deletion, we then iteratively check whether the first rule can destroy occurrences of objects of the condition of the second rule because this might prevent the second rule's application. This check is performed by (iteratively) applying the computation of proto-essences as recalled above. The only difference is that the cospans that we use as input for the computation do not start at the LHSs of two rules but at the LHS L_1 and an object from the condition of the second. Our iteration starts at L_2, the LHS of the second rule over which the condition is defined. In particular, the first step of our iteration is the classical computation of proto-essences for plain rules. We stop the iteration as soon as a non-trivial proto-essence is computed because, if $P_i, P_j, j > i$, are objects from the condition, the destruction of an occurrence of P_i immediately implies the destruction of all occurrences of P_j that encompass the occurrence of P_i. To obtain proto-essences by insertion, we use the same construction but switch the roles of the LHS and the RHS of the first rule. That is, we replace the first rule with its inverse rule and compute proto-essences by deletion.

Definition 6 (Proto-essences for rules with conditions). *Let two rules* $\rho_j = (L_j \overset{l_j}{\hookleftarrow} K_j \overset{r_j}{\hookrightarrow} R_j, ac_j)$, *with* $j = 1,2$ *and* $ac_2 = Q(p_1\colon L_2 \hookrightarrow P_1, \bar{Q}(p_2\colon P_1 \hookrightarrow P_2, Q(\dots p_n\colon P_{n-1} \hookrightarrow P_n)\dots))$, *and a cospan* $s_L\colon L_1 \xrightarrow{i_{L_1}} L_1P_n \xleftarrow{i_{P_n}} P_n$ *be given. The* proto-essence by deletion *of* i_{L_1} *and* i_{P_n} *at level* $1 \le i \le n$, *denoted as* $\mathrm{DPEss}^i(i_{L_1}, i_{P_n})$, *is defined as*

$$\mathrm{DPEss}^0(i_{L_1}, i_{P_n}) := \mathrm{ess}_{l_1}(i_{L_1}, i_{P_1} \circ p_1),$$

$$\mathrm{DPEss}^i(i_{L_1}, i_{P_n}) := \begin{cases} \mathrm{DPEss}^{i-1}(i_{L_1}, i_{P_n}) & \text{if } \mathrm{DPEss}^{i-1}(i_{L_1}, i_{P_n}) \neq (!_{L_1}, !_{P_{i-1}}) \\ \mathrm{ess}_{l_1}(i_{L_1}, i_{P_i}) & \text{if } \mathrm{DPEss}^{i-1}(i_{L_1}, i_{P_n}) = (!_{L_1}, !_{P_{i-1}}) \end{cases},$$

where $i_{P_i} := i_{P_n} \circ p_n \circ \dots \circ p_{i+1}$ *(see Fig. 5). The* set of proto-essences by deletion of ρ_1 in ρ_2, *denoted by* $\mathrm{DPEss}(\rho_1, \rho_2)$, *is the set of all proto-essences by deletion at level* n *of morphisms* i_{L_1} *and* i_{P_n}.

Given a cospan $s_R\colon R_1 \xrightarrow{i_{R_1}} R_1P_n \xleftarrow{i_{P_n}} P_n$, *the* proto-essence by insertion of i_{R_1} and i_{P_n} at level $1 \le i \le n$, *denoted as* $\mathrm{IPEss}^i(i_{R_1}, i_{P_n})$, *is defined completely analogously; we just stop the recursion at* $i_{P_1} := i_{P_n} \circ p_n \circ \dots \circ p_2$. *The* set of

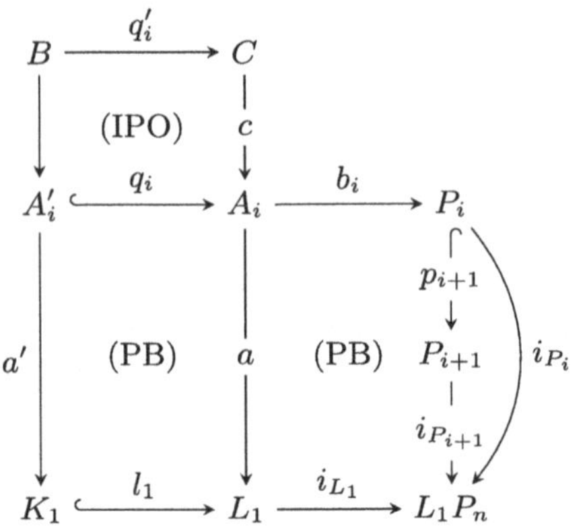

Fig. 5. Construction of proto-essences for rules with conditions

proto-essences by insertion of ρ_1 in ρ_2, *denoted by* $\mathrm{IPEss}(\rho_1, \rho_2)$, *is the set of all proto-essences by insertion at level* n *of morphisms* i_{R_1} *and* i_{P_n}.

By construction, a proto-essence (by deletion or insertion) of a cospan is unique up to isomorphism. Moreover, it is not just a span but both its morphisms come with a decomposition where the essence object is first embedded into the pullback object of the cospan. Consequently, for the rest of the paper, we will often denote proto-essences (by deletion or insertion) as spans $(a \circ c\colon C \to L_1, b_i \circ c\colon C \to P_i)$ of composed morphisms, where a, b_i and c are the morphisms obtained from the computation (Fig. 5). In our proofs, we will regularly use that $\mathrm{DPEss}^{i-1}(i_{L_1}, i_{P_n}) = (!_{L_1}, !_{P_n})$ if and only if the morphism q_i (depicted in Fig. 5) is an isomorphism [2, Lemma 3.6].

Example 1. Fig. 6 shows the construction of proto-essences by deletion of the rules *StopQA* and *StopQA*, starting with the cospan $L_1 \to L_1P_3 \leftarrow P_3$. The mappings of L_1 and P_3 to L_1P_3 are implicitly given by the node names. In the first step, we compute the pullback of i_{L_1} and $i_{P_n} \circ p_3 \circ p_2 \circ p_1$, which results in A_0. Then we compute the pullback of l_1 and the newly created morphism $A_0 \to L_1$, which results in A_0'. Since $q_0\colon A_0' \hookrightarrow A_0$ is an isomorphism, we proceed to the next level and compute the pullbacks of i_{L_1} and $i_{P_n} \circ p_3 \circ p_2$ and of l_1 and the newly created morphism $A_1 \to L_1$. Again, the obtained morphism q_1 is an isomorphism and we proceed with the next level. This shows that the first application of *StopQA* cannot prevent a second application of *StopQA* by destroying the rule match, nor by destroying an occurrence of P_1, if both rules are applied as described by their matches in L_1P_3. In the fourth iteration, where A_3 and A_3' are computed, the morphism q_3 is not an isomorphism, i.e., the initial pushout over q_3 is computed (not depicted but resulting in A_3' and A_3 again) and the span $(a_4 \circ id_{A_3}, b_3 \circ id_{A_3})$ is a proto-essence by deletion of the rules *StopQA* and *StopQA*. This essence shows that *StopQA* might prevent another application of *StopQA* by deleting a QA node of a Member in another Project contained in the same Department, which will destroy an occurrence of the existentially bound

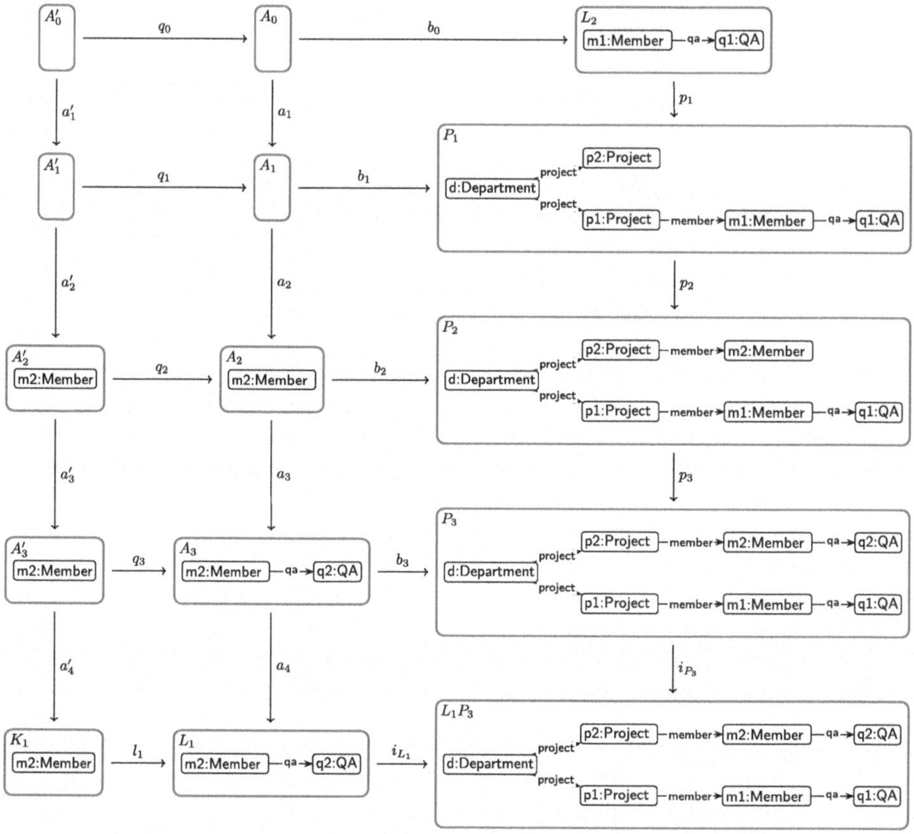

Fig. 6. Example computation of proto-essence by deletion

P_3. Note that an occurrence of this span in a pair of transformations via *StopQA* and *StopQA* does not imply that the second application of *StopQA* is prevented, since the Department may contain a third Project in which all Members do QA.

We now discuss further proto-essences shown in Fig. 7 which were computed using the rules shown in Fig. 1. For all examples, our second rule is *stopQA* with its condition (Fig. 2). We do not show the cospans from which the proto-essences are computed and display the results as spans (concatenating the morphisms).

C_1*:* This proto-essence by deletion of *stopQA* and *stopQA* is a classic disabling essence for plain rules. It shows that the first rule deletes a node of type QA (and its incoming edge) to which the second rule also matches.

C_2*:* This proto-essence by deletion of *stopQA* and *stopQA* at level 3 cannot occur in transformations because the deletion of q2 by the first rule creates a dangling edge (the edge originating in m2).

C_3*:*This proto-essence by insertion of *addMember* and *stopQA* at level 2 indicates that *addMember* can prevent an application of *stopQA* by adding a member

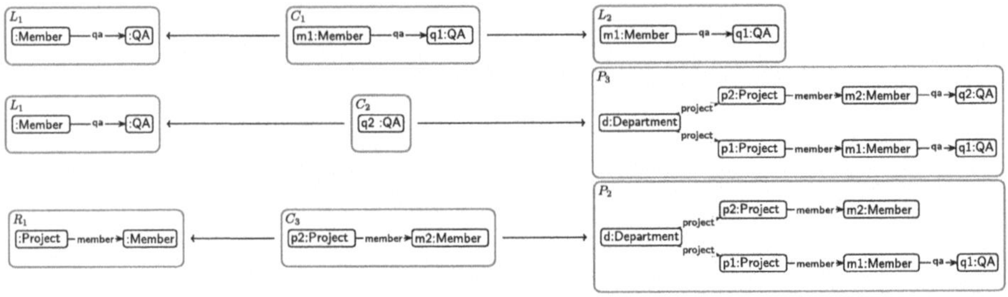

Fig. 7. Examples for proto-essences

that is not connected to a QA node to another Project contained in the same Department. Note that an occurrence of this proto-essence does not imply a conflict if (i) the added user is doing QA or (ii) there is another Project in the same Department so that all of its Members are doing QA.

If we consider the alternative rule *addMemberWithQA* (not shown) which can only add a Member to a project that already has an outgoing edge towards a QA node, we get basically the same proto-essence by insertion (with the difference that R_1 contains an additional QA node that is connected to the Member). However, *addMemberWithQA* cannot cause a conflict for *stopQA* because this insertion will never introduce a violation of the condition of *stopQA*.

While we define proto-essences for arbitrary cospans, at least in categories with all pushouts it is only necessary to consider certain cospans to get an overview of essentially all proto-essences of a pair of rules.

Proposition 1 (Completeness of jointly epic cospans). *If the ambient category has all pushouts, given a pair of rules* ρ_1, ρ_2 *and any proto-essence by deletion* $\mathrm{DPEss}^i(i_{L_1}, i_{P_n}) = (a \circ c, b_i \circ c)$ *at level* i *for them, there exists a pair of jointly epic morphisms* $(\pi_1 \colon L_1 \to X, \pi_2 \colon P_n \to X)$ *such that* $\mathrm{DPEss}^i(i_{L_1}, i_{P_n})$ *and* $\mathrm{DPEss}^i(\pi_1, \pi_2)$ *are isomorphic (as decomposed spans). The same is true for proto-essences by insertion.*

6 Disabling Essences and Embeddability

In this section, we define *disabling essences* and present our results regarding their embedding into transformation pairs. To clarify what it means to embed a proto-essence by deletion into a pair of transformations, we first define when a span is embeddable into a pair of transformations. To later be able to check uniformly at the object G, given a pair of transformations $(t_1 \colon G \Longrightarrow H_1, t_2 \colon G \Longrightarrow H_2)$, for the embeddability of both kinds of essences, we also develop a *shift* construction, which moves proto-essences by insertion from the RHS to the LHS. We then introduce the set of *disabling essences of one rule in another*, which consists of a filtered subset of the proto-essences

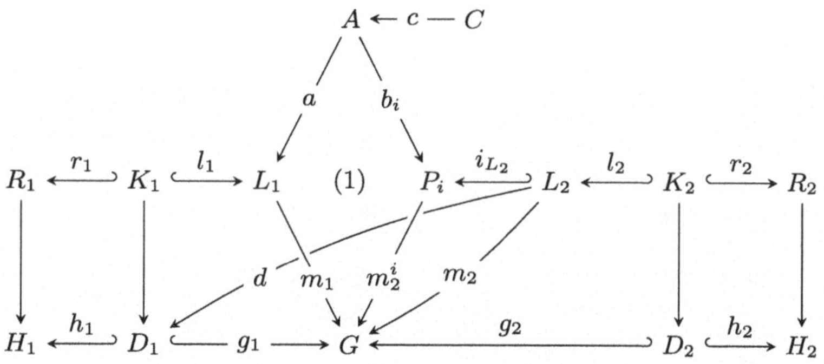

Fig. 8. Embedding of decomposed spans

by deletion and a filtered subset of the shifted proto-essences by insertion, and present our results in terms of the embeddability of these essences.

Now, we will introduce our notion of embeddability. Since a proto-essence for plain rules is of the form $(C \to L_1, C \to L_2)$, it can be embedded in a transformation pair by using only the matches m_1 and m_2 of the applied rules [2]. To embed a proto-essence by deletion (which is of the form $(C \to L_1, C \to P_i)$, where P_i is an object contained in ac_2), we also want to ensure that the embeddings of the LHSs conform to the rule matches. For the embedding of L_1 we can choose the match of the first rule. For P_i, however, we need an additional morphism $i_{L_2}\colon L_2 \hookrightarrow P_i$, which tells us how L_2 is embedded in P_i.

Definition 7 (Embedding of decomposed spans). *Given a pair of transformations* $(t_1, t_2) = (G \Longrightarrow_{\rho_1, m_1} H_1, G \Longrightarrow_{\rho_2, m_2} H_2)$ *via rules* $(L_j \overset{l_j}{\hookleftarrow} K_j \overset{r_j}{\hookrightarrow} R_j, ac_j)$*, with* $j = 1, 2$*, a decomposed span* $(a \circ c\colon C \to A \to L_1, b_i \circ c\colon C \to A \to P_i)$ *and a morphism* $i_{L_2}\colon L_2 \hookrightarrow P_i \in \mathcal{M}$*, a morphism* $m_2^i\colon P_i \to G$ *is an* embedding of $(a \circ c, b_i \circ c)$ in (t_1, t_2) *if* $m_2 = m_2^i \circ i_{L_2}$ *and the square* (1) *in Fig. 8 is a pullback.*

For a proto-essence by deletion $(a \circ c\colon C \to L_1, b_i \circ c\colon C \to P_i)$, the desired embedding of L_2 in P_i is always given by $i_{L_2} := p_i \circ \ldots \circ p_1 \in \mathcal{M}$. For the rest of this paper, we will always use these morphism names when embedding a proto-essence by deletion into a transformation pair.

Since any proto-essence by insertion of a rule ρ_1 in another rule ρ_2 is of the form $(a \circ c\colon C \to R_1, b_i \circ c\colon C \to P_i)$, we cannot embed this proto-essence in a transformation pair $(t_1, t_2) = (G \Longrightarrow_{\rho_1, m_1} H_1, G \Longrightarrow_{\rho_2, m_2} H_2)$. To determine whether the conflict described by $(a \circ c, b_i \circ c)$ occurs in this pair of transformations, we would have to check whether it is embeddable in the pair (t_1^{-1}, t_2'), where t_2' is the transformation via ρ_2 (ignoring ac_2) at match $h \circ d$ (see Fig. 9), which exists if t_1 does not cause a conflict for the plain version of ρ_2.

To still be able to check for an occurrence in (t_1, t_2), we introduce the *shift of proto-essences by insertion* which, given a proto-essence $(a \circ c\colon C \to R_1, b_i \circ c\colon C \to P_i)$, computes a span $(a' \circ c'\colon C' \to L_1, b_i' \circ c'\colon C' \to P_i')$, which is

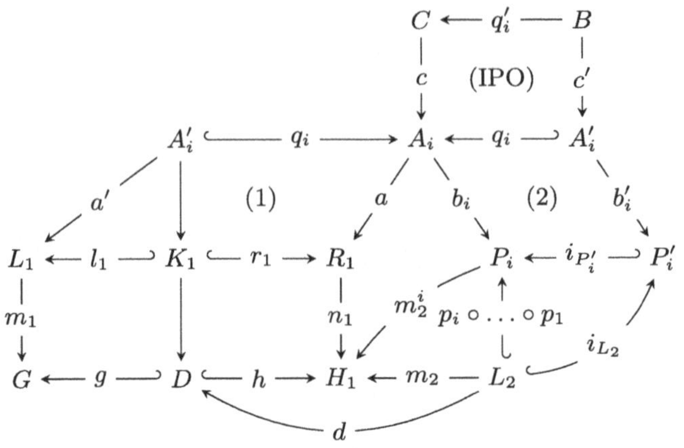

Fig. 9. Shifting of a proto-essence by insertion

equivalent to $(a \circ c, b_i \circ c)$ in the sense that $(a \circ c, b_i \circ c)$ is embeddable in (t_1^{-1}, t'_2) if and only if its shifted version $(a' \circ c', b'_i \circ c')$ is embeddable in (t_1, t_2). Intuitively, we compute the largest subobject P'_i of P_i that is preserved by the transformation t_1^{-1}. To do this, we use an idea similar to that introduced by Habel and Pennemann in [12] for the *shift of nested conditions over rules.* They applied the inverse rule along the nesting structure of the application condition. However, the inverse of the first rule ρ_1 may not be applicable at P_i in the desired way (this is only possible if A_i is isomorphic to R_1), so we apply the largest subrule $A'_i \xleftarrow{id} A'_i \xhookrightarrow{q_i} A_i$ of $K \xleftarrow{id} K \xhookrightarrow{r_1} R_1$ that is applicable at P_i to get P'_i.

Definition 8 (Shift of proto-essence by insertion). *Given two rules* $\rho_j = (L_j \xleftarrow{l_j} K_j \xhookrightarrow{r_j} R_j, ac_j)$, *with* $j = 1, 2$, *and a proto-essence by insertion* $(a \circ c\colon C \to R_1, b_i \circ c\colon C \to P_i) \in \mathrm{IPEss}(\rho_1, \rho_2)$, *the* shift of $(a \circ c, b_i \circ c)$ over ρ_1, *denoted by* $\mathrm{PEShift}_{\rho_1}(a \circ c, b_i \circ c)$, *is defined as*

$$\mathrm{PEShift}_{\rho_1}(a \circ c, b_i \circ c) = (a' \circ c', b'_i \circ c')$$

where (1) *in Fig. 9 is a pullback,* (IPO) *is the initial pushout over* q_i *(both stem from the computation of* $(a \circ c, b_i \circ c)$*, and* A'_i *and* q_i *are depicted twice in Fig. 9) and* (2) *is a pushout constructed by computing the pushout complement of* q_i *and* b_i.

Example 2. We illustrate our shift construction using the case C_3 from Example 1 (see C_3 in Fig. 7). In this case, A'_2 and correspondingly the boundary object B of the initial pushout consist of the Project and the Member nodes, while A_2 additionally contains the members-edge between them. Computing P'_2 as the pushout complement of $A'_2 \to A_i \to P_2$ removes the member-edge from P_2 and finally results in the span shown in Fig. 10. In particular, in this case the inverse of the *addMember* rule is applied to C_3 and P_2.

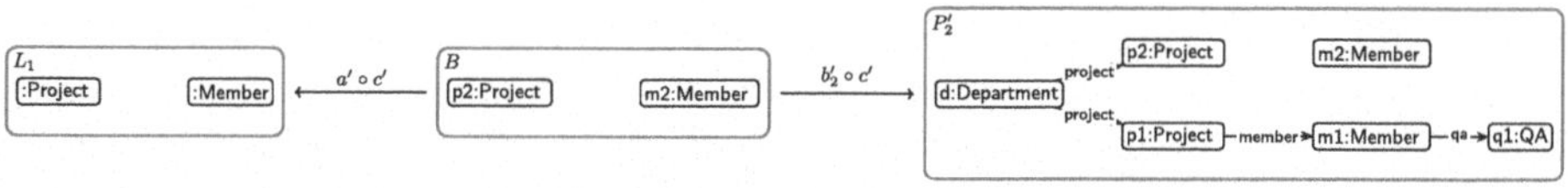

Fig. 10. Example for a shifted proto-essence by insertion

Note that the *shift of proto-essences by insertion* is only partially defined, since the pushout complement of q_i and b_i does not necessarily exist. However, we are only interested in those proto-essences by insertion that can be embedded in a pair of transformations. The following proposition shows that a proto-essence by insertion of a rule ρ_1 in ρ_2 can be shifted over ρ_1 if it can be embedded in a pair of transformations via ρ_1^{-1} and ρ_2.

For a shifted essence $(a' \circ c' \colon B \to L_1, b_i' \circ c' \colon B \to P_i')$, we will also show that there is a morphism $i_{L_2} \colon L_2 \hookrightarrow P_i' \in \mathcal{M}$ with $p_i \circ \ldots \circ p_1 = i_{P_i'} \circ i_{L_2}$ (see Fig. 9), which implies that a shifted essence can also be embedded in a pair of transformations according to Definition 7.

Proposition 2 (Existence of shift). *Let two rules* $\rho_j = (L_j \xleftarrow{l_j} K_j \xrightarrow{r_j} R_j, ac_j)$*, with* $j = 1, 2$*, and a proto-essence by insertion* $(a \circ c, b_i \circ c) \in \mathrm{IPEss}(\rho_1, \rho_2)$ *be given. If there exists an object* H_1 *and morphisms* $n_1 \colon R_1 \to H_1$ *and* $m_2^i \colon P_i \to H_1$ *such that (i) the span* (a, b_i) *is a pullback of* (n_1, m_2^i) *and (ii) the rule* ρ_1^{-1} *(without condition) is applicable at* n_1*, then* $\mathrm{PEShift}_{\rho_1}(a \circ c, b_i \circ c)$ *exists. In particular, there is a pullback complement for* $b_i \circ q_i$ *(see Fig. 9). Moreover, whenever* $\mathrm{PEShift}_{\rho_1}(a \circ c, b_i \circ c)$ *exists, there is a morphism* $i_{L_2} \colon L_2 \hookrightarrow P_i' \in \mathcal{M}$ *with* $p_i \circ \ldots \circ p_1 = i_{P_i'} \circ i_{L_2}$*.*

It remains to show that a shifted essence is embeddable in (t_1^{-1}, t_2') if and only if the original essence is embeddable into (t_1, t_2).

Theorem 1 (Correctness of Shift). *Given a pair of ac-disregarding parallel independent transformations* $(t_1, t_2) = (G \Longrightarrow_{\rho_1, m_1} H_1, G \Longrightarrow_{\rho_2, m_2} H_2)$ *via rules* $\rho_j = (L_j \xleftarrow{l_j} K_j \xrightarrow{r_j} R_j, ac_j)$*, with* $j = 1, 2$*, and a proto-essence by insertion* $(a \circ c, b_i \circ c) \in \mathrm{IPEss}(\rho_1, \rho_2)$*, then* $\mathrm{PEShift}_{\rho_1}(a \circ c, b_i \circ c)$ *is embeddable in* (t_1, t_2) *if and only if* $(a \circ c, b_i \circ c)$ *is embeddable in* (t_1^{-1}, t_2')*, where* $t_2' \colon H_1 \Longrightarrow_{\rho_2, h_1 \circ d} X$ *is the ac-disregarding transformation that exists by parallel independence.*

In the following, we define the set of *disabling essences* for rules with application conditions as a filtered subset of proto-essences by deletion and a subset of filtered shifted proto-essences by insertion. In both cases, we can exclude some proto-essences, since the occurrence of some of them will never cause a conflict. Recall that a condition c can be violated by (i) destroying an occurrence of an existentially bound object or (ii) by inserting an occurrence of a universally bound object. The corresponding proto-essence by deletion describes a situation where an occurrence of object P_i is destroyed so that the embedded occurrence of P_{i-1} is preserved. Since destroying an occurrence of a universally bound P_i

does not affect the satisfaction of a condition, we can exclude all those proto-essences by deletion where P_i is universally bound. Analogously, we can exclude all proto-essences by insertion $(a \circ c\colon C \to R_1, b_i \circ c\colon C \to P_i)$ where P_i is existentially bound. In addition, we exclude all trivial proto-essences and those that cannot be embedded in a transformation pair.

Definition 9 (Disabling essences of rules with conditions). *Given two rules* $\rho_j = (L_j \xleftarrow{l_j} K_j \xrightarrow{r_j} R_j, ac_j)$, *with* $j = 1, 2$, *a proto-essence by deletion* $(a \circ c\colon C \to L_1, b_i \circ c\colon C \to P_i) \in \mathrm{DPEss}(\rho_1, \rho_2)$ *is a* disabling essence by deletion of ρ_1 in ρ_2 *if (i) it is embeddable in a transformation pair, (ii)* P_i *is existentially bound in* ac_2 *or* $P_i = L_2$, *and (iii)* $(a, b_i) \neq (!_{L_1}, !_{P_n})$. *The set of all disabling essences by deletion of* ρ_1 *in* ρ_2 *is denoted by* $\mathrm{DEss}(\rho_1, \rho_2)$.

Given a proto-essence by insertion $(a \circ c\colon C \to R_1, b_i \circ c\colon C \to P_i) \in \mathrm{IPEss}(\rho_1, \rho_2)$, *the span* $\mathrm{PEShift}_{\rho_1}(a \circ c, b_i \circ c)$ *is a* disabling essence by insertion of ρ_1 in ρ_2 *if (i) it is embeddable in a transformation pair, (ii)* P_i *is universally bound in* ac_2, *and (iii)* $(a, b_i) \neq (!_{R_1}, !_{P_n})$. *The set of all disabling essences by insertion of* ρ_1 *in* ρ_2 *is denoted by* $\mathrm{IEss}(\rho_1, \rho_2)$. *The* set of disabling essences of ρ_1 in ρ_2, *denoted by* $\mathrm{Ess}(\rho_1, \rho_2)$, *is defined as* $\mathrm{Ess}(\rho_1, \rho_2) := \mathrm{DEss}(\rho_1, \rho_2) \cup \mathrm{IEss}(\rho_1, \rho_2)$.

The non-trivial disabling essences of plain rules ρ_1 and ρ_2 as introduced by Azzi et al. [2] are a subset of the disabling essences by deletion of ρ_1 in ρ_2.

Proposition 3 (Disabling essences of plain rules as special case). *Given two rules* $\rho_j = (L_j \xleftarrow{l_j} K_j \xrightarrow{r_j} R_j, ac_j)$, *with* $j = 1, 2$, *the morphism* $b_i \circ c$ *of a disabling essence by deletion* $(a \circ c, b_i \circ c) \in \mathrm{DEss}(\rho_1, \rho_2)$ *has the codomain* L_2, *i.e.,* $b_i \circ c\colon C \to L_2$ *if and only if* $(a \circ c, b_i \circ c)$ *is a disabling essence of the plain rule* $L_1 \xleftarrow{l_1} K_1 \xrightarrow{r_1} R_1$ *in the rule* $L_2 \xleftarrow{l_2} K_2 \xrightarrow{r_2} R_2$.

Now we present our results on the embeddability of disabling essences. In contrast to plain rules, where a non-trivial disabling essence is embeddable in a pair of transformations (t_1, t_2) if and only if t_1 and t_2 are parallel dependent, only one direction holds in the presence of application conditions: If t_1 causes a conflict for t_2, a disabling essence of ρ_1 in ρ_2 is embeddable in (t_1, t_2). In Example 1, we already discussed concrete counterexamples to the reverse direction.

Theorem 2 (Embedding of disabling essences). *Given a pair of parallel dependent transformations* $(t_1, t_2) = (G \Longrightarrow_{\rho_1, m_1} H_1, G \Longrightarrow_{\rho_2, m_2} H_2)$ *via rules* $\rho_j = (L_j \xleftarrow{l_j} K_j \xrightarrow{r_j} R_j, ac_j)$, *with* $j = 1, 2$, *so that* t_1 *causes a conflict for* t_2, *there is a disabling essence* $(a \circ c, b_i \circ c) \in \mathrm{Ess}(\rho_1, \rho_2)$ *that is embeddable in* (t_1, t_2).

Note that this implies that if no disabling essence is embeddable in a pair of transformations (t_1, t_2), then t_1 does not cause a conflict for t_2. It also follows that a rule ρ_1 cannot cause a conflict for ρ_2 if the set of disabling essences of ρ_1 in ρ_2 is empty.

Corollary 1 (Conflict-freeness). *Given two rules* $\rho_1 = (L_1 \xleftarrow{l_1} K_1 \xrightarrow{r_1} R_1, ac_1)$ *and* $\rho_2 = (L_2 \xleftarrow{l_2} K_2 \xrightarrow{r_2} R_2, ac_2)$*, if* $\mathrm{Ess}(\rho_1, \rho_2) = \emptyset$*, then* t_1 *does not cause a conflict for* t_2 *for any transformation pair* (t_1, t_2) *via* ρ_1 *and* ρ_2*.*

For rules ρ_1 and ρ_2 where the condition of ρ_2 is a negative application condition (NAC), i.e., ac_2 is of the form $\forall(p_1 \colon L_2 \hookrightarrow P_1, \mathsf{false})$, we get a result that is equivalent to the case of plain rules: A disabling essence for ρ_1 in ρ_2 is embeddable in a transformation pair via ρ_1 and ρ_2 *if and only if* t_1 causes a conflict for t_2.

Theorem 3 (Equivalence result for rules with NACs). *Given two transformations* $t_1 \colon G \Longrightarrow_{\rho_1, m_1} H_1$ *and* $t_2 \colon G \Longrightarrow_{\rho_2, m_2} H_2$ *via rules* $\rho_1 = (L_1 \xleftarrow{l_1} K_1 \xrightarrow{r_1} R_1, ac_1)$ *and* $\rho_2 = (L_2 \xleftarrow{l_2} K_2 \xrightarrow{r_2} R_2, \forall(p_1 \colon L_2 \to P_1, \mathit{false}))$*, then* t_1 *causes a conflict for* t_2 *if and only if a disabling essence of* ρ_1 *in* ρ_2 *can be embedded in* (t_1, t_2)*.*

Note that this result can be extended to the case where the second rule is equipped with a set, i.e., a conjunction of NACs: If a disabling essence of ρ_1 in ρ_2 for one of the NACs can be embedded in a pair of transformations, then, by the above theorem, this NAC is not satisfied after the first transformation. This also implies that the set of NACs is not satisfied after the first transformation. The other direction follows from Theorem 2, by using the fact that a set of NACs is violated if and only if at least one of its elements is.

7 Conclusion

In this paper, we present a construction for disabling essences for rules with application conditions. Such an essence shows the specific reason why one rule can cause a conflict for another, in particular by violating its application condition. We have presented our construction for conditions in ANF, which are nested conditions with alternating quantifiers but without the logical operators 'and' and 'or'. We showed that, given a pair of conflicting rule applications, there is a disabling essence that can be embedded in that pair. If the conditions are just NACs (or sets of NACs), then the converse is also true, i.e., any embedding of a disabling essence indicates a conflict.

While, in this paper, we restrict ourselves to the linear fragment of nested conditions (without equality), we are confident that the approach also applies to the more general case where conditions can branch. Instead of computing disabling essences iteratively along a chain of morphisms, one has to compute them iteratively along the tree structure of the condition. In this case, the second morphism of the cospan that serves as input for the computation then starts at the colimit of the tree (because there does not need to exist a largest object in the condition any longer). Note that this construction could produce duplicates, since the same proto-essence could be computed in two different branches, requiring an additional filtering step to remove these duplicates. Also, filtering becomes more complex because we can no longer assume that all quantifiers at the same nesting level are of the same type. We are confident that Theorem 2 also holds for general conditions because in the case of a conflict it should still be possible to obtain the embedding disabling essence from the required structure that is deleted by the first rule or from the forbidden structure that is created by it.

We use the fact that our conditions are built from $\mathcal{M}$-morphisms in the proof of Theorem 2: Once a disabling interaction is found between a rule and a condition graph P_i, we compute a pushout along an $\mathcal{M}$-morphism to obtain a cospan that leads to the corresponding proto-essence and whose second morphism has the domain $P_n \supseteq P_i$. Thus, to support arbitrary morphisms in conditions, it would suffice to adapt Definition 6 and use cospans based on P_i to compute the proto-essences at each level i. We developed the computation for all levels from the same cospan because we hope that this can serve as a basis for an efficient implementation.

Regarding complexity, we have the following considerations: complexity scales with the size of the largest graph of the condition (since this increases the number of cospans to consider) rather than with the depth of the nesting level. Note that (e.g., by comparing Fig. 6), with the appropriate data structures, the proto-essence can be determined simply by computing the largest pullback, i.e., that of i_{L_1} and i_{P_n}, and then it is checked in which P_i the individual elements of the computed graph A_n occur first.

In future work, we intend to formalize these observations and to study the relationship between disabling essences and initial conflicts, now for rules with conditions. With this theory at hand, we want to consider two applications: In graph repair [24], repair rules often have conditions and can interfere with each other. Computing their disabling essences can be useful to find the right order for applying repair rules. In evolutionary computation, where mutation operators are specified as graph transformation rules [14], the computation of disabling essences can be used to find mutations with non-local consequences.

Acknowledgments. This work was partially funded by the German Research Foundation (DFG), projects "Triple Graph Grammars (TGG) 3.0" and "Model-Driven Optimization in Software Engineering". The authors would like to thank the ICGT reviewers for their insightful comments.

Disclosure of Interests. The authors have no competing interests to declare that are relevant to the content of this article.

A Appendix

For the proof of our first proposition, we recall a general property of the interrelation of pushouts and pullbacks, namely that, generally, the pushout of a pullback is also a pullback.

Lemma 1. *In an arbitrary category* $\mathcal{C}$, *let the square* $P \xrightarrow{p_1} A \xrightarrow{a} C \xleftarrow{b} B \xleftarrow{p_2} P$, *i.e., the outer right square in Fig. 11, be a pullback and the square* (1) *be a* *pushout of the span* $A \xleftarrow{p_1} P \xrightarrow{p_2} B$. *Then, this central square* (1) *is also a pullback.*

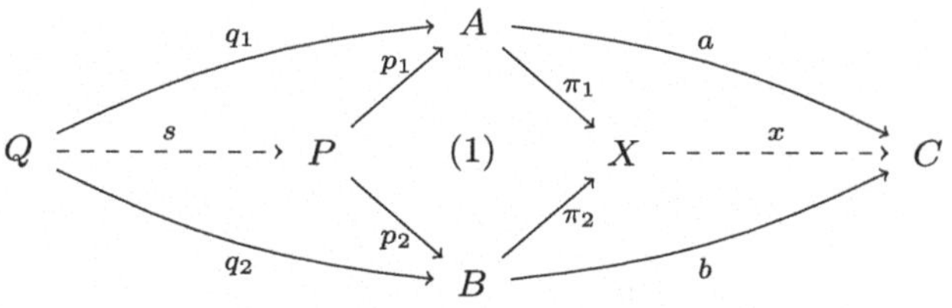

Fig. 11. Proving the pushout of a pullback to constitute a pullback

Proof. Let the square $P \xrightarrow{p_1} A \xrightarrow{a} C \xleftarrow{b} B \xleftarrow{p_2} P$, i.e., the outer right square in Fig. 11, be a pullback and the square (1) be a pushout of the span $A \xleftarrow{p_1} P \xrightarrow{p_2} B$.

To show that (1) is also a pullback, let $q_1 : Q \to A$ and $q_2 : Q \to B$ be morphisms such that $\pi_1 \circ q_1 = \pi_2 \circ q_2$. We have to show that there is a unique morphism $s : Q \to P$ such that $p_i \circ s = q_i$ for $i = 1, 2$. Because $b \circ p_2 = a \circ p_1$, there exists a (unique) morphism $x : X \to C$ with $x \circ \pi_1 = a$ and $x \circ \pi_2 = b$ by the universal property of the pushout (1). Because of $\pi_1 \circ q_1 = \pi_2 \circ q_2$, we also have $x \circ \pi_1 \circ q_1 = x \circ \pi_2 \circ q_2$, which implies $a \circ q_1 = b \circ q_2$. This means that the universal property of the outer right pullback implies the unique existence of the required morphism $s : Q \to P$ such that $p_i \circ s = q_i$ for $i = 1, 2$. □

Proof (of Proposition 1*).* The morphisms π_1 and π_2 are obtained via the pushout of the pullback of i_{L_1} (or i_{R_1}) and i_{P_n}. As pushout coprojections, π_1 and π_2 are jointly epi. As pushout of a pullback, the resulting square is also a pullback (Lemma 1). In particular, up to isomorphism, the same proto-essences by deletion are computed at every level $0 \leq i \leq n$. Moreover, the morphism q_{i-1} constructed during the computation of $\mathrm{DPEss}^{i-1}(\pi_1, \pi_2)$ is an isomorphism and $\mathrm{DPEss}^{i-1}(\pi_1, \pi_2) = (!_{\pi_1}, !_{\pi_2})$. □

Proof (of Proposition 2*).* Let two rules $\rho_j = (L_j \xleftarrow{l_j} K_j \xrightarrow{r_j} R_j, ac_j)$, where $j = 1, 2$, be given and a proto-essence by insertion $(a \circ c, b_i \circ c)$ for them. We first argue for the existence of $\mathrm{PEShift}_{\rho_1}(a \circ c, b_i \circ c)$ (under the mentioned conditions) and then prove the existence of the morphism $i_{L_2} : L_2 \hookrightarrow P_i' \in \mathcal{M}$.

For the first, if there is a cospan $s_R : R_1 \xrightarrow{n_1} H_1 \xleftarrow{m_2^i} P_i$ such that (a, b_i) is its pullback and ρ_1^{-1} is applicable at n_1, we obtain the cube depicted in Fig. 12, where the bottom square is the pushout that exists by applicability of ρ_1^{-1} at n_1, the right front face is the pullback of (m_2^i, n_1) and the rest of the cube is completed by taking pullbacks. By the van Kampen property, the top of the cube is a pushout, which constitutes the desired pushout complement; in particular, $i_{P_i'} \in \mathcal{M}$ because $q_i \in \mathcal{M}$ and $\mathcal{M}$ is stable under pushout.

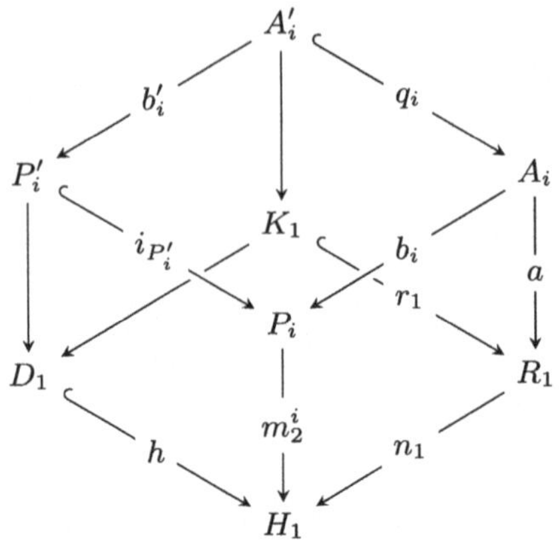

Fig. 12. Existence of the pushout complement computing the shifted proto-essence by insertion

For the existence of the morphism $i_{L_2}\colon L_2 \hookrightarrow P'_i \in \mathcal{M}$, we show the existence of a morphism $j_{P'_i}\colon P_{i-1} \hookrightarrow P'_i \in \mathcal{M}$ such that $i_{P'_i} \circ j_{P'_i} = p_i$. Then we set $i_{L_2} := j_{P'_i} \circ p_{i-1} \circ \ldots \circ p_1\colon L_2 \hookrightarrow P'_i$, which implies $p_i \circ \ldots \circ p_1 = i_{P'_i} \circ j_{P'_i} \circ p_{i-1} \cdots p_1 = i_{P'_i} \circ i_{L_2}$, as desired. Moreover, as composition of $\mathcal{M}$-morphisms, $i_{L_2} \in \mathcal{M}$ holds in that way. The existence of that morphism basically follows from [17, Lemma 2.8]; what follows is a reiteration of the proof of that lemma adapted to our situation.

Whenever $\mathrm{PEShift}_{\rho_1}(a \circ c, b_i \circ c)$ exists, we obtain the cube shown in Fig. 13, where the bottom is the pushout complement that computes $\mathrm{PEShift}_{\rho_1}(a \circ c, b_i \circ c)$, the front right face and the right back face are the pullbacks computed during the construction of $(a \circ c, b_i \circ c)$ (see Definition 6 and Fig. 4), and the cube is completed via pullbacks. To see that the pullback in the left front actually

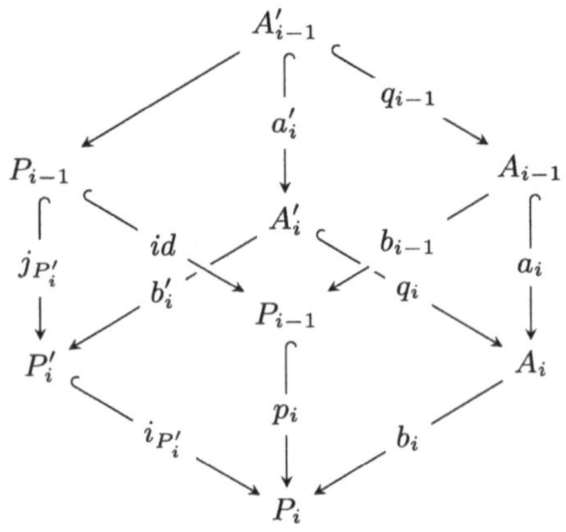

Fig. 13. Proving the existence of a morphism from L_2 to P'_i

computes P_{i-1} and our desired morphism $j_{P'_i}\colon P_{i-1} \hookrightarrow P'_i$, one argues as follows. By the van Kampen property, the top square is a pushout. Because $(a \circ c, b_i \circ c)$ is a proto-essence by insertion with P_i being the codomain of b_i, q_{i-1} is an isomorphism. Therefore, the morphism opposite to it in the top pushout is also an isomorphism, which we can assume, w.l.o.g. to be the identity on P_{i-1}. Because $p_i \in \mathcal{M}$ and $\mathcal{M}$ is stable under pullback, also $j_{P'_i} \in \mathcal{M}$. □

Proof (of Theorem 1). Let a proto-essence by insertion $(a \circ c\colon C \to R_1, b_i \circ c\colon C \to P_i) \in \mathrm{IPEss}(\rho_1, \rho_2)$ for rules ρ_1, ρ_2 be given.

We first prove that $\mathrm{PEShift}_{\rho_1}(a \circ c, b_i \circ c)$ embeds into the transformation pair $(t_1, t_2) = (G \Longrightarrow_{\rho_1, m_1} H_1, G \Longrightarrow_{\rho_2, m_2} H_2)$ if $(a \circ c, b_i \circ c)$ embeds into the transformation pair $(t_1^{-1}, t'_2) = (H_1 \Longrightarrow_{\rho_1^{-1}, n_1} G, H_1 \Longrightarrow_{\rho_2, m'_2} X)$, where $m'_2 = h \circ d$ exists by ac-disregarding parallel independence of t_1 and t_2. Let $n_2^i\colon P_i \to H_1$ be the morphism that embeds $(a \circ c, b_i \circ c)$ in (t_1^{-1}, t'_2) and compare Fig. 14 for the whole proof. We have to find a morphism $m_2^i\colon P'_i \to G$ that embeds $(a' \circ c, b'_i \circ c) = \mathrm{PEShift}_{\rho_1}(a \circ c, b_i \circ c)$ to (t_1, t_2).

Since the squares (1) and (2) in Fig. 14 are pullbacks, (1) + (2) is a pullback and (3)+(4) is pullback, because (3) and (4) are pullbacks (where (4) is a pullback because q_i is a monomorphism). By the *essential characterization of parallel independence* of Corradini et al. [5] (see our Definition 4), the pullback (3) + (4) computing an isomorphism can be interpreted as saying that the transformations t_1 and $H_1 \Longrightarrow_{\mathsf{check}_{P'_i}, n_2^i \circ i_{P'_i}} H_1$ are parallel independent, where $\mathsf{check}_{P'_i}$ denotes the trivial rule $P'_i \overset{id}{\hookleftarrow} P'_i \overset{id}{\hookrightarrow} P'_i$ that just checks for the existence of P'_i. In particular, there is a morphism $d_i\colon P'_i \to D$ with $n_2^i \circ i_{P'_i} = h \circ d_i$. We set $m_2^i := g \circ d_i\colon P'_i \to G$ and prove that this morphism provides an embedding. Since $p_i \circ \ldots \circ p_1 = i_{P'_i} \circ i_{L_2}$ (by Proposition 2) and $m'_2 = n_2^i \circ p_i \circ \ldots \circ p_i$ (by definition of an embedding), we obtain $d = d_i \circ i_{L_2}$ (using the left-cancelability of h) and in particular $g \circ d = g \circ d_i \circ i_{L_2}$. Thus, the first requirement on m_2^i is satisfied (because $m_2 = g \circ d$). It remains to show that the square (5) is a pullback. For this, let (6) denote the square $k \circ a'_i = d_i \circ b'_i$ behind squares (5) and (8) and note that (6) is a pullback as in the proof of Proposition 2 (see Proposition 12). Moreover, we have that

$$
\begin{aligned}
m_1 \circ a' &= m_1 \circ l_1 \circ a'_i \\
&= g \circ k \circ a'_i \\
&= g \circ d_i \circ b'_i \ ,
\end{aligned}
$$

which means that (5) = (6) + (8), which makes (5) a pullback because both (6) and (8) are pullbacks. Summarizing, $m_2^i = g \circ d_i$ embeds $\mathrm{PEShift}_{\rho_1}(a \circ c, b_i \circ c)$ to (t_1, t_2).

Conversely, given an embedding $m_2^i\colon P'_i \to G$ of $(a' \circ c', b'_i \circ c') = \mathrm{PEShift}_{\rho_1}(a \circ c, b_i \circ c)$ in (t_1, t_2), we have to find a morphism $n_2^i\colon P_i \to H_1$ that embeds $(a \circ c, b_i \circ c)$ in (t_1^{-1}, t'_2). For the whole proof compare Proposition 15. First note that, by assumption, $g \circ d = m_2^i \circ i_{L_2}$ and $m_1 \circ a' \circ c' = m_2^i \circ b'_i \circ c'$ and (5) in Fig. 15 is a pullback. Since (3) is a pullback and $a' = l_1 \circ a'_i$, (4) is also a

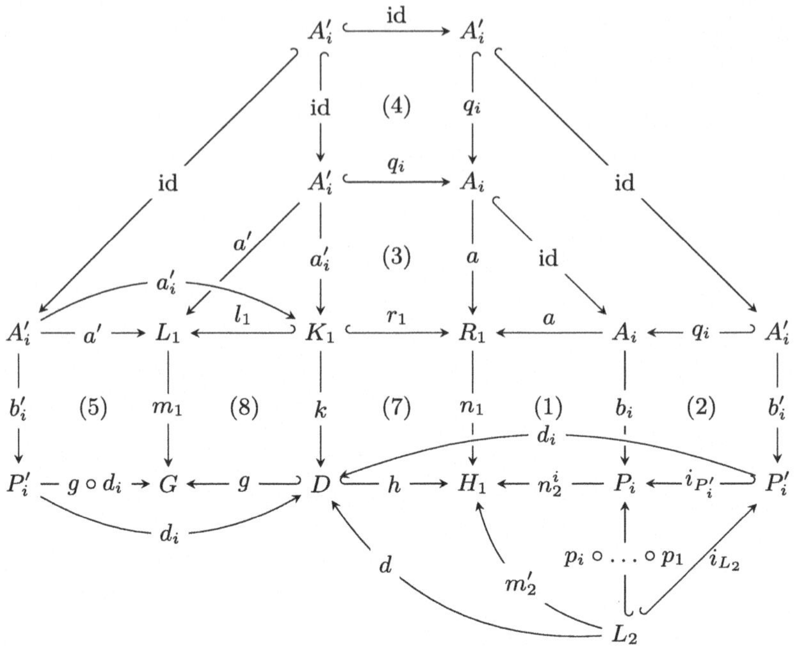

Fig. 14. Embedding of $\text{PEShift}_{\rho_1}(a \circ c, b_i \circ c)$ in the transformation pair (t_1, t_2)

pullback, i.e., there is a morphism $d_i \colon P'_i \to D$ with $m_2^i = g \circ d_i$ and we have $m_2 = h \circ d_1 \circ i_{L_2}$ (because $g, h \in \mathcal{M}$).

By construction, (2) is a pushout, and we have $n_1 \circ a \circ q_i = h \circ k \circ a'_i$ and $h \circ d_i \circ b'_i = h \circ k \circ a'_i$ (since it also holds that $m_2^i = g \circ d_i$, and $g \circ d_i \circ b'_i = g \circ k \circ a'_i$ which implies that $d_i \circ b'_i = k \circ a'_i$), i.e., there is a morphism $n_2^i \colon P_i \to H_1$ with $n_1 \circ a = n_2^i \circ b_i$, i.e., square (1) commutes, and $h \circ d_i = n_2^i \circ i_{P'_i}$ which implies that $m_2 = n_2^1 \circ p_i \circ \ldots \circ p_1$ (as $g \circ d = g \circ d_i \circ i_{L_2}$ implies that $d = d_i \circ i_{L_2}$ and therefore $m_2 = h \circ d = h \circ d_i \circ i_{L_2}$ and $p_i \circ \ldots \circ p_1 = i_{P'_i} \circ i_{L_2}$).

Finally, we need to show that (1) is a pullback. Consider the cube shown in Fig. 16. The bottom is equal to the first pushout of the transformation t_1, the back left face is equal to (3) in Fig. 15 (i.e., a pullback) and the top is equal to (2). The back right face is equal to (6) and also a pullback, because (6) + (7) and (7) are pullbacks. This implies that the front faces, and in particular (1) is a pullback, i.e., n_2^i is an embedding of $(a \circ c, b_i \circ c)$ in (t_1^{-1}, t'_2). □

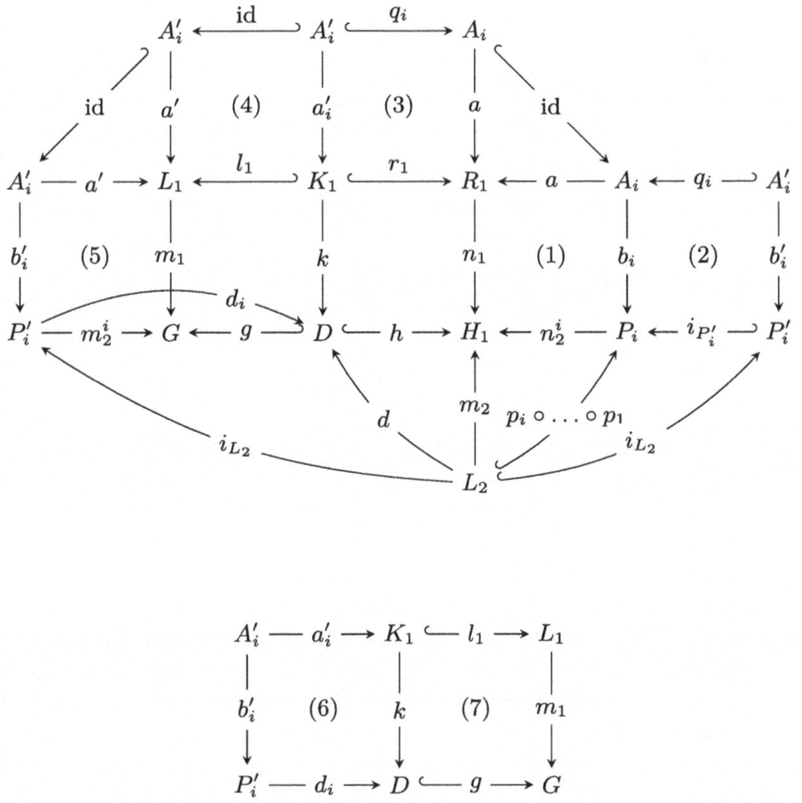

Fig. 15. Embedding of $(a \circ c, b_i \circ c)$ in the transformation pair (t_1^{-1}, t_2')

Proof (of Proposition 3). Let a disabling essence by deletion $(a \circ c\colon C \to L_1, b_i \circ c\colon C \to P_i) \in \mathrm{DEss}(\rho_1, \rho_2)$ of a rule $\rho_1 = (L_1 \xleftarrow{l_1} K_1 \xrightarrow{r_1} R_1, ac_1)$ in $\rho_2 = (L_2 \xleftarrow{l_2} K_2 \xrightarrow{r_2} R_2, ac_2)$ be given. By Definition 6, there is some cospan $s_L\colon L_1 \xrightarrow{i_{L_1}} L_1P_n \xleftarrow{i_{P_n}} P_n$ with $\mathrm{DPEss}^n(i_L, i_{P_n}) = (a \circ c, b_i \circ c)$. Also by Definition 6 (the definition of the recursion), we have that $P_i = L_2$ and $b_i = b_0$ if and only if $\mathrm{DPEss}^0(i_{L_1}, i^L_{P_n}) = \mathrm{ess}_{l_1}(i_{L_1}, i_{P_n} \circ \cdots \circ p_1) \neq (!_{L_1}, !_{L_2})$ (otherwise, we would have continued the recursion). That is, in this case, we compute exactly the disabling essence of Azzi et al. [2, Definition 3.8]. Moreover, any such disabling essence meets our Definitions 6 and 9 because also their definition requires disabling essences to be embbedable in transformation pairs. □

Proof (of Theorem 2). Given a pair of parallel dependent transformations $(t_1, t_2) = (G \Longrightarrow_{\rho_1, m_1} H_1, G \Longrightarrow_{\rho_2, m_2} H_2)$ via rules $\rho_1 = (L_1 \xleftarrow{l_1} K_1 \xrightarrow{r_1} R_1, ac_1)$ and $\rho_2 = (L_2 \xleftarrow{l_2} K_2 \xrightarrow{r_2} R_2, ac_2)$ so that t_1 causes the conflict for t_2, then either the match $m_2\colon L_2 \to G$ is destroyed (i.e., there is no morphism $d_1\colon L_2 \to D_1$ with $m_2 = g_1 \circ d_1$) or $h_1 \circ d_1 \not\models ac_2$ (see Fig. 17).

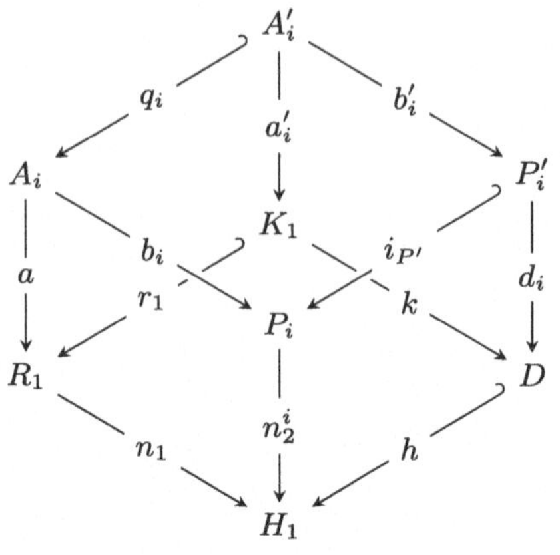

Fig. 16. Proving the pullback property for the embedding of $(a \circ c, b_i \circ c)$ in the transformation pair (t_1^{-1}, t'_2)

If the conflict arises by destroying the match m_2, the claim follows by Proposition 3 because a non-trivial disabling essence embeds into the transformation pair [2, Theorem 3.13] that we also compute as disabling essence.

Therefore, we consider the case that $h_1 \circ d_1 \not\models ac_2$ in the following. In this case, there is an object P_i contained in ac_2 so that either (i) P_i is existentially bound and an occurrence of P_i in G is deleted by t_1 or (ii) P_i is universally bound and an occurrence of P_i in H_1 inserted by t_1 – both in such a way that the contained occurrence of P_{i-1} is not destroyed (inserted). (That this intuitive way of reasoning for conditions in ANF is also formally precise is worked out in [16, Lemmas 3–6].) If P_i is existentially bound, there is an occurrence $m_2^i \colon P_i \to G$ with $m_2 = m_2^i \circ p_i \circ \ldots \circ p_1$ so that there is no morphism $d_i \colon P_i \to D_1$ with $m_2^i = g_1 \circ d_i$. This means, the transformations t_1 and $t_2^i \colon G \Longrightarrow_{\text{check}(P_i), m_2^i} G$ are parallel dependent, where $\text{check}(P_i) = P_i \xleftarrow{id} P_i \xhookrightarrow{id} P_i$ is the plain rule that just checks for the presence of P_i. By using the characterization of parallel dependence presented by Corradini et al. [5], the morphism $q_i \colon A'_i \hookrightarrow A_i$, obtained by constructing the pullbacks (1) and (2), is not an isomorphism. By computing the initial pushout over q_i, we obtain the span $(a \circ c, b_i \circ c)$. It remains to show that this span is a disabling essence and is contained in $\text{DEss}(\rho_1, \rho_2)$. For this, it suffices to show that $(a \circ c, b_i \circ c)$ is a proto-essence by deletion of ρ_1 in ρ_2 because m_2^i is an embedding of $(a \circ c, b_i \circ c)$ in (t_1, t_2) and P_i is existentially bound. This means, we have to find a cospan $s_L \colon L_1 \xhookrightarrow{i_{L_1}} L_1P_n \xhookleftarrow{i_{P_n}} P_n$, i.e., a cospan where the second morphism has domain P_n (and not P_i) that computes $(a \circ c, b_i \circ c)$ as a proto-essence by deletion.

By constructing the pushout (4) in Fig. 18, we obtain the cospan $s_L \colon L_1 \xhookrightarrow{i_{L_1}} L_1P_n \xhookleftarrow{i_{P_n}} P_n$. Since (1) (shown in Figs. 17 and 18) and (4) are pullbacks, (1)+(4) is also a pullback. This means, we can replace the square (1) in Fig. 17 by the square (1)+(4) and still compute the same squares (2) and (3) over l_1. Moreover,

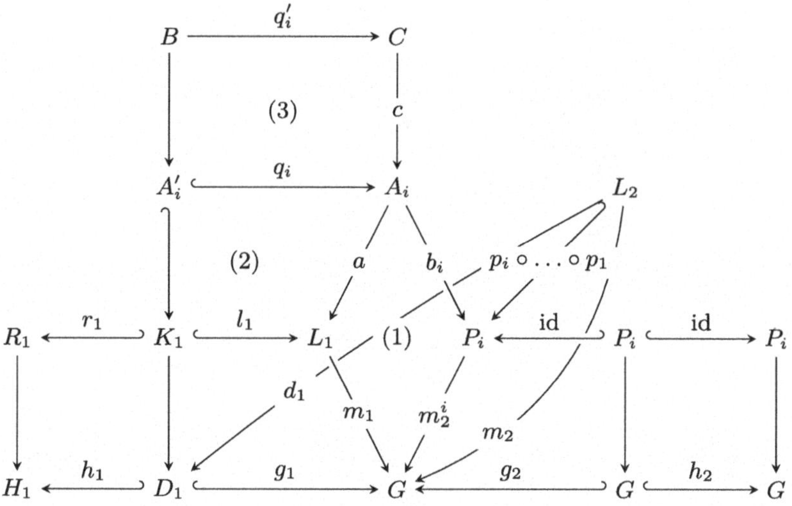

Fig. 17. Construction of proto-essence and conflict essence and transformation condition

since $m_2^i \circ p_i \colon P_{i-1} \to G$ is not destroyed by t_1, the morphism q_{i-1} which is obtained by (i) computing the pullback of $m_2^i \circ p_i$ and m_1 and (ii) computing the pullback of l_1 and $A_{i-1} \to L_1$ (which was computed by the first pullback) is an isomorphism. We also obtain q_{i-1} when computing $\mathrm{DPEss}^{i-1}(i_{L_1}, i_{P_n})$, i.e., $\mathrm{DPEss}^{i-1}(i_{L_1}, i_{P_n}) = (!_{L_1}, !_{P_{i-1}})$ and, as q_i is not an isomorphism, it follows that $\mathrm{DPEss}^n(i_{L_1}, i_{P_n}) = (a \circ c, b_i \circ c)$.

If P_i is universally bound, there is a newly introduced morphism $m_2^i \colon P_i \to H_1$ and the transformations t_1^{-1} and $H_1 \Longrightarrow_{\mathrm{check}_{P_i}, m_2^i} H_1$ are parallel dependent. Analogous to the first part, we can then show that there is a proto-essence by insertion $(a \circ c \colon C \to R_1, b_i \circ c \colon C \to P_i)$ so that m_2^i is an embedding of $(a \circ c \colon C \to R_1, b_i \circ c \colon C \to P_i)$ in $(t_1^{-1}, H_1 \Longrightarrow_{\rho_2, h_1 \circ d_1} X)$. With the correctness of the shift of proto-essences by insertion (Theorem 1), $\mathrm{PEShift}_{\rho_1}(a \circ c \colon C \to R_1, b_i \circ c \colon C \to P_i) \in \mathrm{IEss}(\rho_1, \rho_2)$ can be embedded into the transformation pair (t_1, t_2). □

Proof (of Theorem 3). Let two parallel independent transformations $t_1 \colon G \Longrightarrow_{\rho_1, m_1} H_1$ and $t_2 \colon G \Longrightarrow_{\rho_2, m_2} H_2$ via rules $\rho_1 = (L_1 \xleftarrow{l_1} K_1 \xrightarrow{r_1} R_1, ac_1)$ and $\rho_2 = (L_2 \xleftarrow{l_2} K_2 \xrightarrow{r_2} R_2, \forall(p_1 \colon L_2 \hookrightarrow P_1, \mathsf{false}))$ be given. In case no disabling essence of ρ_1 in ρ_2 is embeddable in (t_1, t_2), Theorem 2 implies that t_1 does not cause a conflict for t_2. For plain rules, the converse statement follows from Proposition 3 and [2, Theorem 3.13]. Thus, we only have to exclude the case that a disabling essence embeds into (t_1, t_2), which stems from the interaction of ρ_1 with the forbidden object P_1.

For this, assume that such a disabling essence embeds into (t_1, t_2). Because P_1 is universally bound, such a disabling essence $(a' \circ c', b_1' \circ c')$ stems from a proto-essence by insertion $(a \circ c \colon C \to R_1, b_1 \circ c \colon C \to P_1)$ and embeds into (t_1, t_2) if

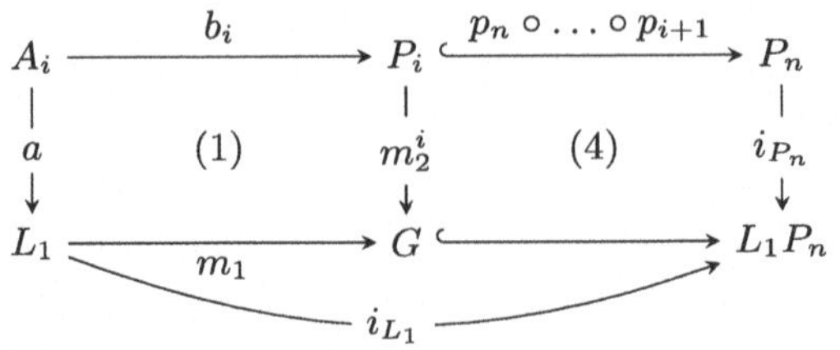

Fig. 18. Construction of proto-essence and conflict essence and transformation condition

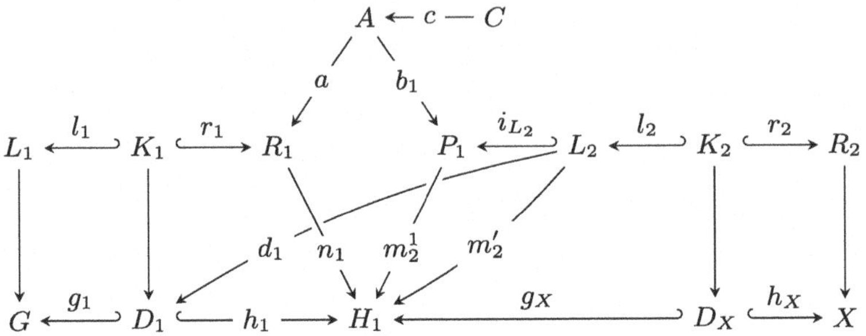

Fig. 19. Embedding of disabling essence caused by a NAC

and only if $(a \circ c, b_1 \circ c)$ embeds into (t_1^{-1}, t_2'), where $t_2' \colon H_1 \Longrightarrow_{\rho_2, m_2'} X$ exists by parallel independence of the underlying plain rules (i.e., $m_2' = h_1 \circ d_1$). Figure 19 shows such an embedding. By definition of the embedding, $m_2^1 \circ i_{L_2} = m_2'$; however, by definition of satisfaction of conditions, this means that $m_2' \not\models ac_2$. By $m_2' = h_1 \circ d_1$ this is a contradiction to the assumption that t_1 and t_2 are parallel independent. □

References

1. Azzi, G.G., Corradini, A., Ribeiro, L.: On the essence and initiality of conflicts. In: Lambers, L., Weber, J. (eds.) ICGT 2018. LNCS, vol. 10887, pp. 99–117. Springer, Cham (2018). https://doi.org/10.1007/978-3-319-92991-0_7
2. Azzi, G.G., Corradini, A., Ribeiro, L.: On the essence and initiality of conflicts in $\mathcal{M}$-adhesive transformation systems. J. Log. Algebraic Methods Program. **109** (2019). https://doi.org/10.1016/J.JLAMP.2019.100482
3. Born, K., Lambers, L., Strüber, D., Taentzer, G.: Granularity of conflicts and dependencies in graph transformation systems. In: de Lara, J., Plump, D. (eds.) ICGT 2017. LNCS, vol. 10373, pp. 125–141. Springer, Cham (2017). https://doi.org/10.1007/978-3-319-61470-0_8
4. Campbell, G., Plump, D.: Confluence up to garbage in graph transformation. Theor. Comput. Sci. **884**, 1–22 (2021). https://doi.org/10.1016/J.TCS.2021.06.010

5. Corradini, A., et al.: On the essence of parallel independence for the double-pushout and sesqui-pushout approaches. In: Heckel, R., Taentzer, G. (eds.) Graph Transformation, Specifications, and Nets. LNCS, vol. 10800, pp. 1–18. Springer, Cham (2018). https://doi.org/10.1007/978-3-319-75396-6_1
6. Ehrig, H., Ehrig, K., Prange, U., Taentzer, G.: Fundamentals of Algebraic Graph Transformation. Monographs in Theoretical Computer Science. An EATCS Series. Springer, Heidelberg (2006). https://doi.org/10.1007/3-540-31188-2
7. Ehrig, H., Ermel, C., Golas, U., Hermann, F.: Graph and Model Transformation – General Framework and Applications. Monographs in Theoretical Computer Science. An EATCS Series. Springer, Heidelberg (2015). https://doi.org/10.1007/978-3-662-47980-3
8. Ehrig, H., Golas, U., Habel, A., Lambers, L., Orejas, F.: $\mathcal{M}$-adhesive transformation systems with nested application conditions. Part 2: embedding, critical pairs and local confluence. Fundam. Inform. **118**(1–2), 35–63 (2012). https://doi.org/10.3233/FI-2012-705
9. Ehrig, H., Golas, U., Habel, A., Lambers, L., Orejas, F.: $\mathcal{M}$-adhesive transformation systems with nested application conditions. Part 1: parallelism, concurrency and amalgamation. Math. Struct. Comput. Sci. **24**(4) (2014). https://doi.org/10.1017/S0960129512000357
10. Ehrig, H., Habel, A., Lambers, L., Orejas, F., Golas, U.: Local confluence for rules with nested application conditions. In: Ehrig, H., Rensink, A., Rozenberg, G., Schürr, A. (eds.) ICGT 2010. LNCS, vol. 6372, pp. 330–345. Springer, Heidelberg (2010). https://doi.org/10.1007/978-3-642-15928-2_22
11. Ehrig, H., Habel, A., Padberg, J., Prange, U.: Adhesive high-level replacement categories and systems. In: Ehrig, H., Engels, G., Parisi-Presicce, F., Rozenberg, G. (eds.) ICGT 2004. LNCS, vol. 3256, pp. 144–160. Springer, Heidelberg (2004). https://doi.org/10.1007/978-3-540-30203-2_12
12. Habel, A., Pennemann, K.: Correctness of high-level transformation systems relative to nested conditions. Math. Struct. Comput. Sci. **19**(2), 245–296 (2009). https://doi.org/10.1017/S0960129508007202
13. Heckel, R., Taentzer, G.: Graph Transformation for Software Engineers - With Applications to Model-Based Development and Domain-Specific Language Engineering. Springer, Heidelberg (2020). https://doi.org/10.1007/978-3-030-43916-3
14. John, S., Kosiol, J., Lambers, L., Taentzer, G.: A graph-based framework for model-driven optimization facilitating impact analysis of mutation operator properties. Softw. Syst. Model. **22**(4), 1281–1318 (2023). https://doi.org/10.1007/S10270-022-01078-X
15. Knuth, D.E., Bendix, P.B.: Simple word problems in universal algebras. In: Leech, J. (ed.) Computational Problems in Abstract Algebra, pp. 263–297. Pergamon (1970). https://doi.org/10.1016/B978-0-08-012975-4.50028-X
16. Kosiol, J., Strüber, D., Taentzer, G., Zschaler, S.: Sustaining and improving graduated graph consistency: a static analysis of graph transformations. Sci. Comput. Program. **214**, 102729 (2022). https://doi.org/10.1016/J.SCICO.2021.102729
17. Lack, S., Sobocinski, P.: Adhesive and quasiadhesive categories. RAIRO Theor. Inform. Appl. **39**(3), 511–545 (2005). https://doi.org/10.1051/ITA:2005028
18. Lambers, L., Born, K., Kosiol, J., Strüber, D., Taentzer, G.: Granularity of conflicts and dependencies in graph transformation systems: a two-dimensional approach. J. Log. Algebraic Methods Program. **103**, 105–129 (2019). https://doi.org/10.1016/J.JLAMP.2018.11.004

19. Lambers, L., Born, K., Orejas, F., Strüber, D., Taentzer, G.: Initial conflicts and dependencies: critical pairs revisited. In: Heckel, R., Taentzer, G. (eds.) Graph Transformation, Specifications, and Nets. LNCS, vol. 10800, pp. 105–123. Springer, Cham (2018). https://doi.org/10.1007/978-3-319-75396-6_6
20. Lambers, L., Ehrig, H., Orejas, F.: Efficient conflict detection in graph transformation systems by essential critical pairs. In: Bruni, R., Varró, D. (eds.) Proceedings of the Fifth International Workshop on Graph Transformation and Visual Modeling Techniques, GT-VMT@ETAPS 2006, Vienna, Austria, 1–2 April 2006. Electronic Notes in Theoretical Computer Science, vol. 211, pp. 17–26. Elsevier (2006). https://doi.org/10.1016/J.ENTCS.2008.04.026
21. Lambers, L., Kosiol, J., Strüber, D., Taentzer, G.: Exploring conflict reasons for graph transformation systems. In: Guerra, E., Orejas, F. (eds.) ICGT 2019. LNCS, vol. 11629, pp. 75–92. Springer, Cham (2019). https://doi.org/10.1007/978-3-030-23611-3_5
22. Lambers, L., Orejas, F.: Transformation rules with nested application conditions: critical pairs, initial conflicts & minimality. Theor. Comput. Sci. **884**, 44–67 (2021). https://doi.org/10.1016/J.TCS.2021.07.023
23. Lambers, L., Strüber, D., Taentzer, G., Born, K., Huebert, J.: Multi-granular conflict and dependency analysis in software engineering based on graph transformation. In: Chaudron, M., Crnkovic, I., Chechik, M., Harman, M. (eds.) Proceedings of the 40th International Conference on Software Engineering, ICSE 2018, Gothenburg, Sweden, 27 May–03 June 2018, pp. 716–727. ACM (2018). https://doi.org/10.1145/3180155.3180258
24. Lauer, A., Kosiol, J., Taentzer, G.: Empowering model repair: a rule-based approach to graph repair without side effects - extended version. Innov. Syst. Softw. Eng. **20**(4), 597–618 (2024). https://doi.org/10.1007/S11334-024-00587-W
25. Plump, D.: Hypergraph rewriting: critical pairs and undecidability of confluence. In: Sleep, M.R., Plasmeijer, M.J., van Eekelen, M.C.J.D. (eds.) Term Graph Rewriting: Theory and Practice, pp. 201–213. John Wiley and Sons Ltd. (1993). https://doi.org/10.5555/167817.167851
26. Rensink, A.: Representing first-order logic using graphs. In: Ehrig, H., Engels, G., Parisi-Presicce, F., Rozenberg, G. (eds.) ICGT 2004. LNCS, vol. 3256, pp. 319–335. Springer, Heidelberg (2004). https://doi.org/10.1007/978-3-540-30203-2_23
27. Sandmann, C., Habel, A.: Rule-based graph repair. In: Echahed, R., Plump, D. (eds.) Proceedings Tenth International Workshop on Graph Computation Models, GCM@STAF 2019, Eindhoven, The Netherlands, 17th July 2019. EPTCS, vol. 309, pp. 87–104 (2019). https://doi.org/10.4204/EPTCS.309.5

Specifying Graph Properties via Automata and Logic

Specifying and Checking Graph Properties with Alternating Graph Automata

Frank Drewes[1], Berthold Hoffmann[2], and Mark Minas[3](✉)

[1] Umeå universitet, 90187 Umeå, Sweden
drewes@cs.umu.se
[2] Universität Bremen, 28334 Bremen, Germany
hof@uni-bremen.de
[3] Universität der Bundeswehr München, 85577 Neubiberg, Germany
mark.minas@unibw.de

Abstract. In many application areas, it is necessary to specify and check properties of graphs. Such properties may be complex, placing structural requirements on graph regions of unbounded size. In this paper, we show that alternating graph automata can check such graph properties, e.g., whether a given input graph is a tree, or whether it contains a Hamiltonian cycle or not. In fact, we show that these automata can accept PSPACE-complete graph languages, and that their uniform membership problem is contained in PSPACE when only automata without permutation cycles are considered.

1 Introduction

Graph-like structures are ubiquitous in computer science and beyond. In many cases, they need to fulfill application-dependent structural properties that have to be specified and checked. In the literature, such properties have been called global [15] if they concern graph regions of unbounded size, like connectedness or the existence of Hamiltonian paths.

Bruggink *et al.* [5] have lifted alternating (finite) string automata [7] to alternating graph automata,[1] which consume input graphs so that in their existential states, at least one onward transition step must eventually lead to acceptance, and in their universal states, all onward transition steps must do so.

We generalize the definition of alternating graph automata by allowing permutation cycles, i.e., cyclic transition sequences that do not make progress in processing the graph. More importantly, we redefine the semantics of these automata so that, whenever a transition step identifies a subgraph Γ of the input, it can be consumed, even if some otherwise dangling edges attached to Γ have to be deleted in the step. As a consequence, our automata can also accept graph languages like the language of all binary trees, which Bruggink *et al.* conjecture to be beyond the capabilities of their automata [5, Conjecture 1].

[1] They also lifted it further to a category-theoretical setting.

J. Endrullis and M. Tichy (Eds.): ICGT 2025, LNCS 15720, pp. 93–111, 2025.
https://doi.org/10.1007/978-3-031-94706-3_5

We show that the *non-uniform* membership problem for alternating graph automata is in PSPACE, and even PSPACE-complete, by presenting a PSPACE-complete graph language that can be recognized by such automata. Furthermore, we show that the *uniform* membership problem is also in PSPACE when only automata without permutation cycles are considered.

The remainder of this paper is structured as follows. First we recall some preliminaries regarding graphs that can be concatenated at their front and rear interfaces. In Sect. 3, we (re-) introduce alternating graph automata, define their semantics, and discuss the differences between our proposal and the one by Bruggink *et al.* [5]. Our PSPACE complexity results are proved in Sect. 4. We conclude with some remarks on related and future work, in Sect. 5.

2 Preliminaries

We let $\mathbb{N}$ denote the set of non-negative integers, and $[n]$ the set $\{1, \dots, n\}$ for all $n \in \mathbb{N}$. A^* denotes the set of all finite sequences over a set A; the empty sequence is denoted by ε. For a sequence $s = a_1 \cdots a_n$ with $n \in \mathbb{N}$ and $a_1, \dots, a_n \in A$, $[s]$ denotes the set $\{a_1, \dots, a_n\}$.

We consider edge-labeled hypergraphs (which we simply call graphs), i.e., edges are attached to sequences of nodes. To be able to concatenate graphs in the way originally proposed by Engelfriet and Vereijken [13], we supply each graph with two sequences of distinguished nodes that constitute its front and rear interfaces. As in [10], we require these sequences to be free of repetitions.

A ranked alphabet (Σ, rank) consists of a finite set Σ of symbols and a function $\mathit{rank} \colon \Sigma \to \mathbb{N}$, which assigns a rank to each symbol $a \in \Sigma$. The pair (Σ, rank) is usually identified with Σ, keeping *rank* implicit. For $k \in \mathbb{N}$, we let $\Sigma^{(k)} = \{a \in \Sigma \mid \mathit{rank}(a) = k\}$ denote the set of symbols of rank k in Σ.

Definition 1 (Graph). A *graph* over a ranked alphabet Σ is a tuple $G = (\dot{G}, \bar{G}, \mathit{att}_G, \mathit{lab}_G, \mathit{front}_G, \mathit{rear}_G)$, where $\dot{G}$ and $\bar{G}$ are disjoint finite sets of *nodes* and *edges*, respectively, $\mathit{att}_G \colon \bar{G} \to \dot{G}^*$ attaches sequences of nodes to edges, $\mathit{lab}_G \colon \bar{G} \to \Sigma$ labels edges with symbols so that $|\mathit{att}_G(e)| = \mathit{rank}(\mathit{lab}_G(e))$ for every edge $e \in \bar{G}$, and the repetition-free node sequences $\mathit{front}_G, \mathit{rear}_G \in \dot{G}^*$ are the *front* and the *rear* interface, respectively.

The front and rear interfaces determine the *type* of each graph G: a graph G with $|\mathit{front}_G| = m$ and $|\mathit{rear}_G| = n$ has $\mathit{type}(G) = (m, n)$. The set of all graphs of type (m, n) is denoted by $\mathbb{G}_\Sigma^{(m,n)}$. Furthermore, $\mathbb{G}_\Sigma^m = \bigcup_{n \in \mathbb{N}} \mathbb{G}_\Sigma^{(m,n)}$ denotes the set of all graphs over Σ whose front is of length m, and $\mathbb{G}_\Sigma = \bigcup_{m,n \in \mathbb{N}} \mathbb{G}_\Sigma^{(m,n)}$ denotes the set of all graphs over Σ, regardless of type.

A *permutation graph* is a graph G with $\bar{G} = \varnothing$ and $\dot{G} = [\mathit{front}_G] = [\mathit{rear}_G]$. If, furthermore, $\mathit{front}_G = \mathit{rear}_G$ and $|\dot{G}| = n$, then G is called an *identity graph* (on n nodes) and is denoted by Id_n.

Example 1. Figure 1 shows four graphs over the ranked alphabet $\Sigma = \{\smile, \sim, \diamond\}$ with $\mathit{rank}(\smile) = \mathit{rank}(\sim) = 2$ and $\mathit{rank}(\diamond) = 3$. As usual, we draw nodes as circles

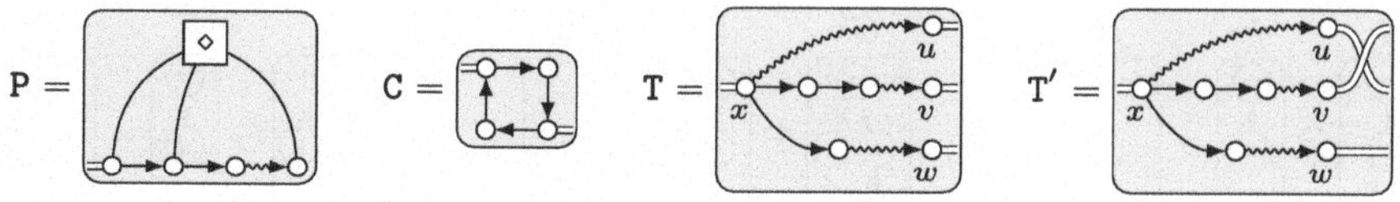

Fig. 1. Four graphs.

and non-binary edges as boxes, the boxes containing the edge label and being connected to the attached nodes by lines.[2] Binary edges like the ones labeled $\llcorner\!\lrcorner$ and $\sim$ in this example are drawn as arrows; in general, their labels are ascribed to the arrows, but here we just distinguish them by drawing edges with the "invisible" label $\llcorner\!\lrcorner$ as straight arrows, and the others as squiggly arrows.

The front and rear interface nodes are represented by double lines that connect to the left and right borders of the graph, respectively. Both node sequences are arranged in order from top to bottom. Here, $\mathtt{P}$ is of type $(1,0)$, $\mathtt{C}$ is of type $(1,1)$, and both $\mathtt{T}$ and $\mathtt{T}'$ are of type $(1,3)$. Their front interfaces are identical, with $\mathit{front}_{\mathtt{T}} = \mathit{front}_{\mathtt{T}'} = x$; however, their rear interfaces differ: $\mathit{rear}_{\mathtt{T}} = uvw$ while $\mathit{rear}_{\mathtt{T}'} = vuw$. Thus, even though $\mathtt{T}$ and $\mathtt{T}'$ would appear identical if the rear interfaces are ignored, they are considered distinct graphs because their rear interfaces are different.

We follow Engelfriet and Vereijken [13] in defining graph concatenation.

Definition 2 (Graph Concatenation). Let $G \in \mathbb{G}_{\Sigma}^{(i,k)}$ and $H \in \mathbb{G}_{\Sigma}^{(k,j)}$. We assume for simplicity that $\mathit{rear}_G = \mathit{front}_H$, $\dot{G} \cap \dot{H} = [\mathit{rear}_G] = [\mathit{front}_H]$, and $\bar{G} \cap \bar{H} = \varnothing$. (Otherwise appropriate isomorphic copies of G and H are used.) The *(typed) concatenation* $G \cdot H$ of G and H is the graph C such that $\dot{C} = \dot{G} \cup \dot{H}$, $\bar{C} = \bar{G} \cup \bar{H}$, $\mathit{att}_C = \mathit{att}_G \cup \mathit{att}_H$, $\mathit{lab}_C = \mathit{lab}_G \cup \mathit{lab}_H$, $\mathit{front}_C = \mathit{front}_G$, and $\mathit{rear}_C = \mathit{rear}_H$. Thus $C \in \mathbb{G}_{\Sigma}^{(i,j)}$.

Note that the concatenation $G \cdot H$ is defined on concrete graphs if the assumptions in Definition 2 are satisfied, but only up to isomorphism if an isomorphic copy of G or H must be taken. To avoid unnecessary technicalities, we usually assume without mentioning that the former is indeed the case.

We will need the notion of frontal subgraphs and how to cut them off in order to define the semantics of alternating graph automata in the next section.

Definition 3 (Frontal Subgraphs). A graph $G \in \mathbb{G}_{\Sigma}^{(i,k)}$ is a *frontal subgraph* of a graph $H \in \mathbb{G}_{\Sigma}^{(i,j)}$ if

- $\dot{G} \subseteq \dot{H}$,
- $\bar{G} \subseteq \bar{H}$ with $\mathit{lab}_G(e) = \mathit{lab}_H(e)$ and $\mathit{att}_G(e) = \mathit{att}_H(e)$ for all $e \in \bar{G}$,
- $\mathit{front}_G = \mathit{front}_H$, and
- for all $v \in \dot{G}$, $v \in [\mathit{rear}_H]$ implies $v \in [\mathit{rear}_G]$.

[2] In our examples, there will be no need to indicate the order of the attached nodes of non-binary edges.

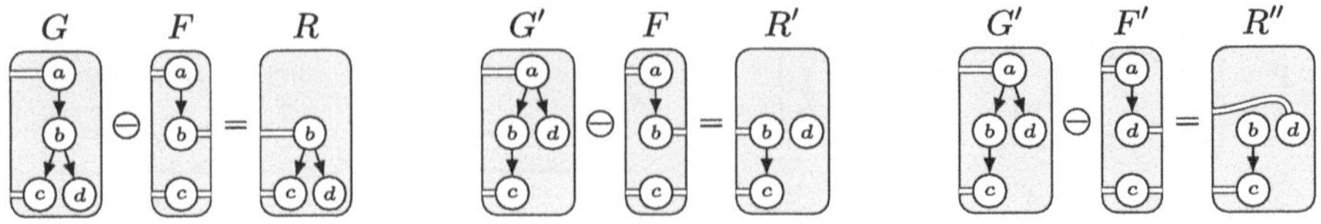

Fig. 2. Cutting off frontal subgraphs.

Then, *cutting G off H* yields the graph $H \ominus G$, which is obtained from H by removing all nodes in $\dot{G}\backslash[rear_G]$, as well as all edges that are either in $\bar{G}$ or attached to a node in $\dot{G}\backslash[rear_G]$. In the resulting graph, $rear_G$ becomes the front interface and $rear_H$ becomes the rear interface.

Example 2 (Cutting off Frontal Subgraphs). Figure 2 shows graphs G and G' of type $(2, 0)$, from which frontal subgraphs F and F' of type $(2, 2)$ are cut off, yielding the results R, R', and R''. Nodes with the same name (written inside the circle) correspond to each other in subgraph relations. Though F and F' are isomorphic, they are different frontal subgraphs of G', leading to different results $R' = G' \ominus F$ and $R'' = G' \ominus F'$. Note that cutting F and F' off G' also removes the other edge attached to node a, which would otherwise dangle. Thus, $\ominus$ is not the inverse of graph concatenation.

3 Alternating Graph Automata

We now define our slightly adapted version of the alternating graph automata by Bruggink *et al.* [5]. They distinguish between two types of states: *existential* and *universal.* Existential states behave similarly to those in classical finite automata: upon reaching such a state, there must *exist* at least one outgoing transition that consumes part of the input graph and eventually leads to acceptance. In contrast, when a universal state is reached, the automaton must simultaneously follow *all* outgoing transitions in *every* possible way, each consuming part of the input graph, and all resulting paths must lead to acceptance.

Definition 4 (Alternating Graph Automata). An *alternating graph automaton* over Σ is given by $\mathfrak{A} = (\Sigma, Q, \Delta, q_0, Q_\forall)$ where

- Σ is a ranked alphabet of edge labels,
- Q is a ranked alphabet of *states*,
- $\Delta \subseteq \bigcup_{m,n\in\mathbb{N}} \left(Q^{(m)} \times \mathbb{G}_\Sigma^{(m,n)} \times Q^{(n)}\right)$ is the set of *transitions*,
- $q_0 \in Q$ is the *initial state*, and
- $Q_\forall \subseteq Q$ is the subset of *universal states.*

The set $Q\backslash Q_\forall$ of *existential* states is denoted by $Q_\exists$. A *permutation cycle* of $\mathfrak{A}$ is a sequence $\delta_1\delta_2\cdots\delta_\ell$ of $\ell \geqslant 1$ transitions where $\delta_i = (q_i, \Gamma_i, q_{i+1}) \in \Delta$ and Γ_i is a permutation graph for each $i \in [\ell]$, and $q_{\ell+1} = q_1$.

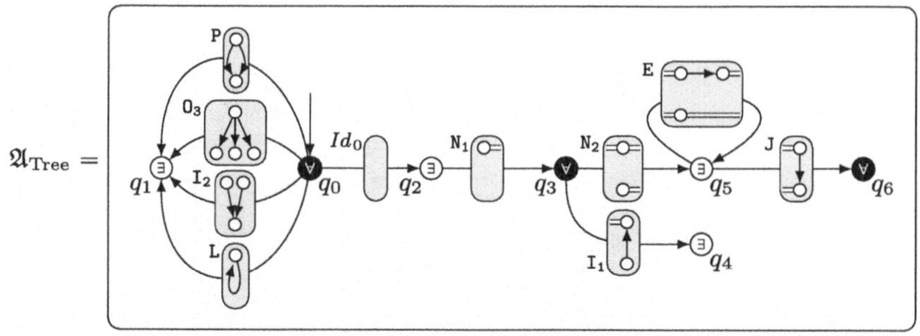

Fig. 3. The automaton accepting the language of binary trees.

In what follows we will sometimes just speak of *automata* if we refer to alternating graph automata according to the preceding definition.

Example 3. Figure 3 shows an example of such an automaton. We will explain in Example 5, after the formal definition of the semantics of our automata, that this one accepts the language of (non-empty) binary trees.

Automata are drawn as transition diagrams. The dangling incoming arrow indicates the initial state q_0, and transitions (q, Γ, q') are drawn as arrows from q to q' with the graph Γ drawn on it. We refer to these graphs by the small labels to the left of them. The interior of existential states is white with a small inscribed $\exists$, and that of universal states is black with a small inscribed $\forall$.

We define how automata accept input graphs. For this, we first define the *configuration graph* of an automaton applied to a given input graph.

Definition 5 (Configuration Graph). Let $\mathfrak{A} = (\Sigma, Q, \Delta, q_0, Q_\forall)$ be an alternating graph automaton.

A *configuration* is a pair $(q, G) \in \bigcup_{m\in\mathbb{N}} \left(Q^{(m)} \times \mathbb{G}_\Sigma^m\right)$. Such a configuration is *universal* if q is, and *existential* otherwise. There is a *(transition) step* $(q, G) \vdash_\Delta (q', G')$ from configuration (q, G) to (q', G') if there is a transition $(q, \Gamma, q') \in \Delta$ such that (an isomorphic copy of) Γ is a frontal subgraph of G and $G' = G \ominus \Gamma$.

The *configuration graph* of $\mathfrak{A}$ for an input graph $G_0 \in \mathbb{G}_\Sigma^{rank(q_0)}$ is the smallest graph $CG_\mathfrak{A}(G_0)$ over configurations containing the *initial configuration* (q_0, G_0) and, for each configuration (q, G) that it contains and each step $(q, G) \vdash_\Delta (q', G')$, the node (q', G') and an edge from (q, G) to (q', G') representing this step.

Note that configuration graphs have empty fronts and rears, and their edges are unlabeled (or all labeled with a special "invisible" label). Note also that every configuration graph is finite since the set of configurations reachable from (q_0, G_0) is finite, and that it may be cyclic if $\mathfrak{A}$ contains permutation cycles.

Example 4. Figure 4 displays the configuration graph $CG_{\mathfrak{A}_{\mathrm{Tree}}}(T)$ of the automaton $\mathfrak{A}_{\mathrm{Tree}}$ shown in Fig. 3. We draw configurations, i.e., the nodes of the configuration graph, as boxes with $\forall$ or $\exists$, indicating whether they are universal or

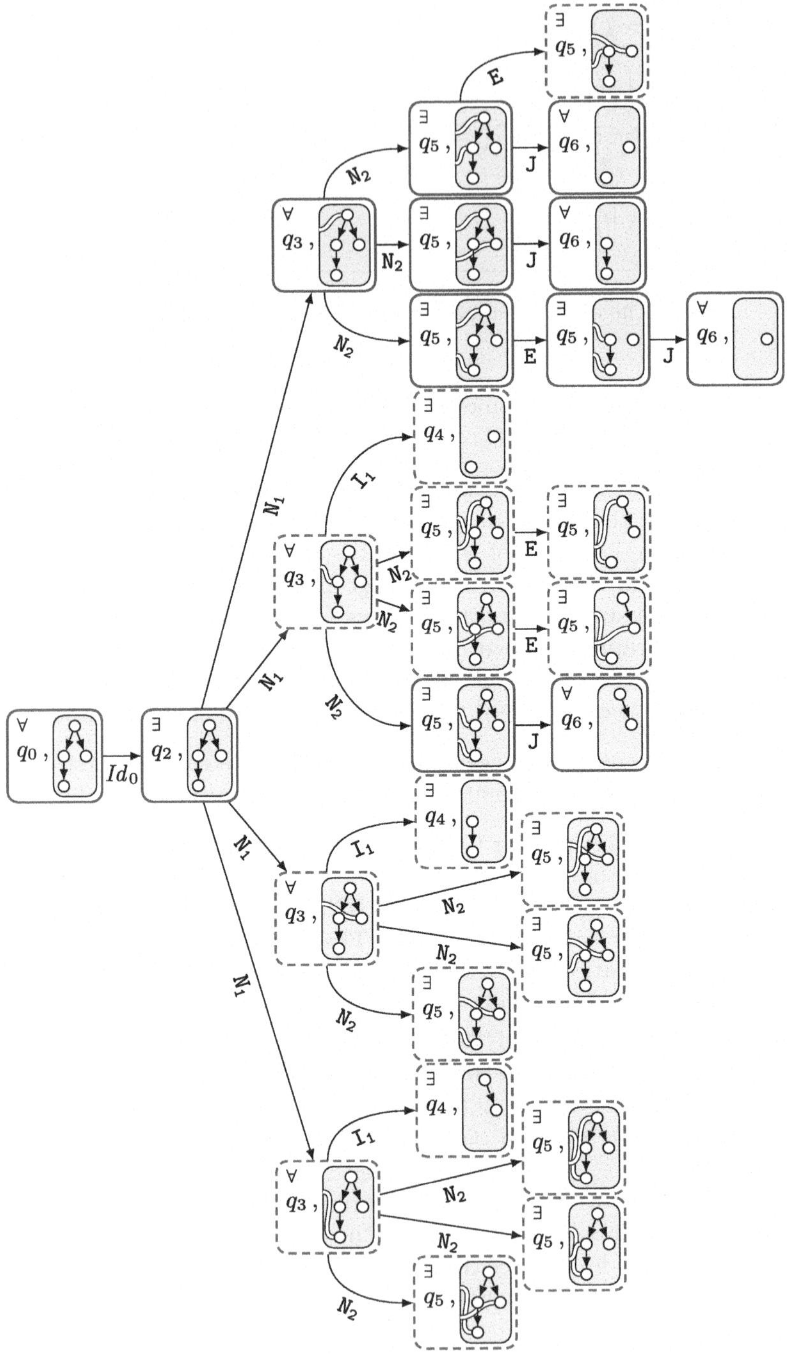

Fig. 4. A configuration graph of $\mathfrak{A}_{\text{Tree}}$ (Color figure online)

existential, and ascribe to the edges the name of the graph consumed in the corresponding transition. The different borders of the boxes will become relevant in Example 5. The initial configuration is represented by the leftmost box; it contains the input graph T, which is a tree with three edges. Note that $CG_{\mathfrak{A}_{\text{Tree}}}(T)$ is acyclic because $\mathfrak{A}_{\text{Tree}}$ does not contain permutation cycles. We will use this configuration graph in Example 5 to demonstrate how $\mathfrak{A}_{\text{Tree}}$ accepts T.

We will define the semantics of automata with respect to their configuration graphs by assigning truth values to configurations based on whether they are accepting or rejecting. Since configuration graphs may contain cycles, this truth assignment cannot, in general, be defined recursively. Instead, we define how this assignment evolves from an initial truth assignment:

Definition 6 (Truth Assignment Evolution). Let $K = CG_{\mathfrak{A}}(G_0)$ be the configuration graph of an alternating graph automaton $\mathfrak{A} = (\Sigma, Q, \Delta, q_0, Q_\forall)$ for an input graph $G_0 \in \mathbb{G}_\Sigma^{rank(q_0)}$. We let Ass_K denote the set of all partial functions $\alpha\colon \dot{K} \to \{true, false\}$ assigning truth values to (some of) the configurations in K. A truth assignment $\alpha \in Ass_K$ *evolves* (directly) into the truth assignment $\alpha' \in Ass_K$ given as follows, for all $c \in \dot{K}$: if

(1) c is universal and
  (1.1) has only outgoing edges to nodes $c' \in \dot{K}$ such that $\alpha(c') = true$ or
  (1.2) has an outgoing edge to some $c' \in \dot{K}$ such that $\alpha(c') = false$, or
(2) c is existential and
  (2.1) has only outgoing edges to nodes $c' \in \dot{K}$ such that $\alpha(c') = false$ or
  (2.2) has an outgoing edge to some $c' \in \dot{K}$ such that $\alpha(c') = true$

then

$$\alpha'(c) = \begin{cases} true \text{ in cases (1.1) and (2.2)} \\ false \text{ in cases (1.2) and (2.1).} \end{cases}$$

If none of the conditions (1.1)–(2.2) is satisfied, then $\alpha'(c) = \alpha(c)$.

The function that maps every truth assignment to the one it evolves into is denoted by Evo_K, i.e., in the case above $\alpha' = Evo_K(\alpha)$.

Given an initial truth assignment $\alpha \in Ass_K$, the sequence $(Evo_K^i(\alpha))_{i\in\mathbb{N}}$ may or may not have a fixed point, i.e., there may or may not be an $i \in \mathbb{N}$ such that $Evo_K^i(\alpha) = Evo_K^{i+1}(\alpha) = \cdots$. If such an i exists, we denote $Evo_K^i(\alpha)$ by $Evo_K^*(\alpha)$. In the following, we also use the notations $\alpha_\perp$, α_{false}, and α_{true} to denote the assignments in Ass_K that are everywhere undefined, everywhere false, and everywhere true, respectively (where K will always be clear from the context).

For $\alpha, \alpha' \in Ass_K$, we let $\alpha \leqslant \alpha'$ if, for all $c \in \dot{K}$, it holds that both $\alpha(c)$ and $\alpha'(c)$ are undefined, or $\alpha(c) = false$ or $\alpha'(c) = true$. Furthermore, $\alpha \sqsubseteq \alpha'$ if, for all $c \in \dot{K}$ such that $\alpha(c)$ is defined, it holds that $\alpha'(c) = \alpha(c)$. Clearly, $\leqslant$ and $\sqsubseteq$ are partial orders.

From here onwards, let $\alpha_b^i = Evo_K^i(\alpha_b)$ for all $i \in \mathbb{N}$ and $b \in \{false, true, \perp\}$.

Lemma 1. *Let $K = CG_{\mathfrak{A}}(G_0)$ be the configuration graph of an automaton $\mathfrak{A} = (\Sigma, Q, \Delta, q_0, Q_\forall)$ for an input graph $G_0 \in \mathbb{G}_\Sigma^{rank(q_0)}$. Then the following hold:*

(1) The fixed point $\alpha^* = Evo_K^*(\alpha_\perp)$ *exists.*
(2) For all $i \in \mathbb{N}$, $b \in \{false, true\}$, *and* $c \in \dot{K}$ *it holds that*

$$\alpha_b^i(c) = \begin{cases} \alpha_\perp^i(c) & \textit{if } \alpha_\perp^i(c) \textit{ is defined} \\ b & \textit{otherwise.} \end{cases}$$

(3) There are unique smallest and largest fixed points $\underline{\alpha}$ *and* $\overline{\alpha}$ *of* Evo_K *with respect to the partial order* $\leqslant$, *namely* $\underline{\alpha} = Evo_K^*(\alpha_{false})$ *and* $\overline{\alpha} = Evo_K^*(\alpha_{true})$. *For all* $c \in \dot{K}$, *these are given by*

$$\underline{\alpha}(c) = \begin{cases} \alpha^*(c) & \textit{if } \alpha^*(c) \textit{ is defined} \\ false & \textit{otherwise} \end{cases} \qquad \overline{\alpha}(c) = \begin{cases} \alpha^*(c) & \textit{if } \alpha^*(c) \textit{ is defined} \\ true & \textit{otherwise.} \end{cases}$$

Thus, $\alpha^* = Evo_K^*(\alpha_\perp)$ is the "most undefined" fixed point of Evo_K, and the smallest and largest fixed points of Evo_K according to $\leqslant$ are obtained from it by turning all undefined truth values into *false* and *true*, respectively.

Proof. (1) We show the following by induction on i: for every $c \in \dot{K}$ and $i \in \mathbb{N}$, if $\alpha_\perp^i(c)$ is defined then $\alpha_\perp^{i+1}(c) = \alpha_\perp^i(c)$. In other words, $\alpha_\perp^i \sqsubseteq \alpha_\perp^{i+1}$ and thus $\alpha^* = Evo_K^*(\alpha_\perp)$ indeed exists because Ass_K is finite (a trivial application of Kleene's fixed-point theorem).
The claim is obviously true for $i = 0$. For the induction step, consider some $c \in \dot{K}$ and $i \in \mathbb{N}$ such that $\alpha_\perp^{i+1}(c)$ is defined and assume as the induction hypothesis that, for every $c' \in \dot{K}$, either $\alpha_\perp^i(c')$ is undefined or $\alpha_\perp^{i+1}(c') = \alpha_\perp^i(c')$. In particular, this holds for all c' such that K contains an edge from c to c'. However, this means in all four cases (1.1)–(2.2) of Definition 6 that $\alpha_\perp^{i+2}(c) = \alpha_\perp^{i+1}(c)$. (For example, if $\alpha_\perp^{i+1}(c) = \mathit{false}$ because c is universal and K contains an edge from c to c' such that $\alpha_\perp^i(c') = \mathit{false}$ (case (1.2)), then $\alpha_\perp^{i+2}(c) = \mathit{false}$ because $\alpha_\perp^{i+1}(c') = \mathit{false}$ by the induction hypothesis.)
(2) This follows by induction on i in a way entirely analogous to the proof of (1), additionally using the monotonicity of $\wedge$ and $\vee$.
(3) This is a direct consequence of (1) and (2). □

In the following, we shall continue to use the notations α^*, $\underline{\alpha}$, and $\overline{\alpha}$ for the most undefined, smallest, and largest fixed points of Evo_K if the alternating graph automaton and the input graph (and thus K) are clear from the context.

Definition 7 (Languages Accepted by an Automaton). Let $\mathfrak{A} = (\Sigma, Q, \Delta, q_0, Q_\forall)$ be an alternating graph automaton. An input graph $G_0 \in \mathbb{G}_\Sigma^{rank(q_0)}$ is *accepted* by $\mathfrak{A}$ if $\alpha^*(q_0, G_0) = \mathit{true}$ and is *rejected* by $\mathfrak{A}$ if $\alpha^*(q_0, G_0) = \mathit{false}$.

The *lower language accepted by* $\mathfrak{A}$ is the set $\underline{L}(\mathfrak{A})$ of all graphs $G_0 \in \mathbb{G}_\Sigma^{rank(q_0)}$ accepted by $\mathfrak{A}$, and the *upper language accepted by* $\mathfrak{A}$ is the set $\overline{L}(\mathfrak{A})$ of all graphs $G_0 \in \mathbb{G}_\Sigma^{rank(q_0)}$ not rejected by $\mathfrak{A}$.

By Lemma 1, an equivalent definition of $\underline{L}(\mathfrak{A})$ and $\overline{L}(\mathfrak{A})$ is

$$\underline{L}(\mathfrak{A}) = \{G_0 \in \mathbb{G}_\Sigma^{rank(q_0)} \mid \underline{\alpha}(q_0, G_0) = \mathit{true}\} \text{ and}$$
$$\overline{L}(\mathfrak{A}) = \{G_0 \in \mathbb{G}_\Sigma^{rank(q_0)} \mid \overline{\alpha}(q_0, G_0) = \mathit{true}\}.$$

In the special case where α^* is a total function for all input graphs in $\mathbb{G}_\Sigma^{rank(q_0)}$, we have $\underline{L}(\mathfrak{A}) = \overline{L}(\mathfrak{A})$. Then we say that $L(\mathfrak{A}) = \underline{L}(\mathfrak{A})$ is the *language accepted by* $\mathfrak{A}$. In particular, this is the case if $\mathfrak{A}$ does not have permutation cycles, because then each configuration graph is a directed acyclic graph (DAG) and the construction of α^* as the fixed point of $Evo_K^0(\alpha_\perp), Evo_K^1(\alpha_\perp), Evo_K^2(\alpha_\perp), \ldots$ can be replaced by a simple recursive evaluation of $\alpha^*(c)$ for all $c \in \dot{K}$. More precisely, let $C = \{c' \in \dot{K} \mid c \vdash_\Delta c'\}$. If $\mathfrak{A}$ does not have permutation cycles then

$$\alpha^*(c) = \begin{cases} \mathit{true} \text{ if } c \text{ is universal and } \alpha^*(c') = \mathit{true} \text{ for all } c' \in C \text{ or} \\ \qquad\qquad c \text{ is existential and } \alpha^*(c') = \mathit{true} \text{ for some } c' \in C \\ \mathit{false} \text{ otherwise.} \end{cases}$$

Lemma 2 (Complementation). *Let* $\mathfrak{A} = (\Sigma, Q, \Delta, q_0, Q_\forall)$ *be an alternating graph automaton and let* $\mathfrak{A}^c = (\Sigma, Q, \Delta, q_0, Q \backslash Q_\forall)$. *Then* $\overline{L}(\mathfrak{A}^c)$ *is the complement of* $\underline{L}(\mathfrak{A})$, *that is,*

$$\overline{L}(\mathfrak{A}^c) = \mathbb{G}_\Sigma^{rank(q_0)} \backslash \underline{L}(\mathfrak{A}).$$

In particular, if $\mathfrak{A}$ *does not have permutation cycles, then* $L(\mathfrak{A}^c)$ *is the complement of* $L(\mathfrak{A})$.

Proof. Straightforward, using the duality principle of Boolean algebra and the fact that all relevant definitions are symmetric with respect to existential and universal states. □

Note that every state q without outgoing transitions can be considered as accepting or rejecting, depending on whether it is universal or not. This is because, for every configuration $c \in \dot{K}$ which does not have outgoing edges in K, by the definition of Evo_K it holds that $Evo_K(\alpha)(c) = \mathit{true}$ if c is universal, and $Evo_K(\alpha)(c) = \mathit{false}$ if c is existential, for every assignment α. In particular, this is the case for $c = (q, G)$ if q has no outgoing transitions at all.

Our definition differs from notions of alternating automata found in the literature as follows:

1. In the definition of alternating string automata of Chandra *et al.* [7], the success of some state q is defined via a Boolean expression over its successor states $q_1, \ldots, q_k$, using conjunction, disjunction, and negation, whereas our definitions just allows conjunctions $q_1 \wedge \cdots \wedge q_k$ for universal states, and disjunctions $q_1 \vee \cdots \vee q_k$ for existential ones. However, nested conjunctions and disjunctions can be simulated by inserting additional states with identical transitions, and negation can be realized by means of complementation according to Lemma 2.
2. Bruggink *et al.* [5] require that alternating graph automata are free of permutation cycles. More importantly, their automata may apply a transition (q, Γ, q') to a graph G only if G is the concatenation $G = \Gamma \cdot G'$ of (an isomorphic copy of) Γ with the remainder G', whereas our definition may apply transitions whenever Γ occurs as a frontal subgraph, and the remainder is $G' = G \ominus \Gamma$, removing edges of G that would be dangling if one of their attached nodes in Γ is removed.

Bruggink *et al.* conjecture that there is no alternating graph automaton with their semantics that can describe the language of binary trees [5, Conjecture 1]. However, the following example shows that this is possible using an automaton with the semantics proposed in this paper. This suggests that the automata defined in this paper may be more powerful than those defined by Bruggink *et al.* [5]. However, a more detailed investigation of this issue is left to future work.

Example 5 (An Alternating Graph Automaton Accepting Binary Trees). For a ranked alphabet Σ consisting of a single "invisible" binary edge label, the automaton $\mathfrak{A}_{\text{Tree}}$ shown in Fig. 3 accepts the language of (non-empty) binary trees. The automaton works as follows: It fails immediately if the input graph contains one of the graphs $\mathtt{L}$, $\mathtt{P}$, $\mathtt{O}_3$, or $\mathtt{I}_2$, because then it can reach the rejecting state q_1 from the universal initial state q_0. So the input graph must contain neither loops, nor parallel edges, nor nodes with three (or more) outgoing edges, nor nodes with two (or more) incoming edges, but there must be a node r as an occurrence of $\mathtt{N}_1$ such that it succeeds at the universal state q_3. It cannot succeed at q_3 if it can reach the rejecting state q_4 from there, i.e., if r has an incoming edge. Consequently, the automaton can succeed at q_3 only if r is a root node of the input graph, and for every other node o as an occurrence of $\mathtt{N}_2$ (together with r) it succeeds at q_5, i.e., if it can reach the accepting state q_6 by finding a path from r to o. Consequently, the input graph is accepted by $\mathfrak{A}_{\text{Tree}}$ if and only if it is a (non-empty) binary tree.

Note that the transitions $(q_0, \mathit{Id}_0, q_2)$ and $(q_2, \mathtt{N}_1, q_3)$ are necessary to specify non-empty trees correctly: If we replaced them by a single "shortcut" transition $(q_0, \mathtt{N}_1, q_3)$, the automaton would accept the empty graph as q_0 is universal; since q_2 is an existential state, the automaton requires the input graph to contain a root in order to succeed.

Figure 4 shows the configuration graph of $\mathfrak{A}_{\text{Tree}}$ for a tree with three edges. Configurations are drawn with a green solid border if they are assigned *true* by α^*, otherwise with a red dashed border. Note that universal leaf configurations are always assigned *true*, while existential leaf configurations are assigned *false*. As a result, only the top $\mathtt{N}_1$ transition leads to acceptance: the corresponding configuration with state q_3 is universal, and it is the only one whose (direct) successor configurations are all assigned *true* by α^*.

$\mathfrak{A}_{\mathtt{HAM}} =$ 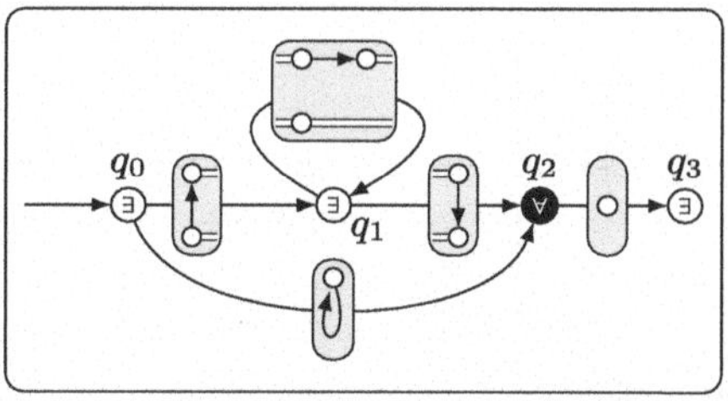

Fig. 5. The automaton accepting graphs containing a Hamiltonian cycle.

This configuration graph illustrates the difference between our notion of transition steps and that of Bruggink *et al.* [5]. Consider the edges $(q_5, G) \longrightarrow_{\mathtt{E}} (q_5, G')$ of the configuration graph, for some graphs G and G'. The corresponding transition steps identify an occurrence of E as a frontal subgraph of G, with its two front nodes, say f_1 and f_2, its edge, say e, and its rear node. The graph G' is then obtained by removing e, its source node f_1, *together with all other edges attached to* f_1, which are outside the occurrence of E. In contrast to this, the automata of [5] *cannot* apply a transition like $(q_5, \mathtt{E}, q_5)$ if the nodes removed in the transitions (f_1 in this case) are attached to any other edge of the input, as it is often the case in this configuration graph. So these steps could not be taken by their automata. As a matter of fact, [5, Conjecture 1] suggests that trees cannot be accepted by their automata altogether.

Example 6 (Alternating Graph Automata for Hamiltonian Cycles). The automaton shown in Fig. 5 detects whether a graph contains a Hamiltonian cycle or not, basically by checking whether there is a cycle in the graph such that there are no more nodes. It does this in the following way: Using the transitions from state q_0 via q_1 to q_2, the automaton finds and follows a cycle of at least two binary edges. Using the transition from state q_0 directly to q_2 instead, it finds a single loop. In both cases the automaton ignores all other edges connected to nodes of the cycle (by cutting them off). The automaton accepts the graph if it finds no more nodes after reaching q_2. Otherwise it reaches the rejecting state q_3, so it rejects the graph.

Example 7 (Alternating Graph Automata with Permutation Cycles). In contrast to the previous examples, which are based on automata without permutation cycles, Fig. 6 shows the automaton $\mathfrak{A}_{\mathtt{P}}$ with a permutation cycle. We will see that its lower language is indeed different from its upper language.

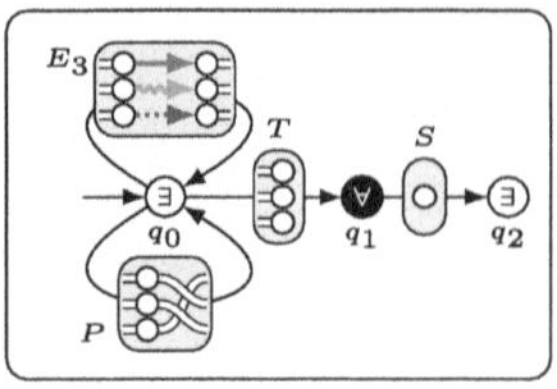

Fig. 6. The automaton $\mathfrak{A}_\mathrm{P}$.

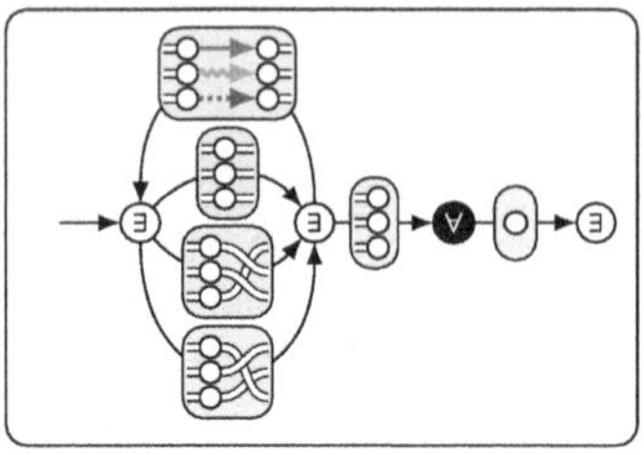

Fig. 7. The automaton $\mathfrak{A}'_\mathrm{P}$.

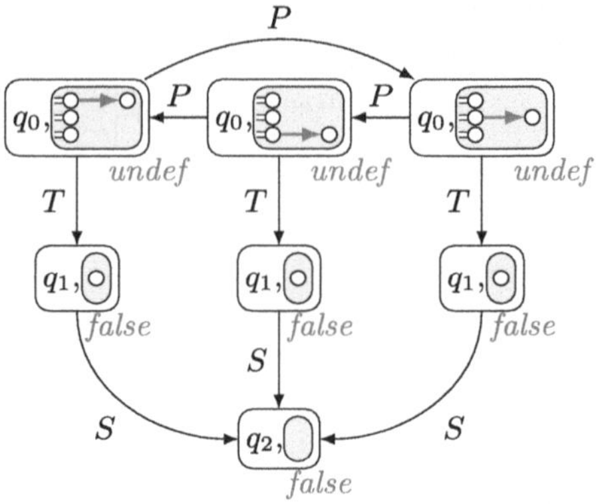

Fig. 8. A configuration graph for $\mathfrak{A}_\mathrm{P}$.

$\mathfrak{A}_\mathrm{P}$ accepts graphs that contain a sequence of edge bundles, where each edge bundle consists of three parallel edges (one red, one squiggly green and one dotted blue), where the three edges may be interchanged cyclically, as is the case for the following graph:

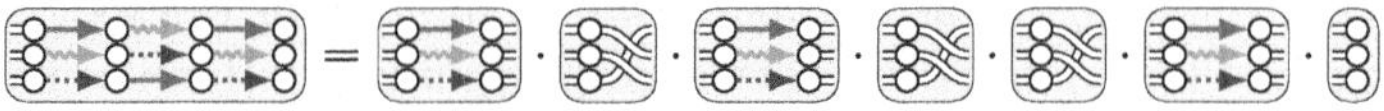

Now consider the graph $G =$ 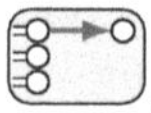, which is not accepted by $\mathfrak{A}_\mathrm{P}$. Figure 8 shows the corresponding configuration graph $CG_{\mathfrak{A}_\mathrm{P}}(G)$ with its initial configuration at the top left. We ascribe to the configurations the truth value assigned by α^*. Because of the permutation cycle, $CG_{\mathfrak{A}_\mathrm{P}}(G)$ also contains a cycle, which means that α^* cannot assign a truth value to the three configurations in the cycle, especially not to the initial configuration. Therefore, $G \notin \underline{L}(\mathfrak{A}_\mathrm{P})$, but $G \in \overline{L}(\mathfrak{A}_\mathrm{P})$, i.e., $\overline{L}(\mathfrak{A}_\mathrm{P}) \neq \underline{L}(\mathfrak{A}_\mathrm{P})$.

Note that the cycle with the permutation graph P in $\mathfrak{A}_\mathrm{P}$ can only produce three different permutation graphs, namely Id_3, a left shift, and a right shift. The automaton $\mathfrak{A}'_\mathrm{P}$ shown in Fig. 7 avoids this cycle by explicitly enumerating these three permutation graphs. So $\mathfrak{A}'_\mathrm{P}$ accepts the same graphs as $\mathfrak{A}_\mathrm{P}$, but rejects all others, i.e., $\overline{L}(\mathfrak{A}'_\mathrm{P}) = \underline{L}(\mathfrak{A}'_\mathrm{P})$.

Resolving the permutation cycle in $\mathfrak{A}_\mathrm{P}$ is easy, since all states in the cycle are existential. We do not yet know whether cycles containing universal and existential states can also be resolved. This is left for future work.

4 Computational Complexity

Next, we want to show that graph languages of the form $\underline{L}(\mathfrak{A})$ and $\overline{L}(\mathfrak{A})$ are in PSPACE. Prior to this, we prove an auxiliary lemma which bounds the number of iterations of Evo_K required to reach a fixed point.

Lemma 3. *Let $K = CG_{\mathfrak{A}}(G_0)$ be the configuration graph of an alternating graph automaton $\mathfrak{A} = (\Sigma, Q, \Delta, q_0, Q_\forall)$ for an input graph $G_0 \in \mathbb{G}_\Sigma^{rank(q_0)}$, and let $b \in \{\bot, false, true\}$. Then, for every configuration $c \in \dot{K}$, it holds that*

$$Evo_K^*(\alpha_b)(c) = \alpha_b^{n_c+1}(c),$$

where n_c is the maximum length of paths in K starting at c and ending at a leaf (and $n_c = 0$ if no such path exists).[3]

Proof. By Lemma 1(2) it suffices to prove the lemma for $b = \bot$. By induction on n, we prove that $\alpha_\bot^{n_c+1}(c) = \alpha^*(c)$ for all configurations c such that $n_c \leqslant n$. For $n = 0$, $n_c \leqslant 0$ means that c is a leaf of K, or no leaf is reachable from c. In the first case, $\alpha_\bot^1(c)$ is defined (it is *true* or *false* depending on whether c is universal or existential), and thus $\alpha_\bot^1(c) = \alpha^*(c)$ because $\alpha_\bot^1 \sqsubseteq \alpha^*$. In the second case, both $\alpha_\bot^1$ and $\alpha^*(c)$ are undefined, as can be shown by a straightforward induction omitted here.

For the inductive step, assume that $n > 0$ and let $n_c \leqslant n$. By the definition of Evo_K, $\alpha_\bot^{n_c+1}(c) = Evo_K(\alpha_\bot^{n_c})(c)$ depends only on the values $\alpha_\bot^{n_c}(c')$ of configurations $c' \in \dot{K}$ such that $c \vdash_\Delta c'$. For these configurations c' it holds that $n_{c'} < n$, and thus $\alpha_\bot^{n_c}(c') = \alpha^*(c')$ by the induction hypothesis. Consequently, $\alpha_\bot^{n_c+1}(c) = Evo_K(\alpha^*)(c) = \alpha^*(c)$. □

Theorem 1. *For every alternating graph automaton $\mathfrak{A}$, there are algorithms which decide the membership problems of $\underline{L}(\mathfrak{A})$ and $\overline{L}(\mathfrak{A})$ in cubic space.*

Proof. We only prove the result for $\underline{L}(\mathfrak{A})$ because this also proves the dual one, using Lemma 2. Consider the function ACCEPT of Algorithm 1. We show that, (1) given an alternating graph automaton $\mathfrak{A}$ and a graph G, we have $G \in \underline{L}(\mathfrak{A})$ if and only if ACCEPT$((q_0, G), \varnothing)$ returns *true* and (2) the algorithm runs in cubic space.

To establish the first claim, assume that ACCEPT$((q_0, G), \varnothing)$ is called. It should be obvious that the tree T of recursive invocations of the function is isomorphic to an unfolding of $K = CG_{\mathfrak{A}}(G_0)$ starting from (q_0, G), where subtrees rooted in configurations already encountered in an enclosing recursive call become leaves, their subtrees being deleted. We call a vertex of the recursion tree *void* if it is such a leaf.[4] The remaining vertices are called *universal* and *existential* depending on the type of the configuration passed as the first parameter. That configuration is called the *configuration* at the vertex, and the *value*

[3] We say that a configuration c is a leaf if there is no configuration c' such that $c \vdash_\Delta c'$.

[4] We speak of vertices of the recursion tree to avoid confusion with the nodes of G.

Algorithm 1. Basic algorithm for the universal membership problem of $\underline{L}(\mathfrak{A})$

1: **function** Accept(configuration c, set P of configurations)
2:     **if** $c \in P$ **then return** *false*
3:     **if** c is universal **then**
4:         *result* $\leftarrow$ *true*
5:         **for each** step $c \vdash_\Delta c'$ **do**
6:             **if** $\neg$Accept$(c', P \cup \{c\})$ **then**
7:                 *result* $\leftarrow$ *false*
8:     **else**
9:         *result* $\leftarrow$ *false*
10:         **for each** step $c \vdash_\Delta c'$ **do**
11:             **if** Accept$(c', P \cup \{c\})$ **then**
12:                 *result* $\leftarrow$ *true*
13:     **return** *result*

of the vertex is the return value of the call. In particular, the value of every void vertex is *false*. The *height* of a vertex is 0 if it is a leaf and otherwise $i+1$, where i is the maximum height of its children. Note that the height i of a vertex with configuration c is the length of the longest path in K from c to a leaf. In other words, Lemma 3 applies to c saying that $Evo_K^{i+1}(\alpha_{false})(c) = \underline{\alpha}(c)$.

Now, consider a vertex with configuration c at height at most i in T. We show by induction on i that the value of the vertex is $Evo_K^{i+1}(\alpha_{false})(c)$.

For $i = 0$, the vertex is a leaf and its value is *true* only if it is universal, as both void and existential leaves have the value *false*, as claimed. For $i > 0$, the configuration c at the vertex under consideration is either universal or existential. In both cases, the value returned depends only on the values of the child vertices, whose height is at most $i-1$. Hence, the induction hypothesis and the fact that lines 3–13 implement precisely the definition of Evo_K, prove that the value of the vertex is indeed $Evo_K^{i+1}(\alpha_{false})(c)$. In particular, the result of Accept$((q_0, G), \varnothing)$ is $\underline{\alpha}(q_0, G)$, proving that the algorithm is correct.

Let us now argue that a standard implementation of the algorithm requires at most cubic space. The key to this is the observation that the height of the recursion tree is linear in the size of G. This is because each step reduces the size of the second component of the configuration unless the transition used is (q, Γ, q') with Γ being a permutation graph. However, there can only be a constant number of such steps in succession (this constant depending on $\mathfrak{A}$ but not on G) before re-encountering a configuration and thus ending in a void vertex of the recursion.

Thus, to implement the recursion in the usual way, a stack of linear depth is sufficient. As a consequence, the parameter P also contains at most a linear number of configurations in every call. Therefore every activation record on this stack has to store at most a linear number of configurations, each at most as large as the input graph (plus some minor bookkeeping information). In total, this proves that the space used is at most cubic. □

The complexity estimation in the preceding proof fails if the automaton $\mathfrak{A} = (\Sigma, Q, \Delta, q_0, Q_\forall)$ is considered to be part of the input (i.e., if we consider the *uniform* membership problem). The reason is that, in this case, the number

$$\tau = \max\{|\mathit{front}_\Gamma| \mid (q, \Gamma, q') \in \Delta, \Gamma \text{ a permutation graph}\},$$

is not a constant anymore, but can become linear in the size n of the input. In this case, even with only two transitions whose graphs are permutation graphs, all $\tau!$ permutations can be generated[5] before a configuration repeats, meaning that the depth of the recursion tree can become superpolynomial.

This effect is ruled out if $\mathfrak{A}$ has no permutation cycles, because there can only ever be a linear number of such transitions which are applied in succession, meaning that the height of the recursion tree becomes at most quadratic (instead of linear, as in the proof above). Hence, we obtain the following corollary.

Corollary 1. *The uniform membership problem of graph languages accepted by alternating graph automata without permutation cycles (given an alternating graph automaton $\mathfrak{A}$ without permutation cycles and a graph G, is G an element of $L(\mathfrak{A})$?) can be decided in space $O(n^4)$.*

Next, we show that even the non-uniform membership problem of alternating graph automata without permutation cycles is PSPACE-complete.

Theorem 2. *There is an alternating graph automaton $\mathfrak{A}$ without permutation cycles such that $L(\mathfrak{A})$ is PSPACE-complete.*

Proof. The proof is by reduction of the well-known PSPACE-complete problem QSAT. The input of QSAT is a quantified Boolean formula of the form $\mathrm{Q}_1 x_1.\mathrm{Q}_2 x_2.\cdots \mathrm{Q}_k x_k.\varphi$ where $\mathrm{Q}_1, \ldots, \mathrm{Q}_k \in \{\exists, \forall\}$ and φ is a Boolean formula over $\wedge$, $\vee$, $\neg$, and the bound variables $x_1, \ldots, x_k$. The question is whether the truth value of the formula is *true*. (Since all variables are bound, the value of the formula is uniquely determined.) Without loss of generality, we may assume that negations are only applied to variables, i.e., φ is a formula over $\wedge$, $\vee$, and literals $x_1, \neg x_1, \ldots, x_k, \neg x_k$. (We do not require φ to be in conjunctive or disjunctive normalform since it would not affect the complexity of the construction.)

An example of such a QSAT formula is $\forall x.\exists y.\forall z.(x \vee \neg y) \wedge (y \vee z)$. The reduction turns the formula into the graph of type $(1, 0)$ shown in Fig. 9. We use straight and squiggly lines for attachments to make it easier to explain the idea behind the reduction. First, consider the right part of the graph, consisting of structures representing the variables. Every variable is represented by a node with two attached edges, labeled T for *true*, and F for *false*. These edges represent the two possible truth values of the variable and are, in turn, connected via squiggly edges to a common node. Now look at the left part of the graph. Every quantifier is represented by an edge of type 3. The straight tentacles arrange the quantifiers in a left-to-right chain according to their nesting. The squiggly

[5] Choose two permutations which are generators of the permutation group on $[\tau]$ and repeat them arbitrarily.

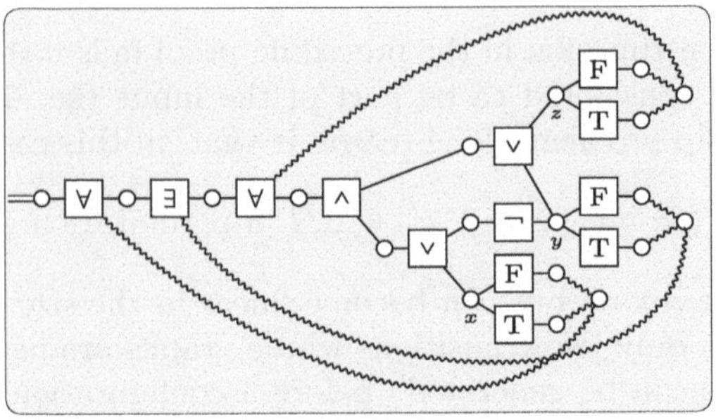

Fig. 9. The image of the formula $\forall x.\exists y.\forall z.(x \vee \neg y) \wedge (y \vee z)$ under the reduction of QSAT to a graph language accepted by an alternating graph automaton.

tentacle connects the quantifier edge to the node through which the edges representing the possible truth values of the variable bound by the quantifier can be accessed (via squiggly binary edges). The rest of the graph, read from left to right, is a straightforward representation of φ as a graph, where $\wedge$, $\vee$, and $\neg$ are represented by edges of type 3, 3, and 2, respectively.

$\mathfrak{A}_{\mathrm{QSAT}} =$ 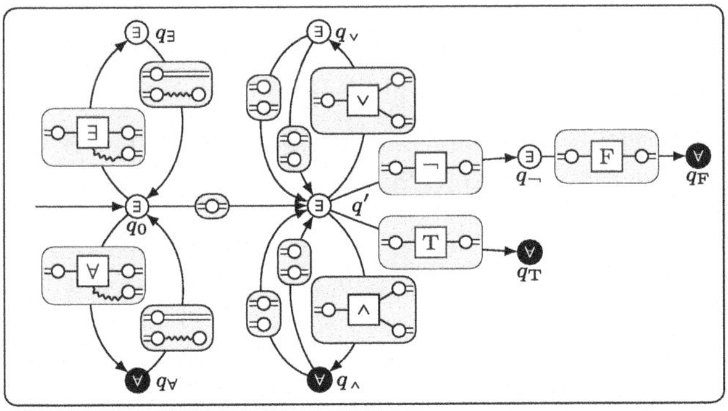

Fig. 10. The automaton accepting an input graph constructed from a QSAT formula φ if and only if the value of φ is *true*.

Figure 10 shows the automaton $\mathfrak{A}_{\mathrm{QSAT}}$ which accepts an input graph obtained in the way just described if and only if the value of the formula is *true*. The most important part is the implementation of the quantifiers. We use an initial existential state q_0. This state "reads" the next quantifier Q (either $\forall$ or $\exists$) and switches to a corresponding state q_{Q}, which is universal or existential, respectively. The transition for q_{Q} selects nondeterministically one of the truth values for the variable bound to it, by deleting the other edge, which is done by choosing one of the squiggly edges and deleting it together with its attached nodes. Note that this, depending on whether $\mathrm{Q} = \forall$ or $\mathrm{Q} = \exists$, implies that either both or at least one of the choices must lead to acceptance. The target state is the initial state q_0, which continues to process the remaining quantifier edges.

After processing all quantifier edges, the automaton enters the state q' to process conjunctions and disjunctions in a rather straightforward way: it switches to state $q_\wedge$ or to state $q_\vee$ which is universal or existential, respectively, and chooses nondeterministically one of the two subexpressions to be evaluated.

Finally, the automaton either processes a *true* edge or a negation followed by a *false* edge, thus checking that a literal with the value *true* has been reached.

We omit the formal correctness proof that a graph of the form exemplified in Fig. 9 is accepted by the automaton if and only if the value of the corresponding formula is *true*, because the manner in which the automaton works should be clear enough. □

5 Conclusions

In this paper, we have studied alternating graph automata, and illustrated that they can specify and check complex properties of graphs, e.g., whether a graph is a tree, or contains a Hamiltonian cycle. They can accept graph languages that are PSPACE-complete, and their uniform membership problem is in PSPACE when only automata without permutation cycles are considered. In a parallel paper [11], we devise graph formulas, a graphical formalism to specify graph properties, and show that they can be translated into equivalent alternating graph automata.

Related Work. While there has been work on finite graph-processing automata [1–3,6,17], the only other work on *alternating* graph automata we know of is that of Bruggink *et al.* [5] and by Brandenburg and Skodinis [4]. As discussed in Example 5, the former automata appear to be weaker than ours. The latter allow to specify graph languages that can be defined by node replacement [12], more precisely languages of undirected node-labelled graphs defined by boundary NCE grammars. In a universal configuration (q, G) of their automata, with ongoing transitions $(q, \Gamma_1, q_1) \ldots (q, \Gamma_k, q_k)$, cutting the occurrences of Γ_1 to Γ_k off G must result in pairwise unconnected remainder graphs $R_1 \ldots R_k$ so that the automaton can proceed with configurations (q_1, R_1) to (q_k, R_k) in parallel. This is rather obscure, since such steps require a global check of the remainder graph, which contradicts the common understanding of automata.

Alternating graph automata provide a means of specifying graph properties and are thus related to nested conditions, first introduced by Rensink [19] and later extended by Habel and Pennemann [16,18]. In particular, the latter introduce nested constraints, which specify local graph conditions using first-order-like logic. In fact, each nested constraint can be translated into an equivalent alternating graph automaton (without permutation cycles) such that a graph is accepted by the nested constraint if and only if it is accepted by the automaton. However, the construction and proof of this equivalence is beyond the scope of this paper. Indeed, alternating graph automata are strictly more expressive than nested constraints. For example, Hamiltonicity can be specified by an alternating graph automaton (see Example 6), but not by any nested constraint.

Flick later extended the framework of nested conditions to recursively nested conditions [14], which can express certain non-local properties such as tree-likeness or acyclicity. However, they still fall short in expressing conditions involving path constraints, such as the existence of disjoint or cycle-free paths. In particular, Hamiltonicity remains inexpressible within this extended formalism.

Future Work. The descriptive power of alternating graph automata should be studied further. In particular, it needs to be investigated whether the notion of alternating graph automata as defined in this paper is really strictly more powerful than that in [5]. Furthermore, we have seen in Example 7 that certain permutation cycles can be resolved quite easily. However, it has to be studied whether this is possible for all permutation cycles, or not. Furthermore, the relation of alternating graph automata to the monadic second-order logic of graphs [8,9] should be investigated.

Also, one may introduce additional features which, if used with care, allow to increase the practical efficiency of alternating graph automata. For example, giving certain transitions evaluation priority over others can decrease the width of a configuration graph considerably. Look, e.g., at the nodes c in the configuration graph in Fig. 4 that represent the universal state q_3. Here the automaton should always try the transition under $\mathtt{I}_1$ first, and, if this transition is applicable, which leads to a rejecting configuration (for state q_4), cancel all other transitions since they cannot make c successful anymore.

Acknowledgments. This paper would not exist without the input of Annegret Habel: she drew the second author's attention to alternating (string) automata. We thank Arend Rensink, and also the anonymous reviewers, for their useful comments, which helped to improve the paper. This work has been supported by the Swedish Research Council under grant no. 2024-05318, and by the Wallenberg AI, Autonomous Systems and Software Program through the NEST project STING—Synthesis and Analysis with Transducers and Invertible Neural Generators.

Disclosure of Interest. The authors have no competing interests to declare that are relevant to the content of this article.

References

1. Blume, C.: Graph Automata and Their Application to the Verification of Dynamic Systems. Ph.D. thesis, University of Duisburg-Essen (2014). http://www.dr.hut-verlag.de/978-3-8439-1881-7.html
2. Blume, C., Bruggink, H.S., Friedrich, M., König, B.: Treewidth, pathwidth and cospan decompositions with applications to graph-accepting tree automata. J. Vis. Lang. Comput. **24**(3), 192–206 (2013). https://doi.org/10.1016/j.jvlc.2012.10.002
3. Bozapalidis, S., Kalampakas, A.: Graph automata. Theor. Comput. Sci. **393**(1–3), 147–165 (2008). https://doi.org/10.1016/j.tcs.2007.11.022
4. Brandenburg, F., Skodinis, K.: Finite graph automata for linear and boundary graph languages. Theor. Comput. Sci. **332**(1–3), 199–232 (2005). https://doi.org/10.1016/j.tcs.2004.09.040

5. Bruggink, H.J.S., Hülsbusch, M., König, B.: Towards alternating automata for graph languages. Electron. Commun. Eur. Assoc. Softw. Sci. Technol. **47** (2012). https://doi.org/10.14279/tuj.eceasst.47.734
6. Bruggink, H., König, B.: Recognizable languages of arrows and cospans. Math. Struct. Comput. Sci. **28**(8), 1290–1332 (2018). https://doi.org/10.1017/S096012951800018X
7. Chandra, A.K., Kozen, D., Stockmeyer, L.J.: Alternation. J. ACM **28**(1), 114–133 (1981). https://doi.org/10.1145/322234.322243
8. Courcelle, B.: The monadic second-order logic of graphs. I. recognizable sets of finite graphs. Inf. Comput. **85**(1), 12–75 (1990). https://doi.org/10.1016/0890-5401(90)90043-H
9. Courcelle, B., Engelfriet, J.: Graph Structure and Monadic Second-Order Logic – A Language-Theoretic Approach, Encyclopedia of Mathematics and its Applications, vol. 138. Cambridge University Press (2012). https://doi.org/10.1017/CBO9780511977619
10. Drewes, F., Habel, A., Kreowski, H.J.: Hyperedge replacement graph grammars. In: Rozenberg, G. (ed.) Handbook of Graph Grammars and Computing by Graph Transformation, chap. 2, vol. I: Foundations, pp. 95–162. World Scientific, Singapore (1997)
11. Drewes, F., Hoffmann, B., Minas, M.: Graph formulas and their translation to alternating graph automata. In: Endrullis, J., Tichy, M. (eds.) Graph Transformation - 18th International Conference, ICGT 2025. LNCS, vol. 15720, pp. xx–yy. Springer, Cham (2025). https://doi.org/10.1007/978-3-031-94706-3_6
12. Engelfriet, J., Rozenberg, G.: Node replacement graph grammars. In: Rozenberg, G. (ed.) Handbook of Graph Grammars and Computing by Graph Transformation, chap. 1, vol. I: Foundations, pp. 1–94. World Scientific, Singapore (1997)
13. Engelfriet, J., Vereijken, J.J.: Context-free graph grammars and concatenation of graphs. Acta Informatica **34**(10), 773–803 (1997). https://doi.org/10.1007/s002360050106
14. Flick, N.E.: Proving correctness of graph programs relative to recursively nested conditions. Ph.D. thesis, Carl-von-Ossietzky-Universität Oldenburg, Germany (2016). https://nbn-resolving.org/urn:nbn:de:gbv:715-oops-29769
15. Gaifman, H.: On local and non-local properties. In: Stern, J. (ed.) Proc. of the Herbrand Symposium, Logic Colloquium. Studies in Logic and the Foundations of Mathematics, vol. 105, pp. 105–135. North-Holland (1982). https://doi.org/10.1016/S0049-237X(08)71879-2
16. Habel, A., Pennemann, K.H.: Nested constraints and application conditions for high-level structures. In: Kreowski, H.J., et al. (eds.) Formal Methods in Software and System Modeling. LNCS, vol. 3393, pp. 293–308. Springer (2005). https://doi.org/10.1007/978-3-540-31847-7_17
17. Kalampakas, A.: Graph automata: the algebraic properties of abelian relational graphoids. In: Kuich, W., Rahonis, G. (eds.) Algebraic Foundations in Computer Science - Essays Dedicated to Symeon Bozapalidis on the Occasion of His Retirement. Lecture Notes in Computer Science, vol. 7020, pp. 168–182. Springer (2011). https://doi.org/10.1007/978-3-642-24897-9_8
18. Pennemann, K.: Development of correct graph transformation systems. Ph.D. thesis, Carl-von-Ossietzky-Universität Oldenburg, Germany (2009). https://nbn-resolving.org/urn:nbn:de:gbv:715-oops-9483
19. Rensink, A.: Representing first-order logic using graphs. In: Ehrig, H., Engels, G., Parisi-Presicce, F., Rozenberg, G. (eds.) Graph Transformations. pp. 319–335. Springer, Heidelberg (2004). https://doi.org/10.1007/978-3-540-30203-2_23

Graph Formulas and Their Translation to Alternating Graph Automata

Frank Drewes[1], Berthold Hoffmann[2], and Mark Minas[3](✉)

[1] Umeå universitet, 90187 Umeå, Sweden
drewes@cs.umu.se
[2] Universität Bremen, 28334 Bremen, Germany
hof@uni-bremen.de
[3] Universität der Bundeswehr München, 85577 Neubiberg, Germany
mark.minas@unibw.de

Abstract. Graphs are widely used in various domains to model complex relationships, often requiring the specification and verification of their properties. These properties may involve complex conditions placing, e.g., structural requirements on subgraphs of unbounded size. In this paper, we propose graph formulas as a new formalism for specifying graph properties, providing a higher level of "graphical" abstraction compared to well-known approaches such as monadic second-order logic. We show how these graph formulas can be translated into alternating graph automata, allowing to check computationally difficult graph properties, such as the existence or non-existence of Hamiltonian paths.

1 Introduction

Graph-like structures are ubiquitous in computer science and beyond. In many application areas, it is necessary to specify and check structural properties of graphs and diagrams. In the literature, such properties have been called global [13] to emphasize that they may concern graph regions of unbounded size, in contrast to conditions that require the existence or non-existence of constant-size subgraphs. The nested graph conditions of Habel and Pennemann [17,20] are well-studied for specifying graph properties in a "graphical" way. However, they only allow to specify local properties. They consist of Boolean formulas (with disjunction, conjunction, and negation as connectives, and constants *true* and *false*) over quantified conditions $\exists(C, \mathit{cond})$ and $\forall(C, \mathit{cond})$, which require that, for some occurrences of (an isomorphic copy of) the graph C in the graph in question (resp., for all of these occurrences), the occurrence also satisfies the sub-condition *cond*.[1] Nested graph conditions are well suited for the specification of local constraints for graph languages and application conditions for graph transformation rules.

[1] The formal definition is based on (mono-) morphisms $p\colon P \to C$ rather than on their images, and is often formulated, more abstractly, in weakly adhesive categories.

J. Endrullis and M. Tichy (Eds.): ICGT 2025, LNCS 15720, pp. 112–132, 2025.
https://doi.org/10.1007/978-3-031-94706-3_6

In this paper, we introduce graph formulas, which borrow the structure of nested graph conditions but use graph expressions ex instead of single graphs C in the quantified conditions $\exists(ex, cond)$ and $\forall(ex, cond)$. This is done in Sect. 3. Graph expressions define graph languages by union, concatenation, and Kleene star. Concatenation of graphs glues the rear interface (a sequence of nodes) of the first graph to the front interface (another sequence of nodes) of the second one. This operation has been proposed by Engelfriet and Verijken [12] to define graph languages similarly to the definition of regular string languages. As graph expressions can define graph languages containing graphs of unbounded size, the graph formulas we propose can express global graph properties.

Bruggink *et al.* [2] have lifted alternating string automata [3] to alternating graph automata.[2] Just like the former, the latter have existential and universal states. A configuration in an existential state accepts an input graph if at least one of its immediate successor configurations does, whereas a configuration in a universal state does this only if all of its immediate successor configurations do. In [11], we have proposed a new semantics for these automata, and studied the computational complexity of the resulting language class. We recall alternating graph automata in Sect. 4.

In Sect. 5, we relate our graph formulas to alternating graph automata by showing that graph formulas can be translated into equivalent alternating graph automata. This provides a more "operational" semantics for graph formulas than theorem proving, which is the method commonly used for nested graph conditions. We also show, in Sect. 6, that every graph formula specifies a graph language which is contained in the polynomial hierarchy PH. Together with a PSPACE-completeness result of [11], this suggests that alternating graph automata are strictly more powerful than graph formulas. Concluding the paper in Sect. 7 with some remarks on related and future work, we hint at a generalization of graph formulas which may close this gap.

2 Preliminaries

We let $\mathbb{N}$ denote the set of non-negative integers, and $[n]$ the set $\{1, \ldots, n\}$ for all $n \in \mathbb{N}$. A^* denotes the set of all finite sequences over a set A; the empty sequence is denoted by ε. For a sequence $s = a_1 \cdots a_n$ with $n \in \mathbb{N}$ and $a_1, \ldots, a_n \in A$, $[s]$ denotes the set $\{a_1, \ldots, a_n\}$. As usual, we denote the transitive reflexive closure of a binary relation $R \subseteq A \times A$ by R^*. For a binary relation $\rightsquigarrow \subseteq A \times B$, we write $\not\rightsquigarrow$ for its complement, i.e., for $a \in A$ and $b \in B$, $a \not\rightsquigarrow b$ holds if and only if $a \rightsquigarrow b$ does not.

We consider edge-labeled hypergraphs (which we simply call graphs), i.e., edges are attached to sequences of nodes. Like Engelfriet and Vereijken [12], we supply each graph with two sequences of distinguished nodes that constitute their front and a rear interfaces. To simplify the model, we require these sequences to be free from repetitions.

[2] They also lifted it further to a category-theoretical setting.

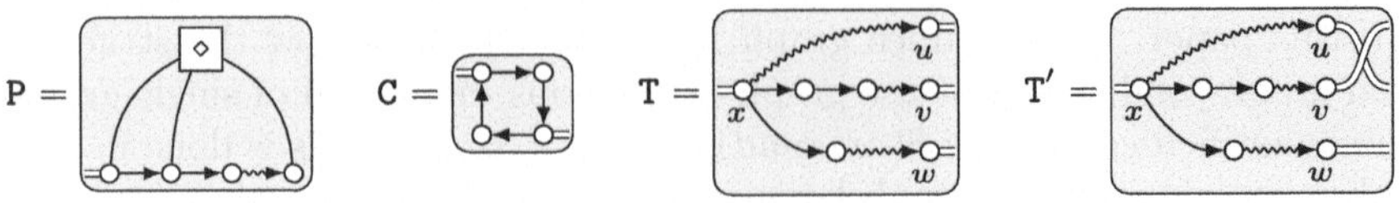

Fig. 1. Four graphs.

A ranked alphabet $(\Sigma, rank)$ consists of a finite set Σ of symbols and a function $rank\colon \Sigma \to \mathbb{N}$, which assigns a rank to each symbol $a \in \Sigma$. The pair $(\Sigma, rank)$ is usually identified with Σ, keeping $rank$ implicit. For $k \in \mathbb{N}$, we let $\Sigma^{(k)} = \{a \in \Sigma \mid rank(a) = k\}$ denote the set of symbols of rank k in Σ.

Definition 1 (Graph). A *graph* over a ranked alphabet Σ is a tuple $G = (\dot{G}, \bar{G}, att_G, lab_G, front_G, rear_G)$, where $\dot{G}$ and $\bar{G}$ are disjoint finite sets of *nodes* and *edges*, respectively, $att_G\colon \bar{G} \to \dot{G}^*$ attaches nodes to edges, $lab_G\colon \bar{G} \to \Sigma$ labels edges with symbols in such a way that $|att_G(e)| = rank(lab_G(e))$ for every edge $e \in \bar{G}$, and the repetition-free node sequences $front_G, rear_G \in \dot{G}^*$ are the *front* and the *rear* interface, respectively.

The *type* of a graph G with $|front_G| = m$ and $|rear_G| = n$ is $type(G) = (m, n)$. $\mathbb{G}_\Sigma^{(m,n)}$ denotes the set of all graphs of type (m, n), and $\mathbb{G}_\Sigma^m = \bigcup_{n\in\mathbb{N}} \mathbb{G}_\Sigma^{(m,n)}$ denotes the set of all graphs over Σ whose front is of length m.

A *permutation graph* is a graph G such that $\bar{G} = \varnothing$ and $\dot{G} = [front_G] = [rear_G]$. If, furthermore, $front_G = rear_G$ and $|\dot{G}| = n$, then G is called an *identity graph* (on n nodes) and is denoted by Id_n.

Example 1 Figure 1 shows four graphs over the ranked alphabet $\Sigma = \{␣, \sim, \diamond\}$ with $rank(␣) = rank(\sim) = 2$ and $rank(\diamond) = 3$. As usual, we draw nodes as circles and non-binary edges as boxes, the boxes containing the edge label and being connected to the attached nodes by lines.[3] Binary edges like the ones labeled ␣ and $\sim$ in this example are drawn as arrows; in general, their labels are ascribed to the arrows, but here we just distinguish them by drawing edges with the "invisible" label ␣ as straight arrows, and the others as squiggly arrows.

The front and rear interface nodes are represented by double lines that connect to the left and right borders of the graph, respectively. Both node sequences are arranged in order from top to bottom. Here, P is of type $(1, 0)$, C is of type $(1, 1)$, and both T and T′ are of type $(1, 3)$. Their front interfaces are identical, with $front_{\mathtt{T}} = front_{\mathtt{T}'} = x$; however, their rear interfaces differ: $rear_{\mathtt{T}} = uvw$ while $rear_{\mathtt{T}'} = vuw$. Thus, even though T and T′ would appear identical if the rear interfaces are ignored, they are considered distinct graphs because their rear interfaces are different.

We follow Engelfriet and Vereijken [12] in defining graph concatenation.

---

[3] In our examples, there will be no need to indicate the order of the attached nodes of non-binary edges.

Definition 2 (Graph Concatenation). Let $G \in \mathbb{G}_\Sigma^{(i,k)}$ and $H \in \mathbb{G}_\Sigma^{(k,j)}$. We assume for simplicity that $rear_G = front_H$, $\dot{G} \cap \dot{H} = [rear_G] = [front_H]$, and $\bar{G} \cap \bar{H} = \varnothing$ (otherwise appropriate isomorphic copies of G and H are used). The (*typed*) *concatenation* $G \cdot H$ of G and H is the graph C such that $\dot{C} = \dot{G} \cup \dot{H}$, $\bar{C} = \bar{G} \cup \bar{H}$, $att_C = att_G \cup att_H$, $lab_C = lab_G \cup lab_H$, $front_C = front_G$, and $rear_C = rear_H$. Thus $C \in \mathbb{G}_\Sigma^{(i,j)}$.

Note that the concatenation $G \cdot H$ is defined on concrete graphs if the assumptions in Definition 2 are satisfied, but only up to isomorphism if an isomorphic copy of G or H must be taken. To avoid unnecessary technicalities, we usually assume without mentioning that the former is indeed the case.

3 Graph Expressions and Graph Formulas

In this section, we define graph expressions and, based on them, graph formulas. Similarly to regular string expressions, graph expressions denote graph languages using singleton languages, union, concatenation, and Kleene star as operators.

Definition 3 (Operations on Graph Languages). Let $i, j, k \in \mathbb{N}$. A subset $\mathcal{L} \subseteq \mathbb{G}_\Sigma^{(i,j)}$ is called a *graph language* of type $type(\mathcal{L}) = (i, j)$.

- For graph languages $\mathcal{L} \subseteq \mathbb{G}_\Sigma^{(i,k)}, \mathcal{M} \subseteq \mathbb{G}_\Sigma^{(k,j)}$, the (*typed*) *concatenation* is the graph language $\mathcal{L} \cdot \mathcal{M} = \{G \cdot H \mid G \in \mathcal{L}, H \in \mathcal{M}\}$, of type (i, j).
- The (*Kleene*) *star* $\mathcal{L}^*$ of $\mathcal{L} \subseteq \mathbb{G}_\Sigma^{(i,i)}$ is the smallest graph language containing Id_i such that $G \cdot H \in \mathcal{L}^*$ for all graphs $G \in \mathcal{L}$ and $H \in \mathcal{L}^*$.

According to [12], graph languages constructed by these operations "could be called regular". We do not call these languages regular here, because it might lead to confusion, since Courcelle and Maneth *et al.* [5,14] use the same term for another class of graph languages.

Definition 4 (Graph Expression). For $i, j \in \mathbb{N}$, the set $\mathbb{E}_\Sigma^{(i,j)}$ of *graph expressions* of type (i, j) over Σ is the smallest set of formal expressions formed according to the following rules:

1. For every graph $G \in \mathbb{G}_\Sigma^{(i,j)}$, $G \in \mathbb{E}_\Sigma^{(i,j)}$, defining the language $L(G) = \{G\}$.
2. If $ex_1, ex_2 \in \mathbb{E}_\Sigma^{(i,j)}$, $ex_1 \oplus ex_2 \in \mathbb{E}_\Sigma^{(i,j)}$, defining the language $L(ex_1 \oplus ex_2) = L(ex_1) \cup L(ex_2)$.
3. If $ex_1 \in \mathbb{E}^{(i,j)}$ and $ex_2 \in \mathbb{E}_\Sigma^{(j,k)}$, $ex_1 \odot ex_2 \in \mathbb{E}_\Sigma^{(i,k)}$, defining the language $L(ex_1 \odot ex_2) = L(ex_1) \cdot L(ex_2)$.
4. If $ex \in \mathbb{E}_\Sigma^{(i,i)}$, $ex^{\circledast} \in \mathbb{E}_\Sigma^{(i,i)}$, defining the language $L(ex^{\circledast}) = L(ex)^*$.

A graph expression ex is called *permutation-free* if all permutation graphs that occur as sub-expressions of ex are identity graphs.

Example 2 (Paths and Cycles). Let $\Sigma = \{\_\}$ contain only the invisible label. Then the following graph expressions specify a path from the front node to a node attached to a unary edge labeled L, and a cycle in a graph (with empty front interface), respectively.

Path = 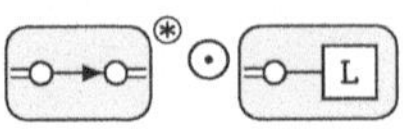   Cycle = 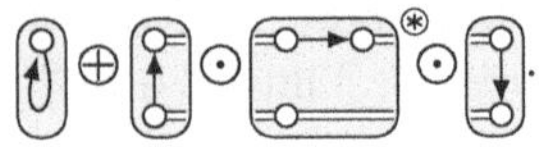.

In examples, $\circledast$ binds stronger than $\odot$, and $\odot$ binds stronger than $\oplus$.

We now turn from graph expressions to graph formulas. The structure of these formulas corresponds to that of nested graph conditions [16,20], but their basic building blocks are graph expressions instead of graph morphisms.

Definition 5 (Graph Formulas). For $m \in \mathbb{N}$, the set $\mathbb{F}_\Sigma^m$ of *graph formulas* of type m (*formulas*, for short) over Σ is defined inductively as follows:

1. $true, false \in \mathbb{F}_\Sigma^m$;
2. if $ex \in \mathbb{E}_\Sigma^{(m,i)}$ and $fo \in \mathbb{F}_\Sigma^i$, then $\exists(ex, fo) \in \mathbb{F}_\Sigma^m$ and $\forall(ex, fo) \in \mathbb{F}_\Sigma^m$;
3. if $fo, fo' \in \mathbb{F}_\Sigma^m$, then $\neg fo \in \mathbb{F}_\Sigma^m$, $fo \wedge fo' \in \mathbb{F}_\Sigma^m$, and $fo \vee fo' \in \mathbb{F}_\Sigma^m$.

We let $\mathbb{F}_\Sigma = \bigcup_{m \in \mathbb{N}} \mathbb{F}_\Sigma^m$ denote the set of all graph formulas over Σ.

A formula fo is called *basic* if $fo = false$, $fo = \neg fo'$, $fo = \exists(ex, fo')$, or $fo = fo' \vee fo''$, and fo', fo'' are basic graph formulas.

Example 3 (Hamiltonian Cycle). Using the graph expression Cycle of Example 2, we can build the formula ∃Cycle, which specifies that a graph is cyclic. Then the following graph formula specifies whether a graph G contains a *Hamiltonian cycle*, i.e., a cyclic path through all nodes of G.

HAM = ∃ 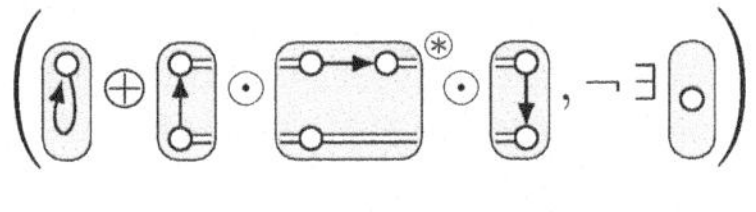

The intended interpretation of this graph formula is "There exists a cycle which contains no additional nodes". In other words, the input graph on which the predicate HAM is to be checked must be decomposed into smaller subgraphs whose concatenation forms a cycle within the original graph. This requires identifying a sequence of edges that together form a cycle. Typically, however, the input graph contains many other edges that do not belong to this cycle and must be ignored during the decomposition. This observation motivates the introduction of the notion of subgraphs that appear at the front of a given graph.

Definition 6 (Frontal Subgraphs). A graph $G \in \mathbb{G}_\Sigma^{(i,k)}$ is a *frontal subgraph* of a graph $H \in \mathbb{G}_\Sigma^{(i,j)}$ if

- $\dot{G} \subseteq \dot{H}$,
- $\bar{G} \subseteq \bar{H}$ with $lab_G(e) = lab_H(e)$ and $att_G(e) = att_H(e)$ for all $e \in \bar{G}$, and

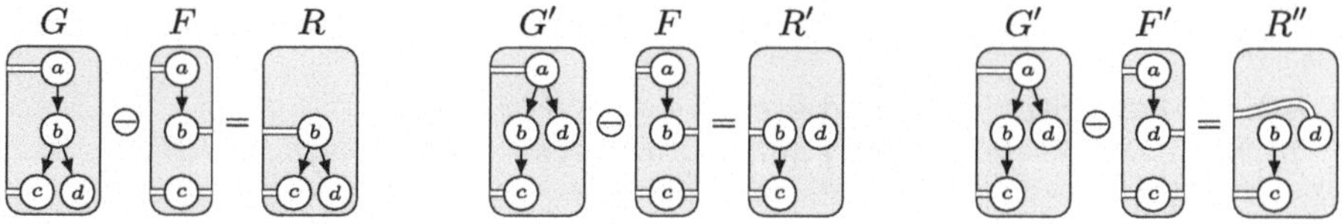

Fig. 2. Cutting off frontal subgraphs.

- $front_G = front_H$, and
- for all $v \in \dot{G}$, $v \in [rear_H]$ implies $v \in [rear_G]$.

Then, *cutting G off H* yields the graph $H \ominus G$, which is obtained from H by removing all nodes in $\dot{G}\backslash[rear_G]$, as well as all edges that are either in $\bar{G}$ or attached to a node in $\dot{G}\backslash[rear_G]$. In the resulting graph, $rear_G$ becomes the front interface and $rear_H$ becomes the rear interface.

Example 4 (Cutting off Frontal Subgraphs). Figure 2 shows graphs G and G' of type $(2, 0)$, from which frontal subgraphs F and F' of type $(2, 2)$ are cut off, yielding the results R, R', and R''. Nodes with the same name (written inside the circle) correspond to each other in subgraph relations. Though F and F' are isomorphic, they are different frontal subgraphs of G', leading to different results $R' = G' \ominus F$ and $R'' = G' \ominus F'$. Note that cutting F and F' off G' also removes the other edge attached to node a, which would otherwise dangle. Thus, $\ominus$ is not the inverse of graph concatenation.

Cutting off frontal subgraphs is compatible with graph concatenation:

Fact 1. For all graphs G, Γ, Γ' of appropriate types, it holds that $(G \ominus \Gamma) \ominus \Gamma' = G \ominus (\Gamma \odot \Gamma')$.

Definition 7 (Satisfaction of Graph Formulas). The satisfaction relation $\models \subseteq \mathbb{G}_\Sigma^m \times \mathbb{F}_\Sigma^m$ is defined by induction as follows. It is the smallest relation such that the following hold for every graph $G \in \mathbb{G}_\Sigma^m$ (where $fo, fo' \in \mathbb{F}_\Sigma$):

- $G \models true$.
- $G \models \exists(ex, fo)$ if $G \ominus P \models fo$ holds for some frontal subgraph $P \in L(ex)$ of G.
- $G \models \forall(ex, fo)$ if $G \ominus P \models fo$ holds for all frontal subgraphs $P \in L(ex)$ of G.
- $G \models \neg fo$ if $G \not\models fo$.
- $G \models fo \wedge fo'$ if $G \models fo$ as well as $G \models fo'$.
- $G \models fo \vee fo'$ if $G \models fo$ or $G \models fo'$ (or both).

$\mathcal{L}(fo)$ denotes the language of all graphs that satisfy the graph formula fo.

Observation 1. Let H be a graph and ex a graph expression. Then $H \models \exists(ex, true)$ if and only if H contains a frontal subgraph G such that $G \in L(ex)$.

Now let $\equiv$ be the equivalence on graph formulas, i.e., $fo \equiv fo'$ denotes that, for all graphs G, $G \models fo$ if and only if $G \models fo'$. As a direct consequence of the preceding definition, we get the following observation.

Observation 2. The following equivalences apply to graph formulas:

1. $\forall(ex, fo) \equiv \neg\exists(ex, \neg fo)$ and $\exists(ex, fo) \equiv \neg\forall(ex, \neg fo)$,
2. $fo_1 \wedge fo_2 \equiv \neg(\neg fo_1 \vee \neg fo_2)$ and $fo_1 \vee fo_2 \equiv \neg(\neg fo_1 \wedge \neg fo_2)$,
3. $\neg\neg fo \equiv fo$,
4. $true \equiv \neg false$.

We abbreviate the basic formula $\exists(ex, \neg false)$ as $\exists\, ex$. We can conclude:

Corollary 1. *For every graph formula there is an equivalent basic formula.*

Example 5 (Binary Trees). Given $\Sigma = \{\_\}$ containing only the invisible label, a graph T is a (non-empty) *binary tree* if it does not contain loops, parallel edges, nodes with two (or more) incoming edges, or nodes with three (or more) outgoing edges, and there is a node r with no incoming edges (the root) such that for every other node o there is a path from r to o. This property is expressed by the graph formula

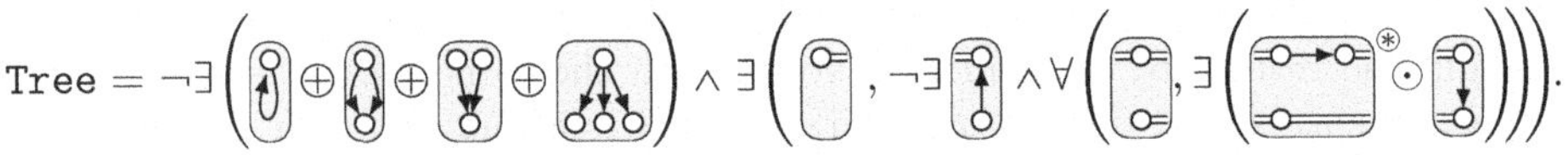

4 Alternating Graph Automata

We recall the notion of alternating graph automata and their semantics from [11] to the extent required for this paper.

Definition 8 (Alternating Graph Automata). An *alternating graph automaton* over Σ without permutation cycles is given by $\mathfrak{A} = (\Sigma, Q, \Delta, q_0, Q_\forall)$ where

- Σ is a ranked alphabet of edge labels,
- Q is a ranked alphabet of *states*,
- $\Delta \subseteq \bigcup_{m,n \in \mathbb{N}} (Q^{(m)} \times \mathbb{G}_\Sigma^{(m,n)} \times Q^{(n)})$ is the set of *transitions*,
- $q_0 \in Q$ is the *initial state*, and
- $Q_\forall \subseteq Q$ is the subset of *universal states*.

The set $Q \backslash Q_\forall$ of *existential* states is denoted by $Q_\exists$.

We require that Δ contains no *permutation cycles*, i.e., no sequence $\delta_1\delta_2\cdots\delta_\ell$ of $\ell \geqslant 1$ transitions such that $\delta_i = (q_i, \Gamma_i, q_{i+1}) \in \Delta$ and Γ_i is a permutation graph for each $i \in [\ell]$, and $q_{\ell+1} = q_1$.

Throughout the rest of the paper, we sometimes use the term *automaton* as a shorthand for *alternating graph automaton*.

Example 6. Figure 8 (in Example 8) shows such an automaton, drawn as a transition diagram. A dangling arrow points to the initial state, and the transitions (q, Γ, q') are drawn as arrows from q to q', with the graph Γ drawn on it or written near it if it is a permutation graph. The interior of existential states is white with a small inscribed $\exists$, and that of universal states is black with a small inscribed $\forall$.

Since, in contrast to [11], we only consider automata without permutation cycles in this paper, we drop this qualification in the following. As discussed in [11], the absence of permutation cycles simplifies the definition and computation of the semantics of automata. We make use of this fact in the definition below.

Definition 9 (Semantics of Automata). A *configuration* of an automaton $\mathfrak{A} = (\Sigma, Q, \Delta, q_0, Q_\forall)$ is a pair $(q, G) \in \bigcup_{m \in \mathbb{N}} (Q^{(m)} \times \mathbb{G}_\Sigma^m)$. It is *universal* if q is, and *existential* otherwise. We write $(q, G) \vdash_\Delta (q', G')$, and call it a *(transition) step*, if there is a transition $(q, \Gamma, q') \in \Delta$ such that (an isomorphic copy of) Γ is a frontal subgraph of G and $G' = G \ominus \Gamma$.

For every configuration c, we let $[\![c]\!]_\mathfrak{A} \in \{\mathit{true}, \mathit{false}\}$ be defined recursively, as follows. Consider the set C of all configurations c' such that $c \vdash_\Delta c'$. Then

$$[\![c]\!]_\mathfrak{A} = \begin{cases} \mathit{true} & \text{if } c \text{ is universal and } [\![c']\!]_\mathfrak{A} = \mathit{true} \text{ for all } c' \in C \text{ or} \\ & c \text{ is existential and } [\![c']\!]_\mathfrak{A} = \mathit{true} \text{ for some } c' \in C \\ \mathit{false} & \text{otherwise.} \end{cases}$$

A graph $G \in \mathbb{G}_\Sigma^{\mathit{rank}(q_0)}$ is *accepted* by $\mathfrak{A}$ if $[\![(q_0, G)]\!]_\mathfrak{A} = \mathit{true}$ and is *rejected* otherwise. The language of all graphs accepted by $\mathfrak{A}$ is denoted by $L(\mathfrak{A})$.

We note here that the absence of permutation cycles in $\mathfrak{A}$ implies that all step sequences are finite (after at most a constant number of steps, the graph component of the configuration decreases in size), which implies that the definition of $[\![c]\!]_\mathfrak{A}$ is sound and can be used directly as an algorithm that decides whether $\mathfrak{A}$ accepts or rejects a given graph. In [11], $[\![c]\!]_\mathfrak{A}$ is denoted by $\alpha^*(c)$.

Informally, a graph G is accepted at an existential state q if it has an outgoing transition (q, Γ, q') such that (an isomorphic copy of) Γ is a frontal subgraph of G, and $G \ominus \Gamma$ is accepted at q'. In contrast, G is accepted at a universal state q if this holds for *every* outgoing transition of q and *every* corresponding frontal subgraph. Consequently, every state with no outgoing transitions can be considered as accepting or rejecting, depending on whether it is universal or existential.

Example 7. The automaton $\mathfrak{A}_{\mathrm{Tree}}$ in Fig. 8 accepts the language of all non-empty binary trees, i.e., the same language as described by the graph formula `Tree` in Example 5. The automaton works as follows: It fails immediately if the input graph contains one of the graphs `L`, `P`, $\mathtt{O}_3$, or $\mathtt{I}_2$, because then it can reach the rejecting state q_2 from the universal initial state q_0. So the input graph must contain neither loops, nor parallel edges, nor nodes with three (or more) outgoing edges, nor nodes with two (or more) incoming edges, but there must be a node r as an occurrence of $\mathtt{N}_1$ such that it succeeds at the universal state q_4. It cannot succeed at q_4 if it can reach the rejecting state q_5 from there, i.e., if r has an incoming edge. Consequently, the automaton can succeed at q_4 only if r is a root node of the input graph, and for every other node o as an occurrence of $\mathtt{N}_2$ (together with r) it succeeds at q_6, i.e., if it can reach the accepting state q_7 by finding a path from r to o.

Given an automaton $\mathfrak{A}$, constructing the *complement automaton* $\mathfrak{A}^c$ that accepts the complement of the language recognized by $\mathfrak{A}$ is straightforward, thanks to the following lemma.

Lemma 1 (Complementation, cf. [11, Lemma 2]**).** *Let* $\mathfrak{A} = (\Sigma, Q, \Delta, q_0, Q_\forall)$ *be an automaton and let* $\mathfrak{A}^c = (\Sigma, Q, \Delta, q_0, Q \backslash Q_\forall)$*. Then* $L(\mathfrak{A}^c)$ *is the complement of* $L(\mathfrak{A})$*, that is,* $L(\mathfrak{A}^c) = \mathbb{G}_\Sigma^{rank(q_0)} \backslash L(\mathfrak{A})$.

5 Translating Graph Formulas to Automata

Let us call an automaton $\mathfrak{A}$ and a graph formula *fo equivalent* if $L(\mathfrak{A}) = \mathcal{L}(fo)$. We will now show how to construct an equivalent automaton $\mathfrak{A}(fo)$ for any graph formula *fo*. Then one can use $\mathfrak{A}(fo)$ to decide whether a graph satisfies *fo*. However, we have to be careful with permutations. In particular, we make sure that no automaton built during the construction contains a permutation cycle. We then know by Lemma 1 that a graph is accepted by such an automaton $\mathfrak{A}$ if and only if it is not accepted by its complement automaton $\mathfrak{A}^c$, which is not the case if $\mathfrak{A}$ contains a permutation cycle [11, Lemma 2]. Thus, we can simply construct the equivalent automaton $\mathfrak{A}(\neg fo)$ for $\neg fo$ by constructing the automaton $\mathfrak{A}(fo)$ for *fo* and taking its complement automaton.

Our construction recursively creates $\mathfrak{A}(fo)$ from a graph formula *fo* by following the inductive definitions of graph expressions in Definition 4 and of graph formulas in Definition 5. To simplify the construction, we assume that the given graph formula is basic. This is not a restriction, because every graph formula can be made basic by Corollary 1. So we also have to consider the case of a graph expression of the form $ex^{\circledast}$ and build an automaton for it. The straightforward approach would be to first create an automaton $\mathfrak{A}(ex)$ for ex, and then construct $\mathfrak{A}(ex^{\circledast})$ by including a cycle that allows looping through $\mathfrak{A}(ex)$. However, this approach would create a permutation cycle in $\mathfrak{A}(ex^{\circledast})$ if $L(ex)$ contains permutation graphs. So we perform a more sophisticated construction, and actually create a triple $(\mathfrak{A}, q_\mathrm{F}, P)$ for any graph expression ex instead of just an automaton $\mathfrak{A}(ex)$. In fact, $\mathfrak{A}$ accepts (by reaching its state q_F) all graphs in $L(ex)$ that are not permutation graphs. And, P consists of all permutation graphs in $L(ex)$. This is possible because $L(ex)$ can contain only finitely many permutation graphs.

We need some additional notation for the following constructions. Given $k \in \mathbb{N}$, we call a bijective function $\pi\colon [k] \to [k]$ a k-permutation and let $\langle\pi\rangle$ be the permutation graph with $front_{\langle\pi\rangle} = 1 \cdots k$ and $rear_{\langle\pi\rangle} = \pi(1) \cdots \pi(k)$. Given sets P and P' of k-permutations, we let $P \circ P' = \{\pi \circ \pi' \mid \pi \in P, \pi' \in P'\}$ where $\pi \circ \pi'$ is the k-permutation defined by $(\pi \circ \pi')(i) = \pi(\pi'(i))$, as usual. Finally, given a set P of k-permutations, let P^* be the smallest set of k-permutations containing the identity function $id_k\colon [k] \to [k]$ and $\pi \circ \pi'$ for every $\pi \in P$ and $\pi' \in P^*$. Note that $|P^*| \leqslant k!$ because every element of P^* is a k-permutation.

In the following, when an automaton is embedded in another, it is represented as a cloud with its initial state on the left edge and, optionally, another state on the right edge.

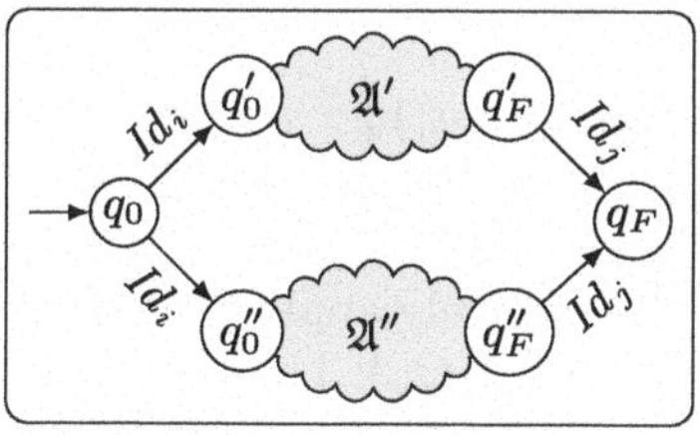

Fig. 3. Automaton for $ex = ex' \oplus ex''$.

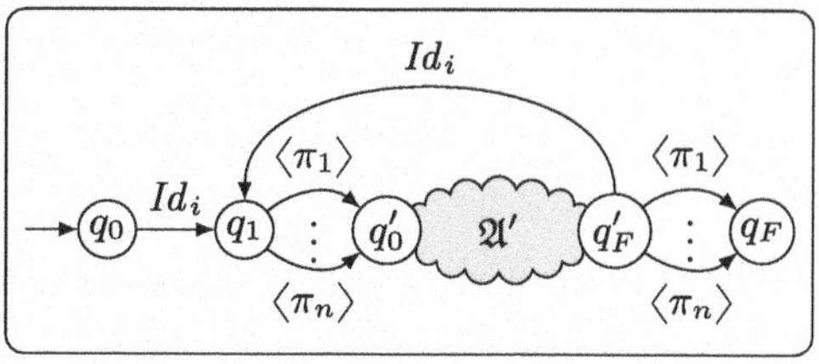

Fig. 4. Automaton for $ex = (ex')^{\circledast}$ where $P = (P')^* = \{\pi_1, \ldots, \pi_n\}$.

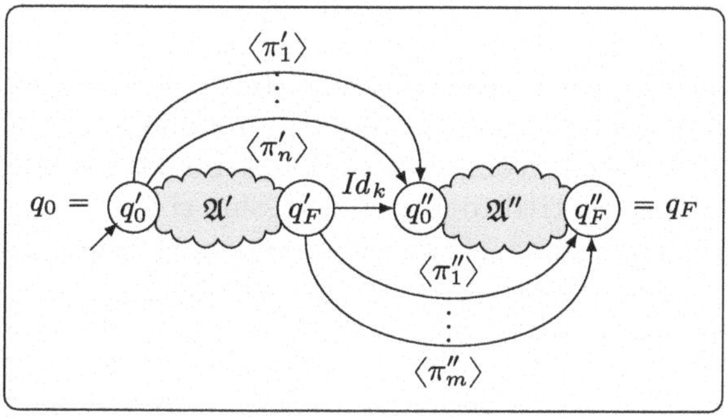

Fig. 5. Automaton for $ex = ex' \odot ex''$ where $P' = \{\pi_1', \ldots, \pi_n'\}$ and $P'' = \{\pi_1'', \ldots, \pi_m''\}$.

Construction 1 (Graph Expression → Automaton). Given $ex \in \mathbb{E}_\Sigma^{(i,j)}$, construct a triple $(\mathfrak{A}, q_\mathrm{F}, P)$ where $\mathfrak{A} = (\Sigma, Q, \Delta, q_0, Q_\forall)$ is an automaton with $q_0 \in Q^{(i)}$, $q_\mathrm{F} \in Q^{(j)}$, $Q_\forall = \varnothing$, and P is a set of i-permutations as follows:

(1) If $ex = G \in \mathbb{G}_\Sigma^{(i,j)}$, and G is no permutation graph: $Q = \{q_0, q_\mathrm{F}\}, \Delta = \{(q_0, G, q_\mathrm{F})\}$, and $P = \varnothing$.
(2) If $ex = G \in \mathbb{G}_\Sigma^{(i,i)}$, and G is a permutation graph: Let π be the i-permutation such that $G \cong \langle\pi\rangle$. Then $Q = \{q_0, q_\mathrm{F}\}$, $\Delta = \varnothing$, and $P = \{\pi\}$.
(3) If $ex = ex' \oplus ex''$ where $ex', ex'' \in \mathbb{E}_\Sigma^{(i,j)}$: Recursively apply Construction 1 to obtain $(\mathfrak{A}', q_\mathrm{F}', P')$ for ex' and $(\mathfrak{A}'', q_\mathrm{F}'', P'')$ for ex'', with $\mathfrak{A}' = (\Sigma, Q', \Delta', q_0', Q_\forall')$ and $\mathfrak{A}'' = (\Sigma, Q'', \Delta'', q_0'', Q_\forall'')$. W.l.o.g, let $Q' \cap Q'' = \varnothing$.

Then let $q_0, q_F \notin Q' \cup Q'''$ be two new states (cf. Fig. 3), $P = P' \cup P''$, $Q = Q' \cup Q'' \cup \{q_0, q_F\}$, and

$$\Delta = \Delta' \cup \Delta'' \cup \{(q_0, Id_i, q_0'), (q_0, Id_i, q_0''), (q_F', Id_j, q_F), (q_F'', Id_j, q_F)\}.$$

(4) If $ex = ex' \odot ex''$ where $ex' \in \mathbb{E}_\Sigma^{(i,k)}$ and $ex'' \in \mathbb{E}_\Sigma^{(k,j)}$: Recursively apply Construction 1 as in the previous case. Then (cf. Fig. 5) $P = P'' \circ P'$, $Q = Q' \cup Q''$, $q_0 = q_0'$, $q_F = q_F''$, and

$$\Delta = \Delta' \cup \Delta'' \cup \{(q_F', Id_k, q_0'')\} \cup \{(q_0, \langle\pi\rangle, q_0'') \mid \pi \in P'\} \cup \{(q_F', \langle\pi\rangle, q_F) \mid \pi \in P''\}$$

(5) If $i = j$ and $ex = (ex')^{\circledast}$ where $ex' \in \mathbb{E}_\Sigma^{(i,i)}$: Recursively apply Construction 1 to obtain $(\mathfrak{A}', q_F', P')$ for ex', with $\mathfrak{A}' = (\Sigma, Q', \Delta', q_0', Q_\forall')$. Then let $q_0, q_1, q_F \notin Q'$ be three new states (cf. Fig. 4), $P = (P')^*$, $Q = Q' \cup \{q_0, q_1, q_F\}$, and

$$\Delta = \Delta' \cup \{(q_0, Id_i, q_1), (q_F', Id_i, q_1)\} \cup \{(q_1, \langle\pi\rangle, q_0') \mid \pi \in P\} \cup \{(q_F', \langle\pi\rangle, q_F) \mid \pi \in P\}$$

Note that many of the identity transitions in this construction could be eliminated by merging their corresponding states. However, we intentionally include these additional transitions to allow the automata to remain structurally unchanged during inductive composition. The straightforward optimization of merging such states is deferred to a postprocessing step.

For proving the correctness of this construction, consider any graph expression $ex \in \mathbb{E}_\Sigma^{(i,j)}$, and let $(\mathfrak{A}, q_F, P)$ with $\mathfrak{A} = (\Sigma, Q, \Delta, q_0, Q_\forall)$ be the result of Construction 1 applied to ex.

Lemma 2. *If $G \in \mathbb{G}_\Sigma^{(i,i)}$ is a permutation graph, there is an i-permutation π such that $G \cong \langle\pi\rangle$, and $G \in L(ex)$ if and only if $\pi \in P$.*

Proof. Let $G \in \mathbb{G}_\Sigma^{(i,i)}$ be a permutation graph. There is apparently an i-permutation π such that $G \cong \langle\pi\rangle$. So it remains to be shown that $\langle\pi\rangle \in L(ex)$ if and only if $\pi \in P$.

For the *forward* direction, let $\langle\pi\rangle \in L(ex)$. We prove $\pi \in P$ by structural induction over ex.

If $ex = H \in \mathbb{G}_\Sigma^{(i,i)}$, $H \cong \langle\pi\rangle$ and $P = \{\pi\}$ by the construction.

If $ex = ex' \oplus ex''$, $ex = ex' \odot ex''$, or $ex = (ex')^{\circledast}$, Construction 1 has been recursively applied to obtain $(\mathfrak{A}', q_F', P')$ for ex' and, in the first two cases, $(\mathfrak{A}'', q_F'', P'')$ for ex''. If $ex = ex' \oplus ex''$, $\langle\pi\rangle \in L(ex')$ or $\langle\pi\rangle \in L(ex'')$. By the induction hypothesis, $\pi \in P'$ or $\pi \in P''$, i.e., $\pi \in P' \cup P'' = P$.

If $ex = ex' \odot ex''$, there must be i-permutations π' and π'' such that $\langle\pi\rangle = \langle\pi''\rangle \cdot \langle\pi'\rangle$, $\langle\pi'\rangle \in L(ex')$, and $\langle\pi''\rangle \in L(ex'')$. By the induction hypothesis, $\pi' \in P'$ and $\pi'' \in P''$, and, therefore, $\pi = \pi'' \circ \pi' \in P'' \circ P' = P$.

If $ex = (ex')^{\circledast}$, $\langle\pi\rangle \in L(ex)$ implies $\langle\pi\rangle = Id_i$, or there exist i-permutations $\pi_1, \ldots, \pi_m \in L(ex')$ $(m \geqslant 1)$ such that $\pi = \pi_m \circ \cdots \circ \pi_1 \in (P')^* = P$.

The other direction can be shown similarly. □

Lemma 3. *Let $G \in \mathbb{G}_\Sigma^{(i,j)}$, $G' \in \mathbb{G}_\Sigma^{(j,k)}$, and $H \in \mathbb{G}_\Sigma^{(i,k)}$ such that $G' = H \ominus G$. It then holds that $(q_0, H) \vdash_\Delta^* (q_F, G')$ if and only if $G \in L(ex)$ and G is not a permutation graph.*

Proof. Let G, G', H be graphs as in the lemma. We first assume $G \in L(ex)$ such that G is not a permutation graph and show $(q_0, H) \vdash_\Delta^* (q_F, G')$ by structural induction over ex.

If $ex = H \in \mathbb{G}_\Sigma^{(i,j)}$, $\Delta = \{(q_0, H, q_F)\}$ and $H \cong G$ because $G \in L(ex)$. Thus $(q_0, H) \vdash_\Delta (q_F, G')$ by Definition 9.

If $ex = ex' \oplus ex''$, $ex = ex' \odot ex''$, or $ex = (ex')^\circledast$, Construction 1 has been recursively applied to obtain $(\mathfrak{A}', q'_F, P')$ for ex' and, in the first two cases, $(\mathfrak{A}'', q''_F, P'')$ for ex''. If $ex = ex' \oplus ex''$, $G \in L(ex')$ or $G \in L(ex'')$. W.l.o.g., $G \in L(ex')$. By the induction hypothesis, $(q'_0, G) \vdash_{\Delta'}^* (q'_F, G')$, and by the construction of $\mathfrak{A}$, $(q_0, G) \vdash_\Delta (q'_0, G) \vdash_{\Delta'}^* (q'_F, G') \vdash_\Delta (q_F, G')$.

If $ex = ex' \odot ex''$, there must be graphs G_1, G_2 such that $G = G_1 \cdot G_2$, $G_1 \in L(ex')$, and $G_2 \in L(ex'')$. By Fact 1, $G' = (H \ominus G_1) \ominus G_2$. G_1 and G_2 cannot both be permutation graphs, i.e., there are three cases:

(i) G_1 and G_2 are no permutation graphs. By the induction hypothesis, $(q'_0, H) \vdash_{\Delta'}^* (q'_F, H \ominus G_1)$ and $(q''_0, H \ominus G_1) \vdash_{\Delta''}^* (q''_F, G')$. By the construction of $\mathfrak{A}$, $(q_0, H) \vdash_\Delta^* (q'_F, H \ominus G_1) \vdash_\Delta (q'_0, H \ominus G_1) \vdash_\Delta^* (q_F, G')$.

(ii) If G_1 is a permutation graph, $G_1 \cong \langle \pi \rangle$ for some $\pi \in P'$ by Lemma 2. Because G_2 is not a permutation graph, $(q''_0, H \ominus G_1) \vdash_{\Delta''}^* (q''_F, G')$ follows by the induction hypothesis, and $(q_0, H) \vdash_\Delta^* (q_F, G')$ by the construction of $\mathfrak{A}$.

The third case (iii), where G_2 is a permutation graph, is handled similarly.

If $ex = (ex')^\circledast$ and because G is not a permutation graph, there must be graphs $G_1, \ldots, G_m \in L(ex')$ such that $m \geqslant 1$, not all of them are permutation graphs, and $G = G_m \cdot \ldots G_2 \cdot G_1$. Let $H_0 = H$ and, for $j \in [m]$, $H_j = H_{j-1} \ominus G_j$. Consequently, $H_m = G'$. For each $j \in [m]$, we can conclude that $G_j = \langle \pi_j \rangle$ for some $\pi_j \in P'$ by Lemma 2 if G_j is a permutation graph. Otherwise, $(q'_0, H_{j-1}) \vdash_{\Delta'}^* (q'_F, H_j)$ follows by the induction hypothesis. Now let $\hat{G}_1, \ldots, \hat{G}_n$ be the sequence obtained from $G_1, \ldots, G_m$ by removing all permutation graphs. Therefore, there are i-permutations $\hat{\pi}_0, \ldots, \hat{\pi}_n \in (P')^* = P$ such that

$$G = \langle \hat{\pi}_0 \rangle \cdot \hat{G}_1 \cdot \langle \hat{\pi}_1 \rangle \cdots \hat{G}_n \cdot \langle \hat{\pi}_n \rangle.$$

If $G_1 = \hat{G}_1$, just choose $\hat{\pi}_0 = id_i \in (P')^* = P$, and if $G_m = \hat{G}_n$, choose $\hat{\pi}_n = id_i$. $(q_0, H) \vdash_\Delta^* (q_F, G')$ then follows by the construction of $\mathfrak{A}$.

The other direction can be shown similarly. □

Lemma 4. *$\mathfrak{A}$ does not have permutation cycles, and Construction 1 can compute $(\mathfrak{A}, q_F, P)$ in polynomial space if ex is permutation-free.*

Proof. First assume that $\mathfrak{A}$ contains a permutation cycle. Only case (5) can produce cycles in Construction 1, i.e., there must be a sub-expression ex' as in case (5) such that either $q'_0 = q'_F$ in $\mathfrak{A}'$, or there is a sequence

$$(p_0, \Gamma_1, p_1), (p_1, \Gamma_2, p_2), \ldots, (p_\ell, \Gamma_\ell, p_\ell)$$

of transitions in $\mathfrak{A}'$ such that $p_0 = q_0'$, $p_\ell = q_F'$, and Γ_i is a permutation graph for each $i \in [\ell]$. Both violate Lemma 3.

To prove the rest of the lemma, assume that ex is permutation-free. Note that every case of Construction 1 produces an automaton of polynomial size in the size of the recursively constructed ones, except for the case (5). In the latter case, the size of the automaton depends on the size of $P = (P')^*$. However, it is easy to see (by a straightforward induction) that P is either empty or consists only of the identity graph Id_i if ex is permutation-free. Consequently, $(\mathfrak{A}, q_F, P)$ can be computed in polynomial space. □

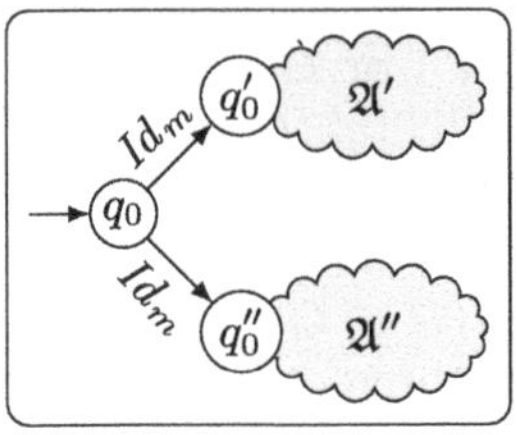

Fig. 6. Automaton for $fo = fo' \vee fo''$.

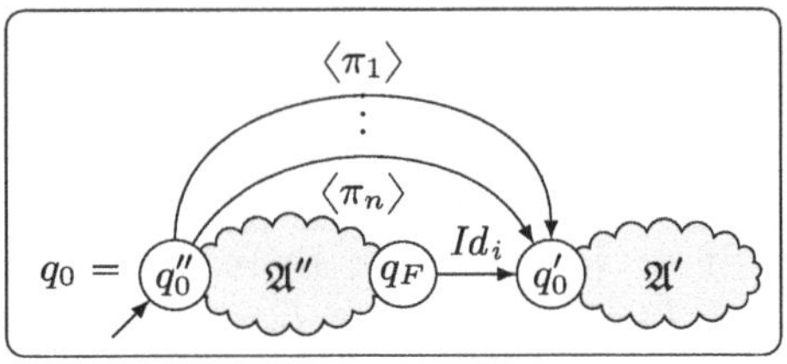

Fig. 7. Automaton for $fo = \exists(ex, fo')$ where $P = \{\pi_1, \ldots, \pi_n\}$.

Note that the size of the automaton produced from a graph expression ex that is not permutation-free can be superpolynomial. To see this, just choose $ex = (\langle \pi_1 \rangle \oplus \langle \pi_2 \rangle)^{\circledast} \odot G$ where π_1 and π_2 are two i-permutations and generators of the permutation group on $[i]$, and $G \in \mathbb{G}_\Sigma^i$ is not a permutation graph, for any Σ and $i \in \mathbb{N}$. The automaton produced from ex by Construction 1 then contains more than $i!$ transitions.

Construction 2. (Graph Formula → Automaton). Given a basic graph formula $fo \in \mathbb{F}_\Sigma^m$, construct an automaton $\mathfrak{A} = (\Sigma, Q, \Delta, q_0, Q_\forall)$ with $q_0 \in Q^{(m)}$ as follows:

(1) If $fo = \mathit{false}$: $Q = \{q_0\}$, $\Delta = \varnothing$, and $Q_\forall = \varnothing$.
(2) If $fo = fo' \vee fo''$ where $fo', fo'' \in \mathbb{F}_\Sigma^m$: Recursively apply Construction 2 to obtain $\mathfrak{A}' = (\Sigma, Q', \Delta', q_0', Q_\forall')$ for fo' and $\mathfrak{A}'' = (\Sigma, Q'', \Delta'', q_0'', Q_\forall'')$ for fo''. W.l.o.g, let $Q' \cap Q'' = \varnothing$. Then let $q_0 \notin Q' \cup Q''$ be a new state (cf. Fig. 6), $Q = Q' \cup Q'' \cup \{q_0\}$, $Q_\forall = Q_\forall' \cup Q_\forall'$, and

$$\Delta = \Delta' \cup \Delta'' \cup \{(q_0, Id_m, q_0'), (q_0, Id_m, q_0'')\}.$$

(3) If $fo = \neg fo'$ where $fo' \in \mathbb{F}_\Sigma^m$: Recursively apply Construction 2 to obtain the automaton $\mathfrak{A}'$ for fo'. $\mathfrak{A}$ is then the complement automaton of $\mathfrak{A}'$.
(4) If $fo = \exists(ex, fo')$ where $ex \in \mathbb{E}^{(m,i)}$ and $fo' \in \mathbb{F}_\Sigma^i$: Recursively apply Construction 2 to obtain $\mathfrak{A}' = (\Sigma, Q', \Delta', q_0', Q_\forall')$ for fo', and Construction 1 to obtain $(\mathfrak{A}'', q_F, P)$ for ex, with $\mathfrak{A}'' = (\Sigma, Q'', \Delta'', q_0'', Q_\forall'')$. W.l.o.g., let $Q' \cap Q'' = \varnothing$. Then (cf. Fig. 7) $Q = Q' \cup Q''$, $Q_\forall = Q_\forall'$, $q_0 = q_0''$, and

$$\Delta = \Delta' \cup \Delta'' \cup \{(q_F, Id_i, q_0')\} \cup \{(q_0, \langle \pi \rangle, q_0') \mid \pi \in P\}.$$

Lemma 5. *Let* $fo \in \mathbb{F}_{\Sigma}^{m}$ *be a basic graph formula and* $\mathfrak{A} = (\Sigma, Q, \Delta, q_0, Q_{\forall})$ *an automaton constructed from* fo *using Construction 2. Then* $\mathfrak{A}$ *does not have permutation cycles, and for each graph* $G \in \mathbb{G}_{\Sigma}^{m}$ *it holds that* G *satisfies* fo *if and only if* $\mathfrak{A}$ *accepts* G*.*

Proof. Let fo and $\mathfrak{A}$ be as in the lemma, and $G \in \mathbb{G}_{\Sigma}^{m}$ any graph. By construction, $\mathfrak{A}$ does not have permutation cycles. We now prove $G \models fo$ iff $[\![(q_0, G)]\!]_{\mathfrak{A}} = true$ by structural induction on fo, which also proves the lemma.

If $fo = false$, $G \not\models fo$, and $[\![(q_0, G)]\!]_{\mathfrak{A}} = false$ follow by Definition 7 and Definition 9.

If $fo = fo' \vee fo''$, $\mathfrak{A}'$ and $\mathfrak{A}''$ have been constructed from fo' and fo'', respectively, and $\mathfrak{A}$ from $\mathfrak{A}'$ and $\mathfrak{A}''$ as described in case (2) of Construction 2. By the induction hypothesis, $G \models fo'$ iff $[\![(q_0', G)]\!]_{\mathfrak{A}'} = true$ and $G \models fo''$ iff $[\![(q_0'', G)]\!]_{\mathfrak{A}''} = true$. Hence, $G \models fo$ iff $[\![(q_0, G)]\!]_{\mathfrak{A}} = true$, by the construction of $\mathfrak{A}$ and Definition 9.

If $fo = \neg fo'$, $\mathfrak{A}'$ has been constructed from fo', and $\mathfrak{A}$ is the complement automaton of $\mathfrak{A}'$. By the induction hypothesis, $G \not\models fo'$ iff $[\![(q_0', G)]\!]_{\mathfrak{A}'} = false$, where q_0' is the start state of $\mathfrak{A}'$. $\mathfrak{A}$ accepts the complement of $L(\mathfrak{A}')$ by Lemma 1. Hence, $G \models fo$ iff $G \not\models fo'$ iff $[\![(q_0', G)]\!]_{\mathfrak{A}'} = false$ iff $[\![(q_0, G)]\!]_{\mathfrak{A}} = true$.

Finally, if $fo = \exists(ex, fo')$, case (4) of Construction 2 has been used to construct $\mathfrak{A}'$ from fo' and $(\mathfrak{A}'', q_{\mathrm{F}}, P)$ from ex. By Definition 7, $G \models fo$ iff there is a frontal subgraph G' of G such that $G' \in L(ex)$ and $G \ominus G' \models fo'$. By Lemma 2, Lemma 3, the induction hypothesis, and the construction of $\mathfrak{A}$, we can conclude that $G \models fo$ iff there is a frontal subgraph G' of G such that $(q_0, G) \vdash_{\Delta}^{*} (q_0', G \ominus G')$ and $[\![(q_0', G \ominus G')]\!]_{\mathfrak{A}} = true$, which is the case iff $[\![(q_0, G)]\!]_{\mathfrak{A}} = true$. □

Using these auxiliary results, we are now ready to express the main result following from the constructions described above:

Theorem 1. *Every graph formula* fo *can be translated into an automaton* $\mathfrak{A}$ *such that* fo *is satisfied by any graph* G *if and only if* G *is accepted by* $\mathfrak{A}$*.* $\mathfrak{A}$ *can be constructed in polynomial space if* fo *is permutation-free.*

Proof. Given any graph formula fo, it can be turned into an equivalent basic graph formula fo' by Corollary 1. By Lemma 5, Construction 2 then produces an automaton $\mathfrak{A}$ from fo' such that, for any graph G, it holds that $G \models fo$ if and only if $G \in \mathcal{L}(\mathfrak{A})$. A straightforward induction using Lemma 4 also shows that $\mathfrak{A}$ can be constructed in polynomial space if fo (and thus fo') is permutation-free. □

Example 8 (Constructing an Alternating Automaton for Trees). In the graph formula `Tree` of Example 5, universal quantification, conjunction, and double negations can be removed with the equivalences of Observation 2, yielding the

equivalent basic graph formula

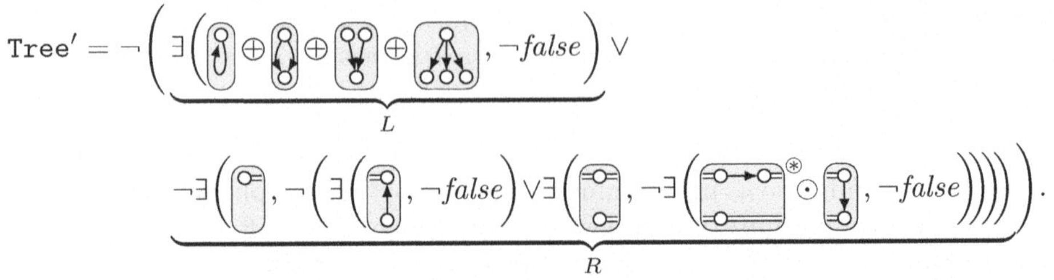

We construct the automaton for Tree′ using Construction 2. In this process, it will build the following (intermediate) automata $\mathfrak{A}_L$ and $\mathfrak{A}_R$ for the sub-expressions L and R of the outer disjunction of Tree′:

$\mathfrak{A}_L =$

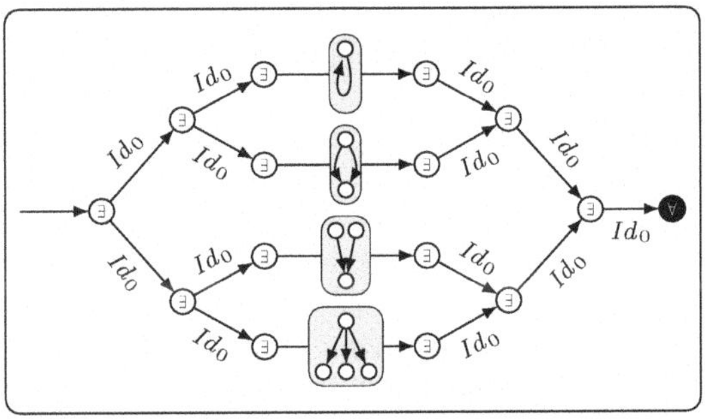

$\mathfrak{A}_R =$

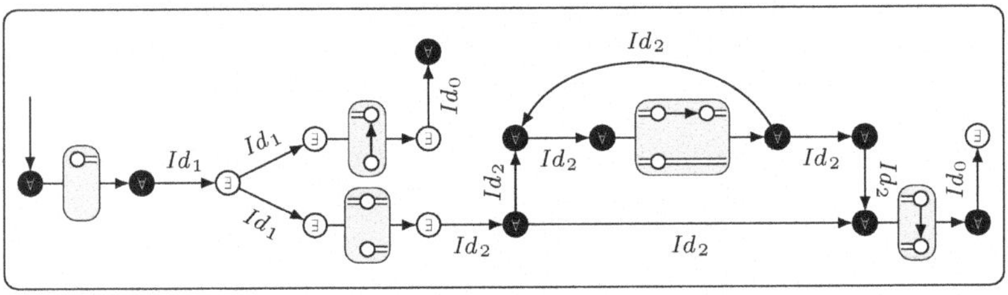

Obviously, the construction generates many unnecessary transitions with identity graphs Id_i. It is easy to see that $\mathfrak{A}_L$ and $\mathfrak{A}_R$ can be simplified. We show their smaller versions for clarity, even though these simplifications are not part of Construction 1 or Construction 2:

$\mathfrak{A}'_L =$

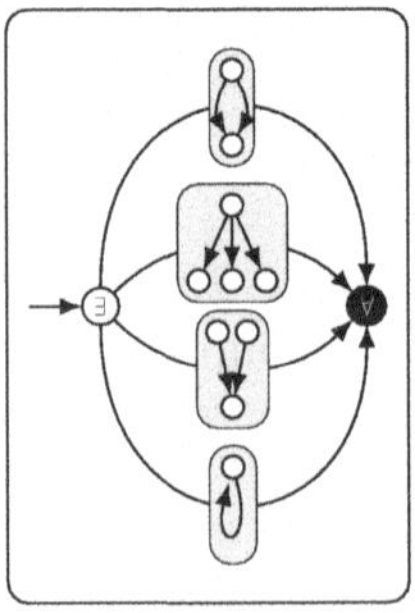

$\mathfrak{A}'_R =$

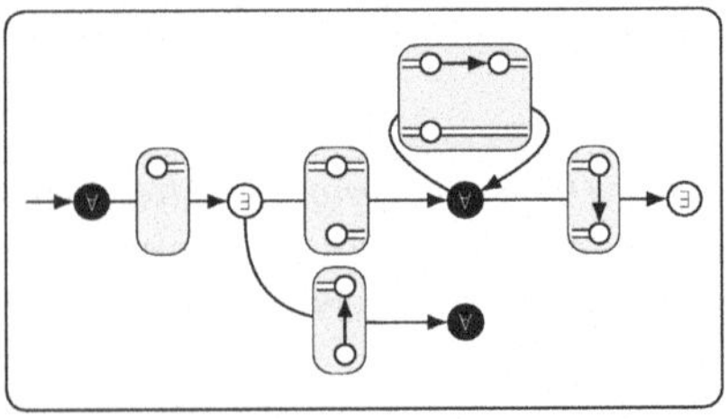

The automaton for the entire formula is obtained from $\mathfrak{A}_L$ and $\mathfrak{A}_R$ by first applying case (2) of Construction 2 and then taking its complement (case (3)) because of the negation. To save space, Fig. 8 instead shows the smaller version obtained from $\mathfrak{A}'_L$ and $\mathfrak{A}'_R$. It corresponds to the even smaller version shown in [11, Figure 3].

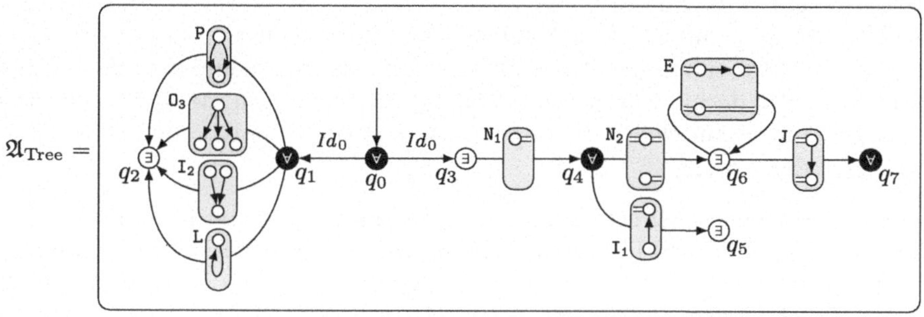

Fig. 8. The automaton constructed from `Tree'`.

6 Computational Complexity

Let us discuss how computationally complex it is to decide satisfaction of graph formulas. For this, recall the polynomial hierarchy: $\mathbf{\Sigma}_0 = \text{P}$ and, for all $i \in \mathbb{N}$, $\mathbf{\Sigma}_{i+1}$ is the set of all problems that can be solved by a polynomial-time nondeterministic Turing machine that has access to an oracle deciding a language in $\mathbf{\Sigma}_i$. Moreover, $\mathbf{\Pi}_i$ is the set of complements of languages in $\mathbf{\Sigma}_i$, for all $i \in \mathbb{N}$. Recall further that $\mathbf{\Sigma}_d \subseteq \mathbf{\Sigma}_{d+1} \cap \mathbf{\Pi}_{d+1}$ for all $d \in \mathbb{N}$.

The following theorem is based on the notion of *alternation depth* for basic formulas, defined as follows:

Definition 10. The *alternation depth* of a basic graph formula is defined inductively. The formula *false* has an alternation depth of 0. If fo_i (for $i \in [2]$) has an alternation depth of d_i, then

- $fo_1 \vee fo_2$ has an alternation depth of $\max(d_1, d_2)$, and
- $\neg fo_1$ and $\exists(ex, fo_1)$ have an alternation depth of $d_1 + 1$.

Theorem 2. *For every basic graph formula fo of alternation depth d, $\mathcal{L}(fo)$ is in $\mathbf{\Sigma}_d$.*

Proof. We proceed by structural induction on *fo*. If *fo* = *false*, then $\mathcal{L}(fo) = \varnothing \in \mathrm{P} = \mathbf{\Sigma}_0$. Now, assume that the claim holds for all subformulas of *fo*.

If $fo = fo_1 \vee fo_2$, then $\mathcal{L}(fo_1), \mathcal{L}(fo_2) \in \mathbf{\Sigma}_d$ because the alternation depths of fo_1 and fo_2 are each at most d, and the claim follows from the fact that $\mathbf{\Sigma}_d$ is closed under union.

If $fo = \neg fo_1$, then $\mathcal{L}(fo_1)$ is in $\mathbf{\Sigma}_{d-1}$ by the induction hypothesis, and thus its complement $\mathcal{L}(fo)$ is in $\mathbf{\Pi}_{d-1} \subseteq \mathbf{\Sigma}_d$.

Finally, if $fo = \exists(ex, fo_1)$, then $\mathcal{L}(fo_1) \in \mathbf{\Sigma}_{d-1}$ according to the induction hypothesis. Let $\mathcal{O}$ be an oracle deciding $\mathcal{L}(fo_1)$. Let M be the nondeterministic oracle Turing machine which first constructs, via "guessing", a graph $G \in L(ex)$ which is a frontal subgraph of the input graph H, then computes $H \ominus G$, and finally uses $\mathcal{O}$ to decide whether $H \ominus G \models fo_1$. By definition, M decides whether $H \in \mathcal{L}(fo)$, and it can clearly be implemented to run in polynomial time, which shows that $\mathcal{L}(fo) \in \mathbf{\Sigma}_d$. □

As a consequence, all graph languages definable by graph formulas are contained in the polynomial hierarchy PH, i.e., the non-uniform membership problem of $\mathcal{L}(fo)$ for any graph formula *fo* is in PH. As shown in Example 3, the NP-complete language of all graphs with a Hamiltonian cycle is in $\mathbf{\Sigma}_1 = \mathrm{NP}$. On the other hand, we have shown in [11, Theorem 2] that there is an alternating graph automaton whose language is PSPACE-complete and thus cannot be the language of any graph formula if we assume that PH $\neq$ PSPACE.

Now consider the uniform decision problem where the graph formula *fo* is part of the input, which turns out to be more computationally complex (again assuming PH $\neq$ PSPACE).

Theorem 3. *The uniform decision problem* "Given a permutation-free graph formula *fo* and a graph G, does G satisfy *fo*?" *is PSPACE-complete.*

Proof Sketch. The problem is in PSPACE because every permutation-free graph formula *fo* can be translated to an automaton $\mathfrak{A}$ such that $G \in L(\mathfrak{A})$ iff $G \models fo$, and this construction can be done in polynomial space in the size of *fo* by Theorem 1. $G \in L(\mathfrak{A})$ can then be decided in polynomial space by [11, Corollary 1].

The proof of PSPACE-hardness is by reduction of the well-known PSPACE-complete problem QSAT to the decision problem of the theorem. The input of QSAT is a quantified Boolean formula of the form $\psi = \mathsf{Q}_1 x_1.\mathsf{Q}_2 x_2.\cdots \mathsf{Q}_k x_k.\varphi$ where $\mathsf{Q}_1, \ldots, \mathsf{Q}_k \in \{\exists, \forall\}$ and φ is a Boolean formula over $\wedge$, $\vee$, $\neg$, and the bound variables $x_1, \ldots, x_k$. The question is whether the truth value of the formula is *true*. (Since all variables are bound, the truth value of the formula is uniquely determined.)

Given any such QSAT formula ψ, the reduction turns it into a graph formula fo_ψ and a graph G_ψ, which only depends on the set $\{x_1, \ldots, x_k\}$ of bound variables in ψ. In fact, G_ψ has $2k$ nodes, two for each of the k bound variables x_i, $i \in [k]$, to each of which a unary x_i-labeled edge is attached. A unary T-labeled edge is also attached to one of these nodes; this node represents the assignment of *true* to the variable x_i, the other represents the assignment of *false*. Figure 9 shows G_ψ for a QSAT formula with $\{x, y, z\}$ as bound variables.

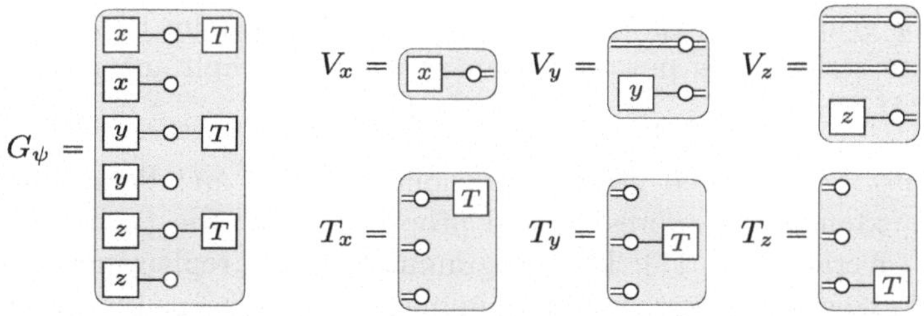

Fig. 9. Graphs for the reduction of ψ with bound variables x, y, and z.

The graph formula fo_ψ is basically a copy of ψ, where each occurrence of a bound variable is replaced by a graph expression or a simple graph formula. This will be demonstrated using the QSAT formula

$$\psi = \forall x.\exists y.\forall z.(x \vee \neg y) \wedge (y \vee z),$$

which is translated into

$$fo_\psi = \forall(V_x, \exists(V_y, \forall(V_z, (f_x \vee \neg f_y) \wedge (f_y \vee f_z))))$$

where $f_v = \exists(T_v, \mathit{true})$ for $v \in \{x, y, z\}$. The graphs T_v and V_v for $v \in \{x, y, z\}$ are shown in Fig. 9. The graph formula fo_ψ is obtained from ψ by converting each occurrence of $\exists u. \ldots$ or $\forall v. \ldots$ in ψ into $\exists(V_u, \ldots)$ and $\forall(V_v, \ldots)$, respectively, and by turning each occurrence of a bound variable $v \in \{x, y, z\}$ in the propositional part φ of ψ into f_v. To check whether fo_ψ is satisfied by G_ψ, the quantifications in fo_ψ select three nodes of G_ψ, one for each of the bound variables x, y, and z. This means that f_v is satisfied if and only if the node selected for $v \in \{x, y, z\}$ has an attached T edge, i.e., *true* is assigned to v. This makes it clear that $G_\psi \models fo_\psi$ if and only if the truth value of ψ is *true*.

The size of G_ψ and fo_ψ is quadratic in the size of ψ, because the size of V_v and T_v ($v \in \{x_1, \ldots, x_k\}$) grows linearly with the number k of bound variables in ψ. This proves PSPACE-hardness, since G_ψ and fo_ψ can be computed from ψ in polynomial time. □

7 Conclusions

In this paper, we have introduced graph formulas as a new formalism for describing graph properties that go beyond local conditions and can also make statements about subgraphs of unbounded size. Graph formulas are similar to the well-known nested graph conditions of Habel and Pennemann [16,20], but use graph expressions similar to regular expressions as building blocks. They can be used, for example, to characterize the language of all binary trees or NP-complete problems such as the existence of Hamiltonian cycles. In order to define an "operational" semantics of graph formulas, we have specified their translation into alternating graph automata. We also proved that the non-uniform satisfaction problem for graph formulas is in the polynomial hierarchy PH, so that graph formulas are strictly less powerful than alternating graph automata (provided PH $\neq$ PSPACE).

Related Work. The nested graph conditions of Habel and Pennemann [16,20] have been extended to express global properties (like the (non-) existence of paths) in several ways: (1) The programmed graph replacement systems of Schürr [24] allow "path expressions" in application conditions for graph transformation rules. (2) Habel and Radke [18,22] allow context-free hyperedge replacement (HR, [7,15]) in conditions; this generalizes graph expressions, which are equivalent to linear HR [12]. (3) Wulandari, Plump, and Poskitt [21,25] propose (textual) conditions expressed in monadic second-order logic for graphs [4]. (4) In the navigational logic of Navarro *et al.* [19], an edge labeled with a regular expression R matches any edge sequence that is labeled with a string in the language of R. All these concepts rely on theorem proving for the actual implementation of conditions, whereas graph formulas are implemented by automata.

Future Work. The descriptive power of graph formulas and automata has to be studied further. In particular, their relation to the monadic second-order logic of graphs [4,6] should be investigated.

Graph formulas can specify *constraints* for a class of graphs and *application conditions* for graph transformation rules. In particular, they could be used to enhance the descriptive power of grammars, like those based on HR [7,15] and contextual HR [10]. Eventually, this could be integrated in the Grappa tool[4] generating parsers for contextual HR grammars [8,9].

We are currently pursuing an approach to close the gap in the expressiveness of graph formulas compared to that of alternating graph automata. We are inspired by the well-known method of determining the regular expression of a given finite automaton by assigning to each state a variable whose value is the language recognized by that state. These values are determined by a linear system of equations that can be solved step by step using Arden's rule [1]. Our approach uses systems of graph formulas which associate each variable with a graph formula that may, in turn, contain variables, thus creating a recursive logic scheme for defining graph languages. We aim to show that these formula systems describe precisely the same graph language as alternating graph automata.

Acknowledgments. This paper would not exist without the contributions of Annegret Habel: not only did she draw the second author's attention to alternating graph automata, but she also sketched a first translation of graph formulas into alternating graph automata. This work has been supported by the Swedish Research Council under grant no. 2024-05318, and by the Wallenberg AI, Autonomous Systems and Software Program through the NEST project STING - Synthesis and analysis with Transducers and Invertible Neural Generators.

Disclosure of Interests. The authors have no competing interests to declare that are relevant to the content of this article.

[4] See https://www.unibw.de/inf2/grappa.

References

1. Arden, D.N.: Delayed-logic and finite-state machines. In: 2nd Annual Symposium on Switching Circuit Theory and Logical Design (SWCT 1961), pp. 133–151 (1961). https://doi.org/10.1109/FOCS.1961.13
2. Bruggink, H.J.S., Hülsbusch, M., König, B.: Towards alternating automata for graph languages. Electron. Commun. Eur. Assoc. Softw. Sci. Technol. **47** (2012). https://doi.org/10.14279/tuj.eceasst.47.734
3. Chandra, A.K., Kozen, D., Stockmeyer, L.J.: Alternation. J. ACM **28**(1), 114–133 (1981). https://doi.org/10.1145/322234.322243
4. Courcelle, B.: The monadic second-order logic of graphs. I. recognizable sets of finite graphs. Inf. Comput. **85**(1), 12–75 (1990). https://doi.org/10.1016/0890-5401(90)90043-H
5. Courcelle, B.: The monadic second-order logic of graphs V: on closing the gap between definability and recognizability. Theor. Comput. Sci. **80**(2), 153–202 (1991). https://doi.org/10.1016/0304-3975(91)90387-H
6. Courcelle, B., Engelfriet, J.: Graph Structure and Monadic Second-Order Logic – A Language-Theoretic Approach, Encyclopedia of Mathematics and its Applications, vol. 138. Cambridge University Press (2012). https://doi.org/10.1017/CBO9780511977619
7. Drewes, F., Habel, A., Kreowski, H.J.: Hyperedge replacement graph grammars. In: Rozenberg [23], chap. 2, pp. 95–162
8. Drewes, F., Hoffmann, B., Minas, M.: Extending predictive shift-reduce parsing to contextual hyperedge replacement grammars. In: Guerra, E., Orejas, F. (eds.) Graph Transformation - 12th International Conference, ICGT 2019. LNCS, vol. 11629, pp. 55–72. Springer (2019). https://doi.org/10.1007/978-3-030-23611-3_4
9. Drewes, F., Hoffmann, B., Minas, M.: Graph parsing as graph transformation - correctness of predictive top-down parsers. In: Gadducci, F., Kehrer, T. (eds.) Graph Transformation - 13th International Conference, ICGT 2020. LNCS, vol. 12150, pp. 221–238. Springer (2020). https://doi.org/10.1007/978-3-030-51372-6_13
10. Drewes, F., Hoffmann, B., Minas, M.: Acyclic contextual hyperedge replacement: Decidability of acyclicity and generative power. In: Behr, N., Strüber, D. (eds.) Graph Transformation - 15th International Conference, ICGT 2022, vol. 13349, pp. 3–19. Springer (2022). https://doi.org/10.1007/978-3-031-09843-7_1
11. Drewes, F., Hoffmann, B., Minas, M.: Specifying and checking graph properties with alternating graph automata. In: Endrullis, J., Tichy, M. (eds.) Graph Transformation - 18th International Conference, ICGT 2025. LNCS, vol. 15720, pp. xx–yy. Springer, Cham(2025). https://doi.org/10.1007/978-3-031-94706-3_5
12. Engelfriet, J., Vereijken, J.J.: Context-free graph grammars and concatenation of graphs. Acta Informatica **34**(10), 773–803 (1997). https://doi.org/10.1007/s002360050106
13. Gaifman, H.: On local and non-local properties. In: Stern, J. (ed.) Proceedings of the Herbrand Symposium, Logic Colloquium. Studies in Logic and the Foundations of Mathematics, vol. 105, pp. 105–135. North-Holland (1982). https://doi.org/10.1016/S0049-237X(08)71879-2
14. Gilroy, S., Lopez, A., Maneth, S., Simonaitis, P.: (re)introducing regular graph languages. In: Kanazawa, M., de Groote, P., Sadrzadeh, M. (eds.) Proceedings of the 15th Meeting on the Mathematics of Language, MOL 2017, London, UK, 13–14 July 2017, pp. 100–113. ACL (2017). https://doi.org/10.18653/v1/w17-3410

15. Habel, A.: Hyperedge Replacement: Grammars and Languages. LNCS, vol. 643. Springer (1992). https://doi.org/10.1007/BFb0013875
16. Habel, A., Pennemann, K.H.: Nested constraints and application conditions for high-level structures. In: Kreowski, H.J., et al. (eds.) Formal Methods in Software and System Modeling. LNCS, vol. 3393, pp. 293–308. Springer (2005). https://doi.org/10.1007/978-3-540-31847-7_17
17. Habel, A., Pennemann, K.H.: Correctness of high-level transformation systems relative to nested conditions. Math. Struct. Comput. Sci. **19**(2), 245–296 (2009). https://doi.org/10.1017/S0960129508007202
18. Habel, A., Radke, H.: Expressiveness of graph conditions with variables. Elect. Comm. of the EASST **30** (2010). https://doi.org/10.14279/tuj.eceasst.30.404. International Colloquium on Graph and Model Transformation (GraMoT'10)
19. Navarro, M., Orejas, F., Pino, E., Lambers, L.: A navigational logic for reasoning about graph properties. J. Log. Algebraic Methods Program. **118**, 100616 (2021). https://doi.org/10.1016/j.jlamp.2020.100616
20. Pennemann, K.: Development of correct graph transformation systems. Ph.D. thesis, Carl-von-Ossietzky-Universität Oldenburg, Germany (2009). https://nbn-resolving.org/urn:nbn:de:gbv:715-oops-9483
21. Poskitt, C.M., Plump, D.: Monadic second-order incorrectness logic for GP 2. J. Log. Algebraic Methods Program. **130**, 100825 (2023). https://doi.org/10.1016/j.jlamp.2022.100825
22. Radke, H.: A Theory of HR* Graph Conditions and their Application to Meta-Modeling. Dissertation, Carl-von-Ossietzky-Universität Oldenburg (2016). http://oops.uni-oldenburg.de/2803
23. Rozenberg, G. (ed.): Handbook of Graph Grammars and Computing by Graph Transformation, Vol. I: Foundations. World Scientific, Singapore (1997)
24. Schürr, A.: Programmed graph replacement systems. In: Rozenberg [23], chap. 7, pp. 479–546. https://doi.org/10.1142/9789812384720-0007
25. Wulandari, G.S., Plump, D.: Verifying graph programs with monadic second-order logic. In: Gadducci, F., Kehrer, T. (eds.) Graph Transformation - 14th International Conference, ICGT 2021. LNCS, vol. 12741, pp. 240–261. Springer (2021). https://doi.org/10.1007/978-3-030-78946-6_13

Applications of Graph Transformation for Program Verification and Testing

Fuzzing Graph Database Applications with Graph Transformations

Stefania Dumbrava[1(✉)], Melchior W. M. Oudemans[2], and Burcu Kulahcioglu Ozkan[2]

[1] ENSIIE & SAMOVAR - Télécom SudParis, Paris, France
stefania.dumbrava@ensiie.fr
[2] Delft University of Technology, Delft, Netherlands

Abstract. Graph databases have surged in popularity, and applications increasingly employ them to store and retrieve interconnected data. However, testing graph database-backed applications has distinctive challenges. Due to the sheer dimension of the graph schema state space, testing applications using naive random graph instances is unlikely to cover a large portion of an application program. We present PGFuzz, a graph transformation-based greybox fuzzer for testing graph database-backed applications, that is, to the best of our knowledge, the first fuzzer to specifically target graph database applications. PGFuzz builds on top of state-of-the-art graph generators and utilizes graph transformations guided by code coverage to produce application test inputs. PGFuzz's graph transformations are schema-aware and support recently introduced graph schema, key, and cardinality constraints. We evaluate PGFuzz on graph database applications that we curate from open-source repositories and show that PGFuzz substantially improves the test coverage of graph database-backed applications compared to the state-of-the-art.

Keywords: graph database applications · graph transformations · graph schema constraints · fuzzing · automated testing

1 Introduction

The popularity of graph database management systems (GDBMSs) has surged in recent years, driven by the increasing need to efficiently store, process, and analyze complex graph data [71]. Due to being custom-built to handle relationship-centric scenarios, GDBMSs have numerous use cases [40,58,70] involving highly connected data, ranging from healthcare [7,14,38,67], finance [15,36,50], to transportation [39,41,65], and various enterprise applications [72].

Prominent examples of commercial GDBMSs include both *native graph* and *multi-model* systems. The former leverage *graph data models*, such as in Neo4j [57], MemGraph [53], or TigerGraph [76], or *triple stores*, as in Amazon Neptune [16], AllegroGraph [8], or BlazeGraph [21]. Multi-model systems mix several models, with the main ones being, for example, *wide-column stores*

J. Endrullis and M. Tichy (Eds.): ICGT 2025, LNCS 15720, pp. 135–156, 2025.
https://doi.org/10.1007/978-3-031-94706-3_7

(JanusGraph [46], DataStax [27], Titan [77], etc.), *key-value stores* (HyperGraphDB [42], DGraph [28], RedisGraph [69], etc.), *document stores* (Azure Cosmos DB [55], ArangoDB [13], OrientDB [61], etc.) or *relational tables* (AgensGraph [20], Db2 Graph [43], Oracle Spatial and Graph [60], etc.). In this work, we target native GDBMSs, represented by the popular Neo4j database, and multi-model ones, represented by JanusGraph, one of the few GDBMSs equipped with not only graph schema, but also advanced integrity constraint mechanisms [11].

The behavior of graph database-backed applications depends on the contents of the input graph database. Therefore, exercising various executions of an application's logic requires the use of different input graph instances, as they can activate different parts of the application logic. Unforeseen input database instances may result in unexpected and erroneous application behaviors.

Ensuring the reliability of graph database applications through testing faces significant challenges in terms of *validity* and *effectiveness*. First, the database instances for the graph database applications must satisfy complicated constraints determined by application semantics, which are nontrivial to synthesize. Second, effectively exploring the application logic requires generating graph database instances that can trigger various parts of the application code.

An intuitive approach to test graph database-backed applications is to stress-test them using graph instance generators, such as the state-of-the-art schema-driven graph generators gMark [18] and pgMark [75]. However, while these schema-driven generators solve the problem of validity, they do not offer effective testing of the applications. They only generate graphs that comply with a given graph data schema, which makes them inherently limited in their coverage of all application behaviors. Although most graph databases do not enforce schema constraints, these are crucial to data quality [22,44], data integration, and data exchange [19,33]. The application programs process the retrieved data and ensure its integrity. As such, they perform operations and substitutions on graph database instances. However, schema-driven graph generators often fail to cover all branches of an application, particularly logic involving application-specific substitution values. Additionally, these generators produce database states independently, without leveraging insights from previously generated states. This limits their ability to cover complex applications with nested branching logic.

Fuzzing [52,82,84] is a popular approach for software testing. Blackbox fuzzing treats the program as a black box, i.e., it does not know the internal structure and branching of the program and generates test inputs randomly, independent of the program logic and the other test inputs. In contrast, whitebox fuzzing [35] analyzes the program's internals to create test inputs. This involves examining the program's branching logic and using satisfiability solvers to generate inputs that can trigger specific branches. While blackbox fuzzing may struggle to effectively exercise rare or deep code branches by creating inputs that can navigate less frequent paths, whitebox fuzzing requires complete knowledge of the system being tested and is computationally intensive.

Greybox fuzzing [82] combines both effectiveness and efficiency by utilizing information gathered from the program being tested, such as execution branch coverage, to generate new test executions. A prominent approach is *coverage-guided* greybox fuzzing, which creates new test inputs by modifying previous inputs that trigger new coverage during execution. Testing the application with these modified test cases allows for more effective exploration of low-frequency or deeply nested program paths and helps with the discovery of new paths. Greybox fuzzing has been shown to be an effective automated approach for testing various application domains [52,84]. However, to our knowledge, there are *no fuzzers designed to generate graph inputs for graph database-backed applications.*

Fuzzing at the input level is often insufficient for graph-backed applications, where control flow depends on the structure and content of the database, such as missing properties, relationship counts, or specific subgraph patterns. PGFuzz addresses this by targeting the database state directly to explore such logic. It is the first greybox fuzzer for graph database-backed applications that combines schema-aware and constraint-aware graph transformations with coverage-guided feedback. To ensure validity, PGFuzz builds on top of state-of-the-art gMark [18] and pgMark [75] graph generators and then mutates the generated instances to uncover new application behaviors. To address scalability, PGFuzz operates on an in-memory graph model and simulates database calls through a lightweight API, avoiding the overhead of actual GDBMS operations and enabling fast, efficient fuzzing across large input spaces.

While PGFuzz is the first greybox fuzzer targeting graph database-backed applications, its primary novelty is its use of *schema-aware mutations*, which produce input graphs that *respect schema-defined structure and types* while deliberately *violating key and cardinality constraints* through targeted property mutations. These mutations, informed by the recent PG-Schema [11] language, contrast grammar-based fuzzers—which strictly enforce grammatical constraints—by intentionally generating graph instances that are syntactically valid, according to the schema types, but that violate key and cardinality constraints. This allows PGFuzz to thoroughly test the application logic responsible for maintaining data integrity and schema compliance.

Testing graph database-backed *applications* is an underexplored area, and there are no openly available sets of such benchmark applications. For the empirical evaluation, we examined open-source applications built on top of Neo4j [57], one of the most popular graph databases [73], and JanusGraph [46], one of the few graph databases to enforce schema constraints [11], and compiled a custom benchmark suite for testing database-backed applications. We evaluate PGFuzz against existing graph generators for test generation and show its superior coverage. Our tests revealed several bugs in the application programs that manifest when using certain graph database instances as input states.

In summary, this paper makes the following contributions:

- PGFuzz, the first greybox fuzzer for testing graph database-backed applications that uses schema and constraint-aware graph transformations to generate input graph database instances.

- A benchmark suite collected from real-world graph database-backed applications that leverage heterogeneous graph structures from various domains.
- An empirical analysis evaluating PGFuzz against the gMark and pgMark state-of-the-art schema-driven graph instance generators.

```
void getTransport(Graph g) {
 int transport_count = 0;
 List<String> trsp = new
      ArrayList<>();
 Node n =
      g.getNodes("Centroid").get(0);
 for (Edge e : n.getEdges()) {
   if (e.getLabel().equals("DRT") ||
        e.getLabel().equals("WALK")) {
     trsp.add(e.getLabel());
     transport_count += 1; }
 }
 String transport;
 if (transport_count == 2) {
   transport= "DRT/WALK"; }
   else {
   transport= trsp.get(0); }
 return transport;
}
```

(a) Excerpt from P2 [39].

```
List<PaperResponse> referers =
g.getConnectedNodes
 (paper, "cites", true)
.stream()
.map(v -> {
 Map<String, String> props =
 v.properties;
 return new PaperResponse(
  (String) (v.id + ""),
  (String) props.getOrDefault
   ("title", "N/A"),
  Integer.parseInt(props.getOrDefault
   ("year", "N/A")),
  Integer.parseInt(props.getOrDefault
   ("numOfPaperReferees", "0")),
  Integer.parseInt(props.getOrDefault
   ("numOfPaperReferers", "0")),
  Double.parseDouble(props.getOrDefault
   ("pagerank", "0.0"))); })
```

(b) Excerpt from P3 [3].

Fig. 1. Simplified code excerpts from our benchmark applications.

2 Motivating Examples

We showcase PGFuzz's key challenges with an example involving branching logic based on the input graph schema and database state, and another handling schema-constrained graph instances where the application ensures data integrity.

Example 1 (Transport Graphs). Graph databases are increasingly used to analyze existing public transit networks and to construct novel models that improve accessibility [29,41,65]. In a recent use-case [39], GTFS[1] schedule data has been used to build and process public transportation graph models in the Neo4j graph database. A code snippet from the application is shown in Fig. 1a. Its purpose is to analyze the connectivity of the `centroid` nodes (centers of a tessellated grid map) using demand-responsive transport (edges labeled "DRT") and pedestrian routes (edges labeled "WALK"). Note that the first branching condition depends on these labels, which are part of the typing schema constraint of the underlying graph database. *Efficiently* producing *valid* test graph instances that *effectively* cover the application's branching structure requires custom techniques.

[1] GTFS is the standard open data format detailing transit schedules.

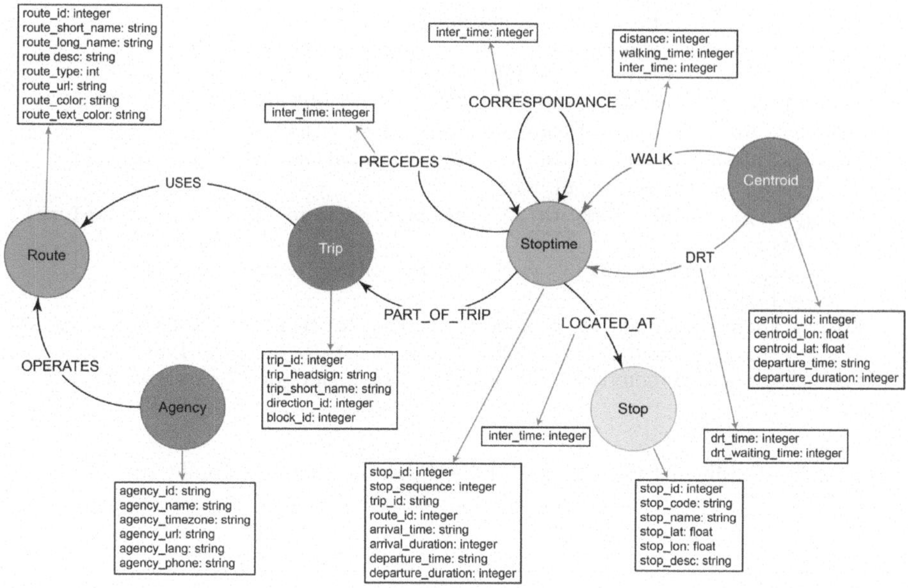

Fig. 2. Property graph schema for the P2 transport application.

Example 2 (Citation Graphs). A recent graph database use-case targets the analysis of bibliographical citation networks [3]. The schema defined by the application requires the nodes representing papers to have string values assigned to the properties that store their title, year of publication, number of referees and referrers, and PageRank. These properties are read by the application logic, and default values are substituted when they are missing, as can be seen in the code snippet in Fig. 1b. Schema-driven generators like gMark or pgMark typically generate *by design* fully schema-compliant instances where all expected properties are present, resulting in test inputs that bypass the application logic meant to handle missing or incomplete data. PGFuzz overcomes this by deliberately mutating such instances to remove or alter expected properties, triggering execution paths that would otherwise remain untested.

3 Preliminaries

Graph databases typically use the labeled property graph data model [10], i.e., a directed multi-labeled multi-graph with properties on nodes and edges.

Example 3 (Transport Graph). We illustrate the property graph model on our running example. In the P2 application, a city's transit line network is represented as a property graph. Figure 2 captures the graph database instance for the city of Royan's transport graph. The node colors correspond to their labeling. `Stoptime` nodes (blue) that represent scheduled passage times are linked

```
CREATE GRAPH TYPE transportationGraphType STRICT {
 (aType : Agency {agency_id STRING,...,agency_phone STRING}),
 (rType : Route {route_id INT32,...,route_text_color STRING}),
 (stimeType : Stoptime {stop_id INT32,..., departure_duration INT32}),
 (sType : Stop {stop_id INT32,..., stop_desc STRING}),
 (cType : Centroid {centroid_id INT32,...departure_duration INT32}), ...
 (:aType)-[:operates]->(:rType),
 (:cType)-[:DRT {drt_time INT32, drt_waiting_time INT32}]->(:stimeType),
 (:cType)
   -[:WALK {distance INT32, walking_time INT32, inter_time INT32}]->
 (:stimeType), ...
 FOR (a:Agency) EXCLUSIVE a.agency_id, FOR (s:Stop) MANDATORY s.stop_id,
 FOR (rt:Route) SINGLETON o WITHIN (a:Agency)-[o:operates]->(rt),
 FOR (c:Centroid) COUNT 2.. OF r WITHIN (s:Stoptime)-[r:DRT|WALK]->(c),
 ... }
```

Fig. 3. Simplified PG-Schema for the P2 transportation application [39].

to specific `Stop` stations (pink) through arcs labeled `LOCATED_AT`. Tessellation `Centroid` nodes (red) are connected to `Stoptime` ones via arcs labeled `WALK` and/or `DRT`, depending on whether these are accessible via pedestrian or, respectively, via demand-responsive routes. Stoptime nodes are inter-connected with arcs labeled `PRECEDES`, denoting temporal ordering, or `CORRESPONDENCE`, denoting pairwise accessibility using a different line. Each `Trip` (orange node) encodes a mobility plan that comprises passage times (`Stoptime` nodes) and employs a `Route` (green node) or several, operated by an `Agency` (brown node) or multiple ones. The property graph model also allows attaching lists of key-value properties on nodes and edges, as can be seen in Fig. 2 discussed next.

Property Graph Schemas and Constraints. While property graph instances represent *data* stored in a GDBMS, graph schemas and constraints describe their *structure*, typing, and integrity requirements to ensure consistency. Although these are crucial for data integration and exchange [19], query optimization [31], and data reliability [32], current GDBMSs provide limited support.

The applications we analyzed are built on top of Neo4j and JanusGraph. We consider these systems, as Neo4j is one of the most popular [73] GDBMSs and JanusGraph is one of the few that can handle not only a priori schema constraints but also rich cardinality ones, as surveyed in [11]. Although these systems provide different levels of support for graph schemas and constraints - Neo4j is schemaless and JanusGraph is schema flexible and allows explicit or implicit schema definitions - both benchmarks use Neo4j and JanusGraph *enforce these at the application level to maintain data integrity.*

PGFuzz supports graph schema constraints, including key constraints and cardinality ones, illustrated with the recent PG-Schema language. *Typing constraints* (PG-Types) specify nodes and edge labels and properties and the relationship types that can connect certain node labels. *Key constraints* (PG-Keys) ensure that graph objects are uniquely identified and referenced. Identification is enforced with the `IDENTIFIER` keyword, comprising `EXCLUSIVE`, `MANDATORY`, and `SINGLETON`, while for referencing, `EXCLUSIVE MANDATORY` is used. PG-KEYS statements are of the form `FOR` $p(x)$ `<qualifier>` $q(x, \bar{y})$, where `<qualifier>` is a combination of `EXCLUSIVE MANDATORY`, `SINGLETON`, and $p(x)$, $q(x, \bar{y})$ are graph queries.

We illustrate these in Fig. 3 on the running example. `EXCLUSIVE` restricts two objects from sharing a key value, e.g., no two agencies have the same `agency_id` (line 28). `MANDATORY` enforces that every object must have at least one key, e.g., every stop should have a `stop_id` (line 30). `SINGLETON` enforces that every object has at most one key, ensuring that two objects with the same key are identical, e.g., every route is operated by a unique agency (lines 32–33).

Cardinality constraints, introduced in PG-SCHEMA as a PG-KEYS extension, enforce bounds on the number of graph object instances, e.g., every centroid should be connected to a stoptime through at least two edges, labeled 'DRT' or 'WALK' (lines 35–36). The constraints are expressed similarly to PG-KEYS, using the qualifier `COUNT <lower bound>?..<upper bound>? OF` to express that the number of distinct results returned by $q(x, \bar{y})$ must be within the specified range.

Among existing GDBMSs, JanusGraph provides one of the most comprehensive supports for cardinality constraints [11]. It allows declaring edge label multiplicity, specifying whether at most one (`SIMPLE`) or multiple (`MULTI`) edges can be defined between any node pair, with `MANY2ONE` and `ONE2MANY` respectively allowing at most one outgoing/incoming edge, without constraining the number of incoming/outgoing ones. It also allows for property key cardinalities, specifying whether a node key can have one or multiple values.

PGFUZZ transformations allow breaking all of the above constraints, as detailed in Sect. 4.3. PGFUZZ considers test inputs for graph database applications to be defined by a core fragment of the previous graph schema constraints, with typing and key constraints being enforced at the property level. As such, each property can be assigned one of the built-in types `String`, `Integer`, `Double`, `Float`, `Boolean`, with `String` being the default when no type is provided. Also, properties can be flagged as `EXCLUSIVE`, `MANDATORY`, or `SINGLETON`. Relationship cardinality constraints are also supported, enabling the specification of the allowed number of edges of a particular type between any two nodes.

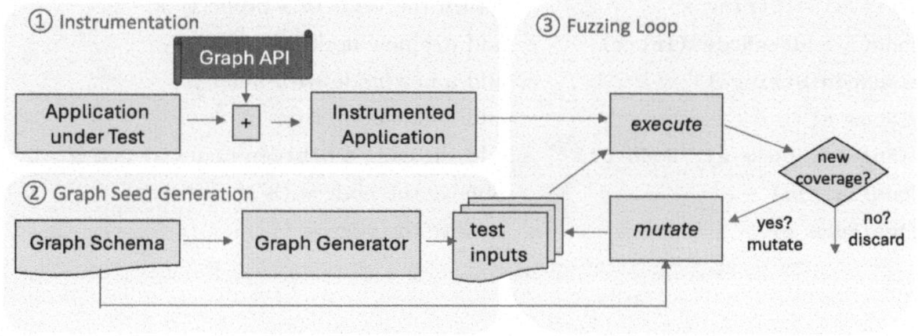

Fig. 4. The workflow of PGFUZZ.

4 The PGFuzz Framework

Figure 4 illustrates the workflow of the PGFuzz testing framework. The main components of PGFuzz are: (1) program instrumentation of the application under test to direct the calls to graph databases to PGFuzz Graph API and track code coverage, (2) leveraging state-of-the-art graph generators to produce graph database input seeds, and (3) using greybox fuzzing to generate new test inputs by graph transformations guided by code coverage.

To avoid the cost of database connection, communication, and state resetting operations at each test execution, PGFuzz maintains an in-memory graph structure for the input graph database state. The model is based on a property graph model (see Sect. 3) and provides a graph processing API modeled after popular graph database query APIs. As shown in Table 1, the PGFuzz API defines the basic methods for accessing and updating graph data.

We instrument the application under test to redirect the calls to graph database operations and instead invoke the API operations on PGFuzz's internal graph. Hence, PGFuzz does not use GDBMS executions but directly explores the behaviors of the graph database application under test.

Table 1. PGFuzz API operations.

Operation	Description
`getNodes(String l, String k, String v)`	return nodes with label `l` and key-value (`k`, `v`)
`getNode(int id)`	return the node with identifier `id`
`getLabel(Node n)`	return the label attached to the node `n`
`getConnectedNodes(Node n, String l)`	return all nodes connected to `n` with edges labeled `l`
`getNodes()`, `getEdges()`	return all nodes/edges
`getNodeEdges(Node n)`	return all edges attached to the node `n`
`getIncomingEdges(Node n)`	return all incoming edges attached to the node `n`
`getOutgoingEdges(Node n)`	return all outgoing edges attached to the node `n`
`getNodePairEdges(Node n1, Node n2)`	return all edges between nodes `n1` and `n2`
`getEdgeProperty(Edge e, String l)`	return a property of an edge `e` labeled `l`
`getNodeProperty(Node n, String l)`	return a property of a node `n` labeled `l`
`getPropertyValue(String k)`	return the value of a property `k`
`addNewNode()`, `addNewNodes(int c)`	add a/c new node(s)
`addLabeledNode(String l)`	add a new node with label `l`
`addNode(Node n)`	add a new node `n`
`addEdge(Edge e, Node n1, Node n2)`	add the edge `e` between nodes `n1` and `n2`
`removeNode(int id)`	remove the node with identifier `id`
`removeEdge(Edge e)`	remove the edge `e`

4.1 Producing Random Graph Instances

PGFuzz constructs initial test seeds using the state-of-the-art gMark [18] and pgMark [75] schema-driven graph generators. gMark produces synthetic edge-labeled graphs based on schema constraints specifying their corresponding typing structure. The system also allows the configuration of the size of the graph, the number of occurrences for each node or edge label, and the in- and out-degree distribution for each relationship type. The pgMark framework builds on top of gMark and adds support for generating attributed graphs with node properties. In the PGFuzz framework, we use an enhanced version of pgMark, which we extended to support the *full expressiveness of the property graph model* by also allowing instances to incorporate edge properties.

4.2 Greybox Fuzzing for Graph Generation

PGFuzz collects the graph database states produced by gMark and pgMark in a set of test inputs and uses them for testing graph database applications. However, these test inputs are generated in a black-box approach, i.e., they are produced independently and randomly, not using any information from the application under test. Therefore, testing the applications using these graphs as test inputs is unlikely to cover all branching logic of the application code.

PGFuzz offers a greybox approach for generating new graph database instances for testing graph database applications. Unlike the graph instance generation by gMark and pgMark, PGFuzz employs a greybox feedback loop to guide the test generation towards more application code coverage.

As given in Fig. 4, the greybox fuzzing loop consists of three main steps: First, it runs the application under test with a test input. During the execution, it collects some greybox information (for PGFuzz, branch code coverage) of the execution. Then, it checks whether the execution covers previously unexplored branches in the application code. Following the intuition that the test inputs similar to the one that triggered new application behavior are also likely to trigger new behaviors, the fuzzer mutates that test input to generate new test inputs. PGFuzz uses graph transformations to generate new test inputs.

PGFuzz uses branch code coverage as greybox feedback information from the test executions, which is commonly used in testing traditional software systems [82]. As the application programs' branching logic is based on the input graph database state, branch code coverage effectively captures the explored parts of the application program. Alternative to code coverage, the framework can be extended to use different feedback information from the program, e.g., defining fitness functions, to evaluate how likely the test inputs generated from an input can trigger new behaviors.

4.3 Schema-Aware Graph Transformations

Given a test input, i.e., test graph database instance, that achieves new coverage PGFuzz mutates that instance to produce a new instance using novel graph transformations that leverage the typing, key, and cardinality constraints of the application under test.

PGFuzz uses graph schema-aware transformations and supports a core fragment of the recent PG-Schema and PG-Keys formalisms, which inform the design of the novel GQL graph query standard [11,12]. PGFuzz integrates information regarding graph typing, key constraints enforced on graph objects, and cardinality restrictions, e.g., on the number of edges between node pairs, as in JanusGraph: `One2One`, `Many2One`, `One2Many`, `Many2Many`. This provides variations of graph instances that comply with application requirements and are likely to cover important states. Rather than adding a random label, the transformation can select one *that is expected to occur*. By design, the transformation can leverage information about graph constraints that might not otherwise be derived from any arbitrary graph state. Schema-aware transformations adhere to the schema when possible but may deviate if no valid candidates are available.

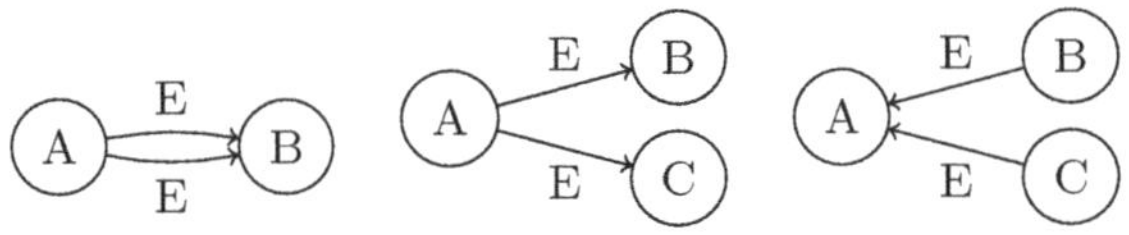

Fig. 5. $\mathbf{M_{14}}$ transformations that violate cardinality constraints. Each subfigure shows the result of a transformation breaking one of: `SIMPLE` (maximum one edge between two nodes), `MANY2ONE` (multiple sources to one target), and `ONE2MANY` (one source to multiple targets). Each graph depicts the input test state *after* the transformation, where an added edge leads to the constraint violation.

PGFuzz transformations are classified into $\mathbf{C_1}$, $\mathbf{C_2}$, $\mathbf{C_3}$, and $\mathbf{C_4}$, depending on whether they add, remove, modify graph objects, or violate graph constraints. Modifications include adding nodes or edges with arbitrary schema labels ($\mathbf{M_{1,2}}$) and assigning arbitrary schema properties ($\mathbf{M_3}$). Nodes and their attached edges can be removed ($\mathbf{M_4}$), as well as edges or properties ($\mathbf{M_{5,6}}$). Node or edge labels can be randomly modified ($\mathbf{M_{7,8}}$), along with the keys or property values ($\mathbf{M_{9,10}}$). Schema constraints can be broken by removing all nodes or edges of a given label ($\mathbf{M_{11,12}}$), altering property typing for arbitrary elements ($\mathbf{M_{13}}$), or cardinality constraints by adding or removing instances ($\mathbf{M_{14}}$ in Fig. 5), breaching exclusivity constraints by assigning values to random elements ($\mathbf{M_{15}}$), and violating mandatory constraints by setting values to `NULL` ($\mathbf{M_{16}}$).

5 Evaluation

We implement PGFuzz[2] on top of the JQF [62] fuzz testing framework for Java applications. JQF allows for instrumenting the application under test to

[2] The source code of PGFuzz and the evaluation data are available at: https://github.com/moudemans/PGFuzz.

track code coverage and implement the coverage-guided testing loop. We redirect graph database operations to PGFuzz's Graph API and then execute the instrumented program with generated graph instances as test inputs. We evaluate PGFuzz's performance compared to state-of-the-art graph instance generators, in terms of test coverage and the number of unique bugs detected, following metrics proposed in [51,54]. We test benchmark graph database-backed applications using baseline graph generators and PGFuzz, measuring branch code coverage and the number of distinct bugs discovered. We collect the test results on a machine running Windows 11 with Intel(R) Core(TM) i7-13700KF 3.40 GHz CPU and 32 GB of memory.

As baselines, we consider the state-of-the-art gMark and pgMark schema-driven graph generators for applications leveraging edge-labeled and property graphs, respectively. These tools follow a black-box testing strategy, producing test inputs independently of each other, as well as of previous test executions. PGFuzz exploits a greybox fuzzing framework to incorporate code coverage feedback into the test generation and uses schema-aware graph data transformations to generate new tests.

Table 2. Benchmark characteristics.

Code	Name	Framework	Language	Characteristics
P1	Medical	Neo4j	Python	Cardinality
P2	Transportation	Neo4j	Python	Properties and node labels
P3	Citation	JanusGraph	Java	Edge properties
P4	Citation Network	JanusGraph	Java	Relationships
P5	Pangenomic	Neo4j	Python	Property values
P6	Pheno4JOut	Neo4j	Java	Property keys and types
P7	Pheno4J	Neo4j	Java	Property types
P8	PanTool1	Neo4j	Java	Property values
P9	PanTool2	Neo4j	Java	Unstructured dependencies

5.1 Benchmarks

We collected graph database-backed applications from *publicly available* repositories. The main criterion is ensuring that graph schema, key, and cardinality constraints are enforced in the branching logic. We curate eight benchmarks (see Table 2) from diverse areas: medical studies (P1), intelligent transportation (P2), bibliographical analysis (P3), and genomics (P5-P8), as described below.

P1 Medical is collected from OpenStudyBuilder [1], an open-source application to manage clinical data standards and study design specifications. It implements CDISC [26] standards[3] by providing additional semantics to prevent

[3] CDISC develops standards for collecting, sharing, and analyzing clinical trial data.

inconsistencies that impede automation scaling. P1 ensures that, when adding or modifying nodes, `EXCLUSIVE` and `MANDATORY` constraints are preserved.

P2 Transportation is collected from a study [2,39] that evaluates how demand-responsive services influence public transportation accessibility. P2 analyzes the connectivity of central locations on a grid with respect to mass transit stations through pedestrian and on-demand routes. Its branching logic depends on *cardinality* constraints on relationships with certain labels and properties.

P3 Citation is collected from CiteGraph [3], an open-source web application for visualizing and analyzing citation networks. It enables discovering connections between papers and authors and evaluating the influence of specific works or researchers. P3 gathers citation, collaboration, PageRank metrics, and bibliographic meta-data through graph traversals. Its logic for processing data is guided by the schema of the underlying property graph.

P4 Citation Network benchmark is collected from the same repository as the P3 Citation. It finds its n-hop references for a given paper by relying on specific relationships to be connected with at least n connected components.

P5 Pangenomic is collected from PanGraph-DB [4], an open-source comparative pangenomics application that analyzes graph structures representing the genomic diversity of species. It inspects genomic family neighborhoods and extracts connectivity properties to identify those with similar characteristics.

P6 Pheno4J1 is collected from Pheno4J [56], an open-source application that integrates various biological data sources, allowing to visualize and analyze relationships between genetic variations and phenotypic traits. Its logic depends on the graph schema, as it computes an export format for data exchange.

P7 Pheno4J2, also collected from Pheno4J, analyses the properties linked to phenotypic traits of genomic variants. Its branching logic is determined by the edge labels of the different variants and by the properties that capture their phenotypic effects. Data integrity is maintained through type conversions.

P8 PanTool1 is collected from PanTool [5], an open-source application that analyzes pangenome graphs to evaluate genetic variations across species. It tracks and organizes phasing information from genomic sequences represented as nodes. Its branching logic depends on `MANDATORY` key constraints on node properties and cardinality constraints on the number of genomes with phasing information.

P9 PanTool2, also collected from PanTool, analyzes the messenger RNA nodes of a genomic graph. Similarly to P8, its branching logic depends on property values and `MANDATORY` key constraints on properties.

Correctness Specification and Test Oracle. We validate the benchmarks by checking whether the application crashes for certain inputs, using this property as our test oracle. We identified several bugs in the applications that led to crashes, with exceptions being thrown when specific graph database instances were used as input.

5.2 Evaluation Results

We evaluate PGFuzz by addressing the following main research questions:

- **RQ1.** Can the test inputs generated by PGFuzz provide a higher test coverage compared to the state-of-the-art graph instance generators?
- **RQ2.** Can the test inputs generated by PGFuzz detect more bugs than the state-of-the-art graph instance generators?
- **RQ3.** How do different categories of graph transformations contribute to the bug detection capability of PGFuzz?

We answer the research questions by testing the benchmark programs using the input graph instances generated by PGFuzz and the state-of-the-art graph instance generators gMark and pgMark with the synthetic graph size of 100 nodes. Due to the randomness of the tests, we conducted each experiment five times for 10 min and reported the average results.

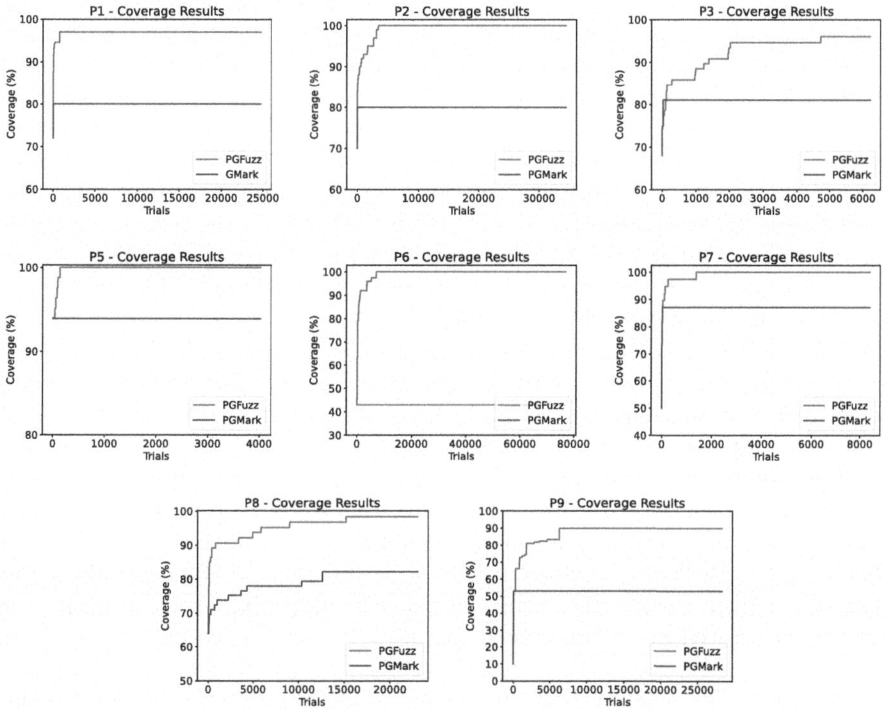

Fig. 6. Test coverage of the benchmarks using PGFuzz and the baseline.

Test Coverage. Figure 6 shows the branch code coverage of testing the benchmark applications using gMark (for P1, which uses an edge-labeled graph), pgMark (for P2-P9, which use property graphs), and PGFuzz. The plots do not include the coverage results for benchmark P4 since all branches in that benchmark program are covered within the first few test cases generated by each method. The results show that testing the benchmarks with the black-box testing approach using pgMark reaches 40%–85% branch coverage. The broad coverage of the application logic achieved with randomly generated instances can be attributed to the diversity of graph instances gMark and pgMark produce by design, as demonstrated by their empirical evaluation [18]. However, the generated graph instances do not cover much of the code in all benchmarks, e.g., in P6 and P9. These benchmarks use various instance nodes, edges, relation labels, and properties in their application logic and use nested branching on these properties. Increasing the test coverage of such benchmarks requires exploring a more extensive set of graph instances. Coverage-guided fuzzing approaches increase the code coverage by generating more graph instances around the instances that hit new code branches, hence exploring deeper branches of the applications.

We answer **RQ1** positively: PGFuzz achieves better coverage results in a smaller number of test executions for all the benchmark applications. PGFuzz's schema-aware test case generation produces test instances that comply with schema types but intentionally break the schema constraints. An example is the branching given in Sect. 3 and Fig. 1b, which requires the exploration of graph instances that do not necessarily satisfy all the constraints in the graph schema. The graph transformation-based mutations of PGFuzz lead to covering the application logic with the substituted values for the graph properties. Moreover, transforming the instances that cover new branches helps in covering a deeper sequence of branching. PGFuzz shows an increase in coverage of up to 57% and 23% on average, with the highest improvement in benchmarks P6 and P8 that have the most complicated branching logic.

While PGFuzz consistently outperforms the baselines, it does not reach 100% coverage for all benchmarks. Further analysis of the applications shows that P1 has an unreachable condition in a private method, as it has already been checked and filtered in an earlier branch. In P3, PGFuzz can reach 100% coverage but cannot consistently cover each branch in all test repetitions. Branches that are not consistently reached depend on the presence of a specific node and require input graph transformations that enable access but make reaching them unlikely. In P9, PGFuzz cannot reach 100% coverage, as doing so would require sequential node property IDs, which are highly improbable to occur in both randomly generated and transformed inputs.

Bug Detection. We answer **RQ2** positively: PGFuzz detects more bugs than testing the applications using the state-of-the-art graph instance generators. In our evaluation, PGFuzz found several bugs in the benchmark applications, while the baseline approaches failed to detect any of them.

Table 3 lists the number of distinct bugs PGFuzz discovered. The most frequent errors, encountered 23 times, are due to invalid data types, as the benchmark tries to parse integers, booleans, or other types while the property value is no longer of that type. Null pointer exceptions occurred 5 times and were triggered by certain graph elements no longer being present. The array index out of bounds occurred 3 times and was caused by removing graph elements or splitting a string that no longer contained a specific character. These errors are

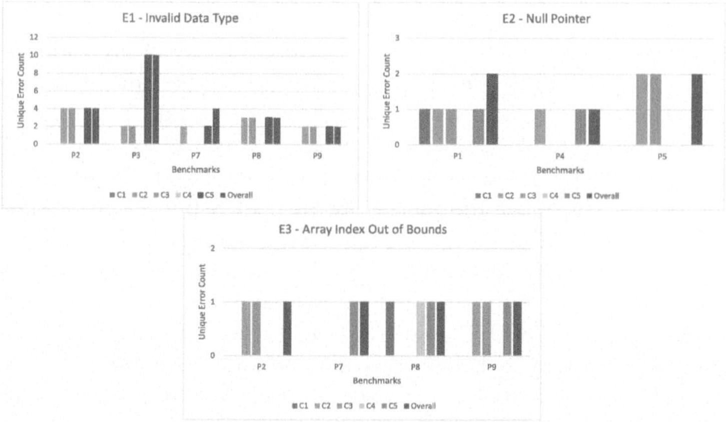

Fig. 7. Contribution of the transformation categories per error type.

thrown by the application under test when the contents of the graph instance trigger parts of the application code that do not anticipate certain nodes, edges, labels, or properties. Although the errors seem simple, they are hard to detect as they manifest only with certain contents of the graph instances to hit certain branches that are neglected by the developers.

Overall, PGFuzz is successful at finding bugs, especially with the graph transformations that break schema constraints. PGFuzz could expose several errors that are caused mainly by changing the graph elements and property types, producing possible test inputs that are neglected in the application logic.

Table 3. Distinct bugs per method causing Null Pointer Exception (NPE), Number Format Exception (NFE), Array Index Out of Bounds Exception (AIOBE).

	PGFuzz	Error(s)
P1	2	NPE
P2	5	AIOBE, NFE
P3	10	NFE
P4	1	NPE
P5	2	NPE
P6	0	-
P7	4	AIOBE, NFE
P8	4	AIOBE, NFE
P9	3	AIOBE, NFE

Contribution of the Transformation Categories for Bug Detection. To answer **RQ3**, we analyzed the effect of the categories of transformations to detect each type of error. Figure 7 shows the number of errors hit by each category of transformations for (1) an invalid data type, (2) a null pointer, and (3) an array index out-of-bounds exception. Breaking graph constraints (C4), which potentially violate the schema constraints of an application, is the most effective strategy, detecting the most unique errors in 6 of 8 benchmarks. This underscores PGFuzz's schema-aware graph transformations, which generate schema-compliant yet semantically unexpected graph instances that violate constraints.

Compound Transformations. The graph transformations in PGFuzz focus on modifying individual graph elements, such as nodes or edges, along with their corresponding labels, properties, or values, as described in Sect. 4.3. To evaluate the effectiveness of making multiple changes to the graph instances, we introduced *compound transformations.* We then repeated the tests by incorporating these into PGFuzz's set of transformations.

Note that compound transformations do not change individual components of a graph; instead, they apply multiple modifications at once. Specifically, we used compound transformations that involve randomly adding multiple nodes and edges in the input graph. This process results in the addition of several new nodes, edges, and relationships that are likely to form more complex structures such as cycles or triangles. We avoid using transformations that remove subsets of graph elements, as such operations can be overly disruptive.

We repeated the evaluation on the set of benchmarks, enabling compound transformations. As given in Fig. 8, the results do not show a significant performance improvement on the application benchmarks. This can be explained by the branching logic in graph database applications, which can be covered by modifying a single element of the test input graph, e.g., labels or properties. Compound mutations can be more beneficial for the applications that branch on more specific patterns, e.g., checking for cycles or triangles.

Threats to Validity. Potential threats to the validity of our findings include the representativeness of the curated set of database-backed application benchmarks and the randomness of the test executions. The generalizability of our results highly depends on how representative our evaluation targets are of real-world application scenarios. As we collect the benchmark suite from open-source repositories, we do not cover enterprise applications. However, our benchmark applications have diverse graph database schemas and schema constraints.

An additional threat to the validity may arise from the decision to redirect database calls to PGFuzz API operations. The effectiveness of this approach depends on the assumption that the API calls accurately represent the behavior of the database methods. Any discrepancies between the two could negatively impact PGFuzz's performance. The API operations model all the database calls in the benchmark programs used in our experiments.

Lastly, the experimental results involve multiple sources of randomness. The input graph instances are randomly generated by gMark and pgMark, which

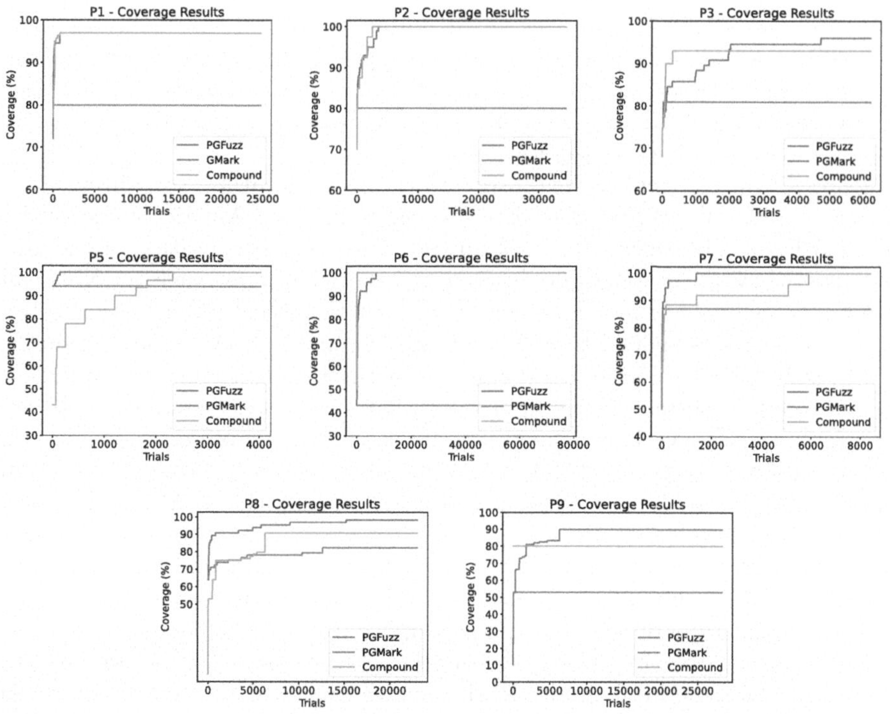

Fig. 8. Test coverage of the benchmarks results using compound transformations.

are later transformed by PGFUZZ. Moreover, the effectiveness of PGFUZZ's tests is highly dependent on the applied transformations, which are also selected randomly as part of the fuzzing framework. To mitigate randomness, we repeat our experiments five times and report averaged results.

6 Related Work

Testing Database Systems. Several recent works focus on testing database management systems [9,17,45] and graph databases [47,49,83]. Different from these works, which target the bugs in the database systems, PGFUZZ targets finding bugs in the graph database-backed applications.

Testing Database-Backed Applications. As the behavior of database-backed applications depends on the state of the database on which they operate, testing them requires generating test database instances. Earlier works [6,30] use constraint solving and static analysis for generating program inputs and input database states. SynDB [63,64] tests database-backed applications using a symbolic database. It generates test inputs and relational database states by considering all program, query, and database constraints and targeting relational databases. Recent work proposes DBGRILLER [81], a greybox testing approach for generating test databases for database-backed applications. Similar to

PGFuzz, DBGRILLER applies small mutations to existing database states to create new test instances. However, it is specifically designed for relational database applications and does not tackle graph-database-backed applications. To our knowledge, PGFuzz is the first greybox fuzzer for generating test inputs for graph database-backed applications.

Fuzz Testing. There is extensive work on generating test input using fuzzing [52,82,84]. While the first fuzzers, such as American Fuzzy Lop (AFL) [79] and libFuzzer [80], work on the bit-level representation of the input seed files, they are not as effective on structured inputs, such as XML and JavaScript, as bit-level mutations either disrupt the structure or are unlikely to produce changes beneficial to code coverage. Smart fuzzing [66] and grammar-based fuzzing [78] operate on test input structures and produce test inputs for a given input structure. These approaches have been successfully used to generate test inputs in various formats [34,48,59,74,78]. GraphFuzz [37] addresses library API testing by modeling executed functions as a dataflow graph and applying graph-based mutations to generate new test cases. Although GraphFuzz also uses graph mutations to modify test cases, it is tailored to dataflows and fundamentally differs from PGFuzz. Similar to smart and grammar-based fuzzers, PGFuzz generates syntactically valid test inputs, satisfying the syntactical schema type of the underlying graph database. Unlike grammar-based fuzzers that aim to cover the space of the *syntactic* structure, enforcing constraints in the generated input, PGFuzz can generate test inputs that may intentionally break *semantic* constraints that capture the intended logic of the data, such as key and cardinality constraints. This enables targeted testing of graph database applications where such constraints play a critical role in ensuring data integrity.

Graph Databases. Graph transformations have been extensively applied to schema evolution, data migration, and interoperability in graph databases. Bonifati et al. [23] propose a framework for supporting evolving graph schemas, which leverages graph rewriting operations to maintain consistency. More recent works [24,25] adopt a high-level, declarative approach aligned with the GQL standard, employing graph pattern matching to support complex migration and cleaning tasks. Similarly, schema-aware transformations ensure semantic interoperability between diverse graph models such as property graphs and RDF [68]. In contrast, PGFuzz leverages a lightweight, *heuristic* use of schema-aware transformations aimed at generating varied test instances for evaluating the robustness of graph database-backed applications.

7 Conclusion

The rise of graph database management systems emphasizes the importance of testing graph database-backed applications that rely on complex, highly interconnected datasets. Our work introduces PGFuzz, the first greybox fuzz-testing framework designed to specifically address the intricacies of testing graph database-backed applications. By incorporating schema-aware transformations and leveraging code coverage, PGFuzz offers a promising solution to testing graph database applications. Our empirical evaluations demonstrate that PGFuzz significantly outperforms existing state-of-the-art baselines, uncovering more distinct bugs and achieving higher test coverage in graph database applications from various domains. PGFuzz opens the promising perspective of also incorporating other recently introduced types of graph constraints [31].

Acknowledgments. This work was partially supported by the grant ANR-24-CE25-1109 (Dumbrava).

Disclosure of Interests. The authors have no competing interests to declare that are relevant to the content of this article.

References

1. OpenStudyBuilder (2022). https://gitlab.com/Novo-Nordisk/nn-public/openstudybuilder/project-description
2. Transport Network Graph Database (2022). https://github.com/CathiaLH/GraphDatabaseCombinedTransportNetwork
3. CiteGraph (2023). https://github.com/citegraph/citegraph
4. PanGraph-DB (2023). https://github.com/jpjarnoux/PanGraph-DB
5. PanTool (2023). https://git.wur.nl/bioinformatics/pantools/
6. Agrawal, P., Chandra, B., Emani, K.V., Garg, N., Sudarshan, S.: Test data generation for database applications. In: ICDE, pp. 1621–1624. IEEE Computer Society (2018)
7. Al-Saleem, J., et al.: Knowledge graph-based approaches to drug repurposing for COVID-19. J. Chem. Inf. Model. **61**(8), 4058–4067 (2021)
8. AllegroGraph: AllegroGraph. https://allegrograph.com/. Visited 2024
9. Alvaro, P., Rigger, M.: Automatically testing database systems: DBMS testing with test oracles, transaction history, and fuzzing. ACM Queue **21**(6), 128–135 (2024)
10. Angles, R.: The property graph database model. In: AMW. CEUR Workshop Proceedings, vol. 2100. CEUR-WS.org (2018)
11. Angles, R., et al.: PG-schema: schemas for property graphs. Proc. ACM Manag. Data **1**(2), 198:1–198:25 (2023)
12. Angles, R., et al.: PG-keys: keys for property graphs. In: SIGMOD Conference, pp. 2423–2436. ACM (2021)
13. ArangoDB: ArangoDB. https://arangodb.com/. Visited 2024
14. Arnoux, J., Bonifati, A., Calteau, A., Dumbrava, S., Gautreau, G.: Integrating complex pangenome graphs. In: ICDEW, pp. 350–354. IEEE (2024)
15. Atzeni, P., Bellomarini, L., Iezzi, M., Sallinger, E., Vlad, A.: Weaving enterprise knowledge graphs: the case of company ownership graphs. In: EDBT, pp. 555–566. OpenProceedings.org (2020)
16. AWS: Amazon Neptune. https://aws.amazon.com/fr/neptune/. Visited 2024
17. Ba, J., Rigger, M.: Testing database engines via query plan guidance. In: ICSE, pp. 2060–2071. IEEE (2023)
18. Bagan, G., Bonifati, A., Ciucanu, R., Fletcher, G.H., Lemay, A., Advokaat, N.: gMark: schema-driven generation of graphs and queries. IEEE Trans. Knowl. Data Eng. **29**(4), 856–869 (2016)

19. Barceló, P., Pérez, J., Reutter, J.L.: Schema mappings and data exchange for graph databases. In: ICDT, pp. 189–200. ACM (2013)
20. Bitnine. Co., Ltd.: AgensGraph. https://bitnine.net/agensgraph. Visited 2024
21. BlazeGraph: BlazeGraph. https://blazegraph.com/. Visited 2024
22. Bohannon, P., Fan, W., Geerts, F., Jia, X., Kementsietsidis, A.: Conditional functional dependencies for data cleaning. In: ICDE, pp. 746–755. IEEE Computer Society (2007)
23. Bonifati, A., Furniss, P., Green, A., Harmer, R., Oshurko, E., Voigt, H.: Schema validation and evolution for graph databases. In: Laender, A., Pernici, B., Lim, E.-P., de Oliveira, J. (eds.) ER 2019. LNCS, vol. 11788, pp. 448–456. Springer, Cham (2019). https://doi.org/10.1007/978-3-030-33223-5_37
24. Bonifati, A., Murlak, F., Ramusat, Y.: Transforming property graphs. Proc. VLDB Endow. **17**(11), 2906–2918 (2024)
25. Bonifati, A., Ramusat, Y., Murlak, F., Fejza, A., Echahed, R.: DTGraph: declarative transformations of property graphs. Proc. VLDB Endow. **17**(12), 4265–4268 (2024)
26. Clinical Data Interchange Standards Consortium: CDISC (2022). https://www.cdisc.org/. Visited 2024
27. DataStax: DataStax Enterprise Graph. https://www.datastax.com/products/datastax-graph. Visited 2024
28. DGraph: DGraph. https://dgraph.io/. Visited 2024
29. Elayam, M.M., Ray, C., Claramunt, C.: A hierarchical graph-based model for mobility data representation and analysis. Data Knowl. Eng. **141**, 102054 (2022)
30. Emmi, M., Majumdar, R., Sen, K.: Dynamic test input generation for database applications. In: ISSTA, pp. 151–162. ACM (2007)
31. Fan, W.: Dependencies for graphs: challenges and opportunities. ACM J. Data Inf. Qual. **11**(2), 5:1–5:12 (2019)
32. Fan, W., Geerts, F.: Foundations of Data Quality Management. Synthesis Lectures on Data Management. Morgan & Claypool Publishers (2012)
33. Francis, N., Libkin, L.: Schema mappings for data graphs. In: PODS, pp. 389–401. ACM (2017)
34. Godefroid, P., Kiezun, A., Levin, M.Y.: Grammar-based Whitebox fuzzing. In: PLDI, pp. 206–215. ACM (2008)
35. Godefroid, P., Levin, M.Y., Molnar, D.A.: Automated Whitebox fuzz testing. In: NDSS. The Internet Society (2008)
36. Gosnell, D., Broecheler, M.: The practitioner's guide to graph data. https://www.oreilly.com/library/view/the-practitioners-guide/9781492044062/. Visited 2024
37. Green, H., Avgerinos, T.: GraphFuzz: library API fuzzing with lifetime-aware dataflow graphs, vol. 2022, pp. 1070–1081. IEEE Computer Society (2022). https://doi.org/10.1145/3510003.3510228
38. Gütebier, L., et al.: CovidGraph: a graph to fight COVID-19. Bioinform. **38**(20), 4843–4845 (2022)
39. Hasif, C.L., Araldo, A., Dumbrava, S., Watel, D.: A graph-database approach to assess the impact of demand-responsive services on public transit accessibility. In: IWCTS@SIGSPATIAL, pp. 2:1–2:4. ACM (2022)
40. Hegeman, T., Iosup, A.: Survey of graph analysis applications. CoRR abs/1807.00382 (2018)
41. Huang, H., Bucher, D., Kissling, J., Weibel, R., Raubal, M.: Multimodal route planning with public transport and carpooling. IEEE Trans. Intell. Transp. Syst. **20**(9), 3513–3525 (2019)

42. HyperGraphDB: HyperGraphDB. https://hypergraphdb.org/. Visited 2024
43. IBM: DB2 Graph. https://www.ibm.com/docs/en/db2-warehouse?topic=applications-db2-graph. Visited 2024
44. Ilyas, I.F., Chu, X.: Data Cleaning. ACM Books, vol. 28. ACM (2019)
45. Ba, J., Rigger, M.: Keep it simple: testing databases via differential query plans. Proc. ACM Manag. Data **2**(3), 188 (2024)
46. JanusGraph: JanusGraph. https://janusgraph.org/. Visited 2024
47. Jiang, Y., Liu, J., Ba, J., Yap, R.H.C., Liang, Z., Rigger, M.: Detecting logic bugs in graph database management systems via injective and surjective graph query transformation. In: ICSE, pp. 46:1–46:12. ACM (2024)
48. Borges Jr., N.P., Havrikov, N., Zeller, A.: Generating tests that cover input structure. In: Software Engineering. LNI, vol. P-310, pp. 85–86. Gesellschaft für Informatik e.V. (2021)
49. Kamm, M., Rigger, M., Zhang, C., Su, Z.: Testing graph database engines via query partitioning. In: ISSTA, pp. 140–149. ACM (2023)
50. Kertkeidkachorn, N., Nararatwong, R., Xu, Z., Ichise, R.: FinKG: a core financial knowledge graph for financial analysis. In: ICSC, pp. 90–93. IEEE (2023)
51. Klees, G., Ruef, A., Cooper, B., Wei, S., Hicks, M.: Evaluating fuzz testing. In: CCS, pp. 2123–2138. ACM (2018)
52. Li, J., Zhao, B., Zhang, C.: Fuzzing: a survey. Cybersecurity **1**(1), 1–13 (2018). https://doi.org/10.1186/s42400-018-0002-y
53. MemGraph: MemGraph. https://memgraph.com/. Visited 2024
54. Metzman, J., Szekeres, L., Simon, L., Sprabery, R., Arya, A.: FuzzBench: an open fuzzer benchmarking platform and service. In: ESEC/SIGSOFT FSE, pp. 1393–1403. ACM (2021)
55. Microsoft: Azure Cosmos DB. https://azure.microsoft.com/fr-fr/products/cosmos-db. Visited 2024
56. Mughal, S., Moghul, I., Yu, J., Clark, T., Gregory, D.S., Pontikos, N.: Pheno4J: a gene to phenotype graph database. Bioinformatics **33**(20), 3317–3319 (2017)
57. Neo4J: Neo4J. https://neo4j.com/. Visited 2024
58. Noy, N.F., Gao, Y., Jain, A., Narayanan, A., Patterson, A., Taylor, J.: Industry-scale knowledge graphs: lessons and challenges. Commun. ACM **62**(8), 36–43 (2019)
59. Olsthoorn, M., van Deursen, A., Panichella, A.: Generating highly-structured input data by combining search-based testing and grammar-based fuzzing. In: ASE, pp. 1224–1228. IEEE (2020)
60. Oracle: Oracle Big Data Spatial and Graph. https://www.oracle.com/database/technologies/bigdata-spatialandgraph.html. Visited 2024
61. OrientDB: OrientDB. http://orientdb.org/. Visited 2024
62. Padhye, R., Lemieux, C., Sen, K., Papadakis, M., Traon, Y.L.: Semantic fuzzing with zest. In: ISSTA, pp. 329–340. ACM (2019)
63. Pan, K., Wu, X., Xie, T.: Automatic test generation for mutation testing on database applications. In: AST, pp. 111–117. IEEE Computer Society (2013)
64. Pan, K., Wu, X., Xie, T.: Guided test generation for database applications via synthesized database interactions. ACM Trans. Softw. Eng. Methodol. **23**(2), 12:1–12:27 (2014)
65. Park, S., Cheng, T.: Framework for constructing multimodal transport networks and routing using a graph database: a case study in London. Trans. GIS **27**(5), 1391–1417 (2023)
66. Pham, V., Böhme, M., Santosa, A.E., Caciulescu, A.R., Roychoudhury, A.: Smart greybox fuzzing. IEEE Trans. Software Eng. **47**(9), 1980–1997 (2021)

67. Preusse, M., et al.: COVIDGraph: connecting biomedical COVID-19 resources and computational biology models. In: SEA-Data@VLDB. CEUR Workshop Proceedings, vol. 2929, pp. 34–37. CEUR-WS.org (2021)
68. Rabbani, K., Lissandrini, M., Bonifati, A., Hose, K.: Transforming RDF graphs to property graphs using standardized schemas. Proc. ACM Manag. Data **2**(6), 242:1–242:25 (2024)
69. RedisGraph: RedisGraph. https://redis.io/. Visited 2024
70. Sahu, S., Mhedhbi, A., Salihoglu, S., Lin, J., Özsu, M.T.: The ubiquity of large graphs and surprising challenges of graph processing: extended survey. VLDB J. **29**(2–3), 595–618 (2020)
71. Sakr, S., et al.: The future is big graphs: a community view on graph processing systems. Commun. ACM **64**(9), 62–71 (2021)
72. Sequeda, J., Lassila, O.: Designing and building enterprise knowledge graphs. In: Synthesis Lectures on Data Semantics, and Knowledge. Morgan & Claypool Publishers (2021)
73. Redgate Software: Neo4J (2024). https://db-engines.com/en/ranking/graph+dbms. Visited 2024
74. Steinhöfel, D., Zeller, A.: Input invariants. In: ESEC/SIGSOFT FSE, pp. 583–594. ACM (2022)
75. Thom Hurks: PGMark: a domain-independent tool for generating property graphs based on a user-defined schema. https://github.com/ThomHurks/pgMark. Visited 2024
76. TigerGraph: TigerGraph. https://www.tigergraph.com/. Visited 2024
77. Titan: Titan. http://espeed.github.io/titandb/. Visited 2024
78. Wang, J., Chen, B., Wei, L., Liu, Y.: Superion: grammar-aware greybox fuzzing. In: ICSE, pp. 724–735. IEEE/ACM (2019)
79. Website, A.: American Fuzzy Loop. http://lcamtuf.coredump.cx/afl/. Accessed 2024
80. AFL Website: libFuzzer: A library for coverage-guided fuzz testing. http://llvm.org/docs/LibFuzzer.html. Accessed 2024
81. Yan, C., Nath, S., Lu, S.: Generating test databases for database-backed applications. In: ICSE, pp. 2048–2059. IEEE (2023)
82. Zeller, A., Gopinath, R., Böhme, M., Fraser, G., Holler, C.: The fuzzing book (2019)
83. Zheng, Y., et al.: Differential optimization testing of Gremlin-based graph database systems. In: ICST, pp. 25–36. IEEE (2024)
84. Zhu, X., Wen, S., Camtepe, S., Xiang, Y.: Fuzzing: a survey for roadmap. ACM Comput. Surv. (CSUR) **54**(11s), 1–36 (2022)

Counterexample-Guided Abstraction Refinement for Generalized Graph Transformation Systems

Barbara König[1], Arend Rensink[2], Lara Stoltenow[1](✉), and Fabian Urrigshardt[1]

[1] University of Duisburg-Essen, Duisburg, Germany
{barbara_koenig,lara.stoltenow}@uni-due.de
[2] University of Twente, Enschede, Netherlands
arend.rensink@utwente.nl

Abstract. This paper addresses the following verification task: Given a graph transformation system and a class of initial graphs, can we guarantee (non-)reachability of a given other class of graphs that characterizes bad or erroneous states? Both initial and bad states are characterized by nested conditions (having first-order expressive power). Such systems typically have an infinite state space, causing the problem to be undecidable. We use abstract interpretation to obtain a finite approximation of that state space, and employ counter-example guided abstraction refinement to iteratively obtain suitable predicates for automated verification. Although our primary application is the analysis of graph transformation systems, we state our result in the general setting of reactive systems.

Keywords: reachability analysis · CEGAR · reactive systems · abstract interpretation · graph transformation

1 Introduction

One of the successful techniques to analyze systems with very large or infinite state spaces is *abstract interpretation*. This either under- or over-approximates the possible behaviours, reducing the state space to a manageable size (and also covering a set of potential initial states rather than just one) at the cost of precision: essentially, different states are regarded as the same not if they are for all intents and purposes equivalent, but if they are, in some precise sense, alike. If we want to check certain properties about the behaviour of the original system, such as the absence of errors in reachable states or the satisfaction of more refined temporal logic properties, this can be done on the abstracted state space instead; however, due to the imprecision, the answers obtained in this way may not be correct for the full state space. In particular, if the abstraction was an under-approximation, analysing whether an error state is reachable may

B. König, A. Rensink, L. Stoltenow and F. Urrigshardt—Equal contribution.

J. Endrullis and M. Tichy (Eds.): ICGT 2025, LNCS 15720, pp. 157–176, 2025.
https://doi.org/10.1007/978-3-031-94706-3_8

yield *false negatives* (the answer *no* may be incorrect for the full, concrete state space); if it was an over-approximation, the analysis may yield *false positives* (the answer *yes* may be incorrect).

Of these two, false positives can be detected more easily, since the answer *yes* to a question of the type "can an error state be reached" or "is there a trace with a certain temporal property" comes with a *witness*, being an actual trace in the abstract state space that reaches that error state or has that property. For historical reasons, such a witness is usually called a *counterexample*, even though in our narrative it is rather an *example*. We can then typically check whether such a counterexample actually exists in the full state space—a task that is feasible even if the full state space is infinite. If the counterexample does exist, the original *yes* answer was correct after all; if it does *not* exist (in which case it is called *spurious*), we know that the original *yes* answer cannot be trusted; it may still be correct, but we do not know.

The method of *counter-example guided abstraction refinement* (CEGAR) [9, 19] builds upon this principle by relying on a notion of abstraction that can be *tuned*. In particular, two states are considered to be "alike" if they satisfy the same predicates. A spurious counterexample contains concrete evidence of where the original abstraction gives rise to "harmful" over-approximation: namely a trace in the abstract transition system leading to a state that does not rule out that a certain property Bad (encoding some unwanted feature) holds, but which does not exist in the concrete state space. This check provides additional predicates, leading to a stronger notion of "alike", which results in a new, *refined* abstraction in which at least this particular (spurious) counterexample no longer exists. We can then start over again using the new, refined, abstract state space, until we get either a *no* answer or a *yes* answer for which the counterexample is real, or we run out of time. This leads to the so-called CEGAR loop (see Fig. 1). In the general case there is no guarantee that the loop will ever terminate (e.g. due to undecidability of the verification problem), but good results have still been achieved in practice with the development of verification tools based on CEGAR [1,2,7].

A related technique for proving correctness relies on finding an *invariant* I, which is a property that holds in the initial state and is preserved by all transitions, such that I entails $\neg\mathsf{Bad}$. In fact, the abstraction refinement procedure outlined above may be seen as a recipe for generating such an invariant.

The contribution of this paper is to propose a variant of CEGAR in the spirit of *reactive systems*. This is a framework developed by Milner et al. [31] that subsumes graph transformation via double-pushout. We assume that the states are arrows in a suitable category, and transitions are generated by conditional rules that modify those objects using categorical operations. Our abstraction is based on so-called *nested conditions* (related to first-order logic) [16,40], again defined relative to the category of choice, with a notion of satisfaction over the states. The predicates mentioned above take the shape of such nested conditions; the abstraction is driven by a finite set P of them. The abstraction of a state, or a set of states, is defined as the subset of elements of P that provably hold in that

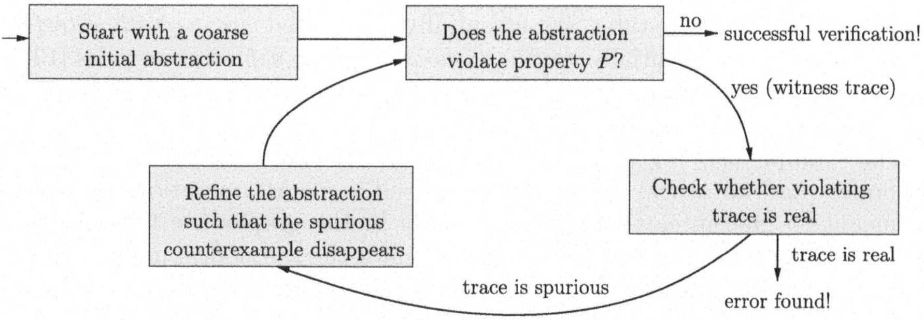

Fig. 1. A schematic depiction of the CEGAR loop

(set of) state(s). In previous work [4], we have shown how to compute strongest postconditions and weakest preconditions for rules with application conditions, and this allows us to compute the abstract (over-approximating) transitions, as well as to check whether a counterexample found on the abstract level fails to exist on the concrete level, i.e., is spurious. Given a spurious counterexample, we can then refine the abstraction by augmenting P with the characteristic properties of all the (sets of) concrete states traversed by the counterexample. This refined abstraction is certain not to include that counterexample any more.

A complication lies in the fact that the method outlined above at several points requires the computation of entailment among nested conditions. Depending on the underlying category, this is equivalent to entailment of First-Order Logic and hence undecidable; in practice, we are forced to rely on *provable* entailment (i.e., using available tooling), which is necessarily a weaker relation—or in the terminology above, may yield false negatives. Fortunately, though this introduces further imprecision in our abstraction, we show that this does not invalidate the method.

A further contribution of this paper consists in a prototype implementation of this CEGAR method for a concrete base category, viz. that of *graphs*. Rules in this category are (essentially) double-pushout transformation rules, and the nested conditions correspond to ones studied before by [16,40]. We report some experiments. Unfortunately, the performance is such that only very small examples can be analysed successfully. The biggest obstacle is the procedure for (semi-)deciding entailment. We believe there is a lot of room for improvement, but this is outside the scope of the current paper.

A full version of the paper, including all the proofs and further material, is available [27].

2 Motivation

This section presents a running example that shows the principal steps of the CEGAR method and gives an idea of its potential benefits. It is based on a particular instantiation of our general framework (presented formally in the next sections): the concrete states are essentially unlabelled *graphs with interfaces* [3,24], the rules are essentially *double-pushout graph transformation rules* [13,14] and our nested conditions essentially correspond to those introduced by Habel and Pennemann [16] and Rensink [40].

The example uses a graph-based representation of a set of lists, in which the tail nodes (and no other) are marked by a self-loop. All operations on the list are meant to guarantee that the only edges in the graphs are those between successive list nodes and the self-loops that mark the tail nodes; in particular, it is an error for any node to have both an outgoing edge to a successor node and a self-loop. This erroneous structure is captured by a condition Bad; in other words, Bad is meant to be unreachable.

We consider a scenario consisting of an initial condition Init_1 that is meant to capture precisely those graphs in which there are only empty lists (i.e., no edges between distinct nodes), and a rule append that specifies the extension of a list (at its tail) with a single element. Using the notation introduced in the next section, this scenario is captured by the following conditions and rules that are based on cospans of graphs:

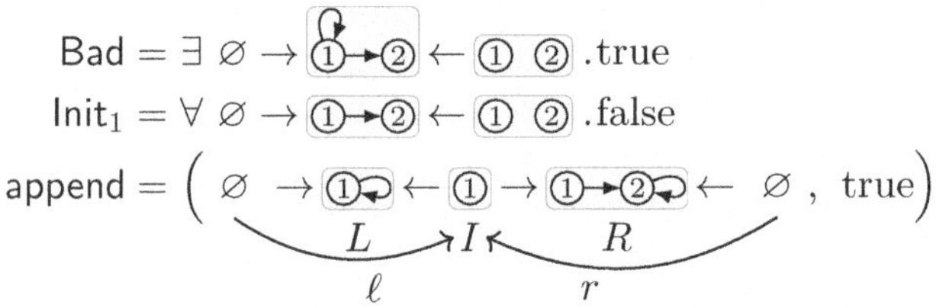

The three graphs L, I, R in rule append correspond to left-hand side, interface, and right-hand side in double-pushout rewriting. It should be noted that $\mathsf{Init}_1 \models \neg\mathsf{Bad}$.

We want to show that Bad does not hold in any graph reachable from any initial graph that satisfies Init_1.

The method works by initially assuming that the only knowledge we have about a state is whether Init_1, Bad or the negation of either holds there, and checking if that knowledge is sufficient to show that $\neg\mathsf{Bad}$ holds everywhere. (In fact, CEGAR can be seen as a way to automatically generate invariants or – more generally – conditions that are guaranteed to hold at certain execution points.)

Let us denote $P = \{\mathsf{Init}_1, \mathsf{Bad}\}$; then abstract states are elements of $\{\text{true}, \text{false}, \text{unknown}\}^P$—or, equivalently, subsets of $P \cup \{\neg\mathcal{P} \mid \mathcal{P} \in P\}$ (or conjunctions of such predicates) that do not include both $\mathcal{P}$ and $\neg\mathcal{P}$ for any $\mathcal{P} \in P$. The method involves the following steps.

1. Compute, for every rule, the strongest postcondition (sp) for the next unexplored abstract state. In our example, the initial state consists of[1] $\{\mathsf{Init}_1, \neg\mathsf{Bad}\}$ and hence we start by computing $\text{sp}(\mathsf{Init}_1 \wedge \neg\mathsf{Bad}, \mathsf{append})$, which yields $\mathcal{A} = \exists\ \varnothing \rightarrow$ ①→②⟲ $\leftarrow$ ① $.\mathcal{A}'$, where the subcondition $\mathcal{A}'$ guarantees

[1] Note that $\mathsf{Init}_1 \models \neg\mathsf{Bad}$ implies $\mathsf{Init}_1 \equiv \mathsf{Init}_1 \wedge \neg\mathsf{Bad}$, hence we could start with either formula.

that $\mathsf{Init}_1 \wedge \neg\mathsf{Bad}$ holds for nodes 1 and 2, as well any additional nodes that already existed in the graph.

2. Infer which of the elements of P or their negations are entailed by $\mathcal{A}$. In our example, we only have $\mathcal{A} \models \neg\mathsf{Init}_1$: we cannot infer either $\mathcal{A} \models \mathsf{Bad}$ or $\mathcal{A} \models \neg\mathsf{Bad}$. It follows that the successor state is $\{\neg\mathsf{Init}_1\}$. This gives rise to the following transition system:

$$\{\mathsf{Init}_1, \neg\mathsf{Bad}\} \xRightarrow{\mathsf{append}} \{\neg\mathsf{Init}_1\}$$

3. Repeat the previous two steps for every new state, until there are no more states to be found (note that the number of reachable states is bounded by $3^{|P|}$) or we find a state s for which $\neg\mathsf{Bad} \notin s$. In the former case we are done: the system is guaranteed to be error-free. In the latter case, the system has a potentially faulty behaviour: namely, the trace from the initial state to s. Such a trace is called a *counterexample*. In our running example, we already have such a counterexample, viz. the trace consisting of a single application of append.
4. Check whether this trace really represents faulty behaviour, by computing the weakest precondition $\mathcal{W}_1$ for $\neg\mathsf{Bad}$. If $\mathsf{Init}_1 \models \mathcal{W}_1$, the counterexample is *spurious*; if not, then any graph satisfying $\mathsf{Init}_1 \wedge \neg\mathcal{W}_1$, when subjected to the successive rules of the counterexample, will indeed give rise to a concrete state satisfying Bad—hence we have a real error. In our running example, the weakest precondition wp(append, $\neg$Bad) is (simplified and in abbreviated notation):

$\mathcal{W}_1 \equiv \forall$  .false $\wedge\ \forall$  .false $\wedge\ \forall$ 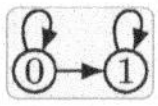 .false ,

which is not entailed by Init_1: for instance,  satisfies $\mathsf{Init}_1 \wedge \neg\mathcal{W}_1$. From it we can reach a bad state in one append-step, hence the system is erroneous.

We see that the method has uncovered the fact that our initial condition Init_1 is not strong enough to guarantee $\neg$Bad: we had not considered that nodes may have more than one self-loop. We can repair this by strengthening Init_1 so as to rule this out. Let us redo the analysis on the basis of

$\mathsf{Init}_2 = \mathsf{Init}_1 \wedge \forall\ \varnothing \rightarrow$  $\leftarrow$  .false

Note that, on its own, $\neg$Bad is not an invariant: it is *not* the case that the result of applying append to a graph satisfying $\neg$Bad will certainly also satisfy $\neg$Bad—otherwise the problem would be easier. Similarly, Init_2 is not an invariant.

After repeating the first two steps above, we once more find that there is a reachable state not containing $\neg$Bad, identifying a counterexample, this time consisting of two rule applications:

$$\{\mathsf{Init}_2, \neg\mathsf{Bad}\} \xRightarrow{\mathsf{append}} \{\neg\mathsf{Init}_2, \neg\mathsf{Bad}\} \xRightarrow{\mathsf{append}} \{\neg\mathsf{Init}_2\}$$

The weakest precondition computation gives us two conditions $\mathcal{W}_1 =$ wp(append, $\neg$Bad) (identical to the one above) and $\mathcal{W}_2 =$ wp(append, $\mathcal{W}_1$) (in this particular case $\mathcal{W}_2 \equiv \mathcal{W}_1$). Now $\mathsf{Init}_2 \models \mathcal{W}_2$, and hence the counterexample is spurious. When that happens, the method continues as follows:

5. Refine P by adding predicates that ensure the counterexample no longer occurs on the abstract states. In particular, augmenting P with the weakest preconditions that we just computed in order to check for spuriousness—in our case $\mathcal{W}_1$—will do the trick. (We could also choose to add individual conjuncts $\mathcal{W}_{ij}$ if $\mathcal{W}_i = \mathcal{W}_{i1} \wedge \cdots \wedge \mathcal{W}_{in}$, to possibly obtain a better abstraction. In our running example, the three subconditions of $\mathcal{W}_1$ could be turned into three new predicates.) We have the guarantee that this will indeed eliminate the spurious counterexample.
6. Restart the analysis at Step 1 on the basis of the refined P. Repeat the process until either a real counterexample is found (as happened in our first iteration), all reachable (abstract) states have been computed and they all entail $\neg\mathsf{Bad}$, or time is up. For our running example, the second case occurs:

$$\{\mathsf{Init}_2, \neg\mathsf{Bad}, \mathcal{W}_1\} \xrightarrow{\mathsf{append}} \{\neg\mathsf{Init}_2, \neg\mathsf{Bad}, \mathcal{W}_1\} \circlearrowleft \mathsf{append}$$

The running example shows the power of the analysis method: it allows us to prove that, for all graphs satisfying a given initial condition (of which there are infinitely many), for all possible sequences of rule applications (of arbitrary length, and generating graphs of unbounded size), the outcome satisfies the well-formedness condition embodied in $\neg\mathsf{Bad}$; or, if that is not the case, to find a counterexample. Of course, there are practical limitations, which we will discuss in Sect. 5.3.

3 Preliminaries

3.1 Abstract Interpretation

We rely on the principles of the theory of abstract interpretation [10,11], based on lattices and Galois connections. We first recall the definitions.

A *complete lattice* $(\mathbb{C}, \sqsubseteq)$ consists of a set $\mathbb{C}$ with a partial order $\sqsubseteq$ such that each $Y \subseteq \mathbb{C}$ has a least upper bound $\bigsqcup Y$ (also called supremum, join) and a greatest lower bound $\bigsqcap Y$ (also called infimum, meet).

Let $\mathbb{C}$, $\mathbb{A}$ be two lattices. A *Galois connection* from $\mathbb{C}$ to $\mathbb{A}$ is a pair $\alpha\colon \mathbb{C} \to \mathbb{A}$, $\gamma\colon \mathbb{A} \to \mathbb{C}$ of monotone functions, such that for all $\ell \in \mathbb{C}$: $\ell \sqsubseteq \gamma(\alpha(\ell))$ and for all $m \in \mathbb{A}$: $\alpha(\gamma(m)) \sqsubseteq m$. Intuitively $\mathbb{C}$ represents (more) concrete values and $\mathbb{A}$ (more) abstract values, which are connected by the abstraction α and concretization γ. The order indicates whether the values are more or less precise: i.e., whenever $a \sqsubseteq b$, then a is supposed to provide higher precision than b. The function α (resp. γ) is also called the *left* (resp. *right*) *adjoint*.

Given a function $f\colon \mathbb{C} \to \mathbb{C}$ on the concrete values, we say that $f^\# : \mathbb{A} \to \mathbb{A}$ is an over-approximation of f whenever $\alpha \circ f \circ \gamma \sqsubseteq f^\#$ (pointwise). Whenever equality holds $f^\#$ is the induced over-approximation of f.

3.2 Categories

We will use standard concepts from category theory. Given an arrow $f\colon A \to B$, we write $\mathsf{dom}(f) = A$, $\mathsf{cod}(f) = B$. For two arrows $f\colon A \to B$, $g\colon B \to C$ we denote their composition by $f;g\colon A \to C$.

We will state our results in a general framework, allowing for easy generalization of our results to other applications. An important type of category that we will focus on are cospan categories, which are particularly useful for reactive systems (to be defined later). Given a base category $\mathbf{D}$ with pushouts, the category $\mathbf{Cospan}(\mathbf{D})$ has as objects the objects of $\mathbf{D}$ and as arrows *cospans*, which are equivalence classes of pairs of arrows of the form $A \xrightarrow{f_L} X \xleftarrow{f_R} B$, where the middle object is considered up to isomorphism. Cospan composition is performed via pushouts.

A cospan is *left-linear* if its left leg f_L is a mono. For adhesive categories [28], the composition of left-linear cospans again yields a left-linear cospan. $\mathbf{ILC}(\mathbf{D})$ will denote the subcategory of $\mathbf{Cospan}(\mathbf{D})$ where the arrows are restricted to left-linear cospans (historically called *input*-linear; hence $\mathbf{ILC}$.)

Our running example is based on the category $\mathbf{D} = \mathbf{Graph_{fin}}$, which has finite graphs as objects and graph morphisms as arrows.

3.3 Generalized Conditions

As in previous work [44] we consider nested conditions—from here on just called *conditions*—over an arbitrary category $\mathbf{C}$ in the spirit of reactive systems [30,31]. Following [16,40], we define conditions as finite tree-like structures, where nodes are annotated with quantifiers and objects, and edges are annotated with arrows.

Definition 1 (Condition). *Let $\mathbf{C}$ be a category. A condition $\mathcal{A}$ over an object A in $\mathbf{C}$ is defined inductively as follows: it is either*

- *a finite conjunction of universals $\bigwedge_{i\in\{1,\ldots,n\}} \forall f_i.\mathcal{A}_i = \forall f_1.\mathcal{A}_1 \wedge \ldots \wedge \forall f_n.\mathcal{A}_n$, or*
- *a finite disjunction of existentials $\bigvee_{i\in\{1,\ldots,n\}} \exists f_i.\mathcal{A}_i = \exists f_1.\mathcal{A}_1 \vee \ldots \vee \exists f_n.\mathcal{A}_n$*

where $f_i\colon A \to A_i$ are arrows in $\mathbf{C}$ and $\mathcal{A}_i$ are conditions over A_i. We call $A = \mathrm{RO}(\mathcal{A})$ the root object *of the condition $\mathcal{A}$. Each subcondition $\mathcal{Q}f_i.\mathcal{A}_i$ ($\mathcal{Q} \in \{\forall, \exists\}$) is called a* child *of $\mathcal{A}$. The constants* true_A *(empty conjunction) and* false_A *(empty disjunction) serve as the base cases. We will omit subscripts in* true_A *and* false_A *when clear from the context. The set of all conditions over A is denoted by $\mathbf{Cond}_A$, and $\mathbf{Arr}_A$ refers to the A-sourced arrows (i.e., potential models) of $\mathbf{C}$.*

Instantiated with $\mathbf{C} = \mathbf{Graph_{fin}}$, conditions are equivalent to graph conditions as defined in [16], and equivalence to first-order logic has been shown in [40]. Cospan conditions (with $\mathbf{C} = \mathbf{ILC}(\mathbf{Graph_{fin}})$) have previously been used [22,44]. Standard graph conditions can trivially be encoded into cospan conditions, and cospan conditions can be translated to equivalent graph conditions.

Intuitively, conditions check for the occurrence of certain subgraphs or patterns for which the context satisfies a child condition. For instance, the cospan condition $\forall\, \varnothing \to$ [①→②] $\leftarrow$ [① ②] $.\, \exists$ [① ②] $\to$ [①←②] $\leftarrow$ [① ②] $.\mathrm{true}$ requires that for every edge, a second edge in the reverse direction also exists. For additional examples and discussion we refer to [22].

To be consistent with [44], Definition 1 restricts conjunction to universal and disjunction to existential subformulas; e.g., $\exists f.\mathcal{A} \wedge \exists g.\mathcal{B}$ is excluded. However, conditions that violate this syntactic restriction can easily be rewritten—e.g., to $\forall \mathrm{id}.\exists f.\mathcal{A} \wedge \forall \mathrm{id}.\exists g.\mathcal{B}$ for the above example. Hence, in examples we sometimes write $\mathcal{A} \wedge \mathcal{B}$ or $\mathcal{A} \vee \mathcal{B}$ for arbitrary conditions.

Definition 2 (Satisfaction). *Let $\mathcal{A} \in \mathbf{Cond}_A$ and let $a\colon A \to B$ be an arrow. We define the satisfaction relation $a \models \mathcal{A}$ as follows:*

- *$a \models \bigwedge_{i\in I} \forall f_i.\mathcal{A}_i$ iff for every $i \in I$ and every arrow $g\colon \mathrm{RO}(\mathcal{A}_i) \to B$ we have: if $a = f_i; g$, then $g \models \mathcal{A}_i$.*
- *$a \models \bigvee_i \exists f_i.\mathcal{A}_i$ iff there exists $i \in I$ and an arrow $g\colon \mathrm{RO}(\mathcal{A}_i) \to B$ such that $a = f_i; g$ and $g \models \mathcal{A}_i$.*

We define a concretization $[\![\_]\!]\colon \mathbf{Cond}_A \to \mathcal{P}(\mathbf{Arr}_A)$ (for arbitrary A) via $[\![\mathcal{A}]\!] = \{x \in \mathbf{Arr}_{\mathrm{RO}(\mathcal{A})} \mid x \models \mathcal{A}\}$, mapping conditions to the set of arrows that satisfy them. From the above it follows that $[\![\mathrm{true}_A]\!] = \mathbf{Arr}_A$ and $[\![\mathrm{false}_A]\!] = \varnothing$.

We write $\mathcal{A} \models \mathcal{B}$ ($\mathcal{A}$ entails $\mathcal{B}$) if $\mathrm{RO}(\mathcal{A}) = \mathrm{RO}(\mathcal{B})$ and for every arrow $a \in \mathbf{Arr}_{\mathrm{RO}(\mathcal{A})}$ we have: if $a \models \mathcal{A}$, then $a \models \mathcal{B}$. We write $\mathcal{A} \equiv \mathcal{B}$ ($\mathcal{A}$ and $\mathcal{B}$ are equivalent) if $\mathcal{A} \models \mathcal{B}$ and $\mathcal{B} \models \mathcal{A}$.

Since conditions are equivalent to first-order logic [40] for $\mathbf{C} = \mathbf{Graph_{fin}}$, the satisfiability, entailment and equivalence problems are undecidable, but semi-decidable. In fact, in [44] we have provided a semi-decision procedure for satisfiability in the general case, based on a predecessor technique for graph conditions [29].

3.4 Conditional Reactive Systems

We now define conditional reactive systems, which were introduced in [31] and extended with application conditions in [21]. In our definition, we closely follow [22]. We fix a distinguished object 0 (not necessarily the initial object in the category).

Definition 3 (Reactive system rules). *Let $\mathbf{C}$ be a category with a distinguished object* 0 *(not necessarily initial). A* rule *$\mathcal{R} = (\ell, r, \mathcal{C})$ consists of arrows $\ell, r\colon 0 \to I$ (called left-hand side and right-hand side) and a condition $\mathcal{C}$ with root object I. A* reactive system *is a set of rules.*

Let $\mathcal{S}$ be a reactive system and $a, a'\colon 0 \to J$ be arrows. We say that a reduces *to a' ($a \rightsquigarrow a'$) in $\mathcal{S}$ whenever there exists a rule $(\ell, r, \mathcal{C}) \in \mathcal{S}$ and an arrow $c\colon I \to J$ (the* reactive context*) such that $a = \ell; c$, $a' = r; c$ and $c \models \mathcal{C}$.*

Note that $\mathcal{C}$ is not a pre- or post-application condition, but is specified over the context in which the reaction takes place. Reactive systems instantiated with cospans (where 0 is the empty graphs) [22,42,43] yield exactly double-pushout rewriting [14], hence reactive systems over $\mathbf{ILC}(\mathbf{Graph_{fin}})$ essentially describe DPO graph transformation systems with monic matching.

Example 4. The reactive system over $\mathbf{ILC}(\mathbf{Graph_{fin}})$ having a single rule $\mathcal{R}$ (defined below) adds edges between arbitrary nodes, but only if such an edge does not already exist:

$$\mathcal{R} = \Big(\overbrace{\varnothing \xrightarrow{} \boxed{① \;\; ②}}^{\ell\;=\;L} \leftarrow \boxed{① \;\; ②} \overbrace{\rightarrow \boxed{① \!\rightarrow\! ②} \leftarrow \varnothing}^{r\;=\;R} , \mathcal{C} \Big)$$

$$\mathcal{C} = \forall \boxed{① \;\; ②} \rightarrow \boxed{① \!\rightarrow\! ②} \leftarrow \boxed{① \;\; ②} \,.\,\text{false}$$

3.5 Shift Operation and Hoare Logic

Nested conditions are equipped with a shift operation. More concretely, given $\mathcal{A} \in \mathbf{Cond}_A$, $c\colon A \to B$, $\mathcal{A}_{\downarrow c} \in \mathbf{Cond}_B$ can be understood as a partial evaluation of $\mathcal{A}$ under the assumption that an arrow c is already "present". In particular, it is defined as $d \models \mathcal{A}_{\downarrow c} :\iff c; d \models \mathcal{A}$. Here we do not delve into details, but just remark that a shift can be computed via so-called representative squares.

As an example in $\mathbf{Graph_{fin}^{inj}}$ (the subcategory of $\mathbf{Graph_{fin}}$ with only injective graph morphisms), shifting $\forall\, \varnothing \rightarrow \boxed{①} \,.\, \exists\, \boxed{①} \rightarrow \boxed{① \!\rightarrow\! ②}\,.\text{true}$ (every node has an outgoing edge) over $\varnothing \rightarrow \boxed{◎}$ (a node exists) yields

$$\begin{aligned} &\forall \boxed{◎} \rightarrow \boxed{◎} \,.\, \exists \boxed{◎} \rightarrow \boxed{◎ \!\rightarrow\! ②}\,.\text{true} \\ \wedge\, &\forall \boxed{◎} \rightarrow \boxed{◎ \;\; ①} \,.\, (\exists \boxed{◎ \;\; ①} \rightarrow \boxed{◎ \;\; ① \!\rightarrow\! ②}\,.\text{true} \vee \exists \boxed{◎ \;\; ①} \rightarrow \boxed{◎ \!\leftarrow\! ①}\,.\text{true}) \end{aligned}$$

(the designated node has an outgoing edge, and so does every other node, possibly to the designated node). This example can be lifted to $\mathbf{ILC}(\mathbf{Graph_{fin}})$ by replacing all graph morphisms $A \xrightarrow{m} B$ with cospans $A \xrightarrow{m} B \xleftarrow{\text{id}} B$.

In order to compute successor states for graph conditions, we need the concepts of Hoare triple, (strongest) postconditions and (weakest) preconditions that is based on the shift operation.

Definition 5 (Hoare triple, weakest precondition, strongest postcondition [4]). *Let $\mathcal{R} = (\ell, r, \mathcal{C})$ be a rule and let $\mathcal{A}, \mathcal{B}$ be conditions. We say that $\mathcal{A}, \mathcal{R}, \mathcal{B}$ form a* Hoare triple *– written as $\{\mathcal{A}\}\mathcal{R}\{\mathcal{B}\}$ – if for all $a, b\colon 0 \to J$ with $a \models \mathcal{A}$ and $a \rightsquigarrow_{\mathcal{R}} b$ we have that $b \models \mathcal{B}$.*

$\mathcal{A}$ is a precondition *for $\mathcal{R}$ and $\mathcal{B}$ whenever $\{\mathcal{A}\}\mathcal{R}\{\mathcal{B}\}$. Similarly, $\mathcal{B}$ is called a* postcondition *for $\mathcal{A}$ and $\mathcal{R}$.*

$\mathcal{A}$ is the weakest precondition *for $\mathcal{R}$ and $\mathcal{B}$ (written $\text{wp}(\mathcal{R}, \mathcal{B})$) whenever it is a precondition and for every other precondition $\mathcal{A}'$ we have that $\mathcal{A}' \models \mathcal{A}$.*

$\mathcal{B}$ is the strongest postcondition *for $\mathcal{A}$ and $\mathcal{R}$ (written $\text{sp}(\mathcal{A}, \mathcal{R})$) whenever it is a postcondition and for every other postcondition $\mathcal{B}'$ we have that $\mathcal{B} \models \mathcal{B}'$.*

It is easy to see that $\{\mathcal{A}\}\mathcal{R}\{\mathcal{B}\}$ iff $\mathcal{A} \models \text{wp}(\mathcal{R}, \mathcal{B})$ iff $\text{sp}(\mathcal{A}, \mathcal{R}) \models \mathcal{B}$. Furthermore all notions can be generalized to traces, i.e., sequences of rules, instead of single rules $\mathcal{R}$.

Proposition 6 (Computing weakest preconditions and strongest postconditions [4]). *Let* $\mathcal{R} = (\ell, r, \mathcal{C})$ *be a rule and let* $\mathcal{A}, \mathcal{B}$ *be conditions. Then* $\text{wp}(\mathcal{R}, \mathcal{B}) \equiv \forall \ell.(\mathcal{C} \to \mathcal{B}_{\downarrow r})$ *and* $\text{sp}(\mathcal{A}, \mathcal{R}) \equiv \exists r.(\mathcal{C} \wedge \mathcal{A}_{\downarrow \ell})$

For instance, for the motivating example in Sect. 2, the strongest postcondition of the first step is the following.

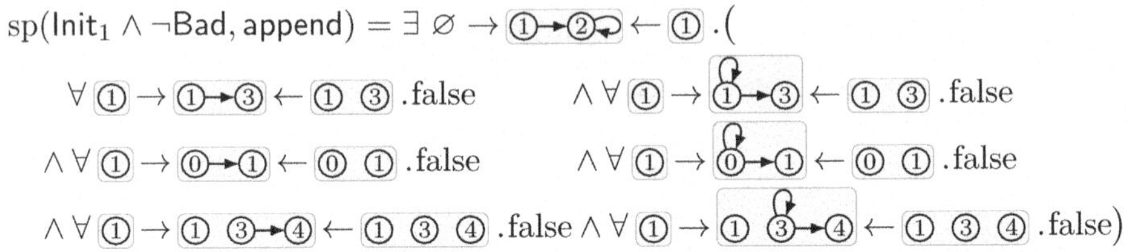

Essentially, this states that a list with one element must exist, and condition $\mathsf{Init}_1 \wedge \neg\mathsf{Bad}$ has to hold for both the second-to-last list element that has just been added, and any other list elements that might exist. Note that the three subconditions in the right column are already "covered" by the three ones on the left, and could be removed to obtain a smaller but equivalent condition.

4 GTS Verification Using Predicate Abstraction

We are now in a position to formalize the method outlined and illustrated in Sect. 2. To reiterate: we want to answer verification questions of the form "from a given initial system state, is it possible to reach a state where a given (undesirable) property holds?"—where, for us, states are elements of $\mathbf{Arr}_0$ in an arbitrary base category $\mathbf{C}$ with distinguished object 0.

4.1 Concrete Transition Systems

As a first observation, in practice we are interested in answering the verification question for a *family* of initial system states, and not just a single one. We therefore immediately generalise the formal notion of states and transitions to sets of arrows, with a disjunctive interpretation: a system being "in" a state means that it is described by *one* of the elements of that state.

Definition 7 (Set-based transition system, rule(_,$\mathcal{R}$), correctness). *Given a reactive system* $\mathcal{S}$, *a* set-based transition system *is a tuple* $T = \langle Q, \to, X_0 \rangle$, *where* $Q \subseteq \mathcal{P}(\mathbf{Arr}_0)$ *is the set of states,* $X_0 \in Q$ *is the initial state, and* $\to \subseteq Q \times \mathcal{S} \times Q$ *is the transition relation, defined by* $X \xrightarrow{\mathcal{R}} \text{rule}(X, \mathcal{R})$ *where* $\text{rule}(X, \mathcal{R}) := \{y \mid x \in X,\ x \leadsto_{\mathcal{R}} y\}$ *for arbitrary* $X \in Q$ *and* $\mathcal{R} \in \mathcal{S}$.

T is called correct *with respect to a given condition* $\mathsf{Bad} \in \mathbf{Cond}_0$ *if* $Y \cap [\![\mathsf{Bad}]\!] = \varnothing$ *for all states* Y *reachable from* X_0.

For our running example we obtain the following set-based transition system, starting from the set of arrows that satisfy Init_2 (rule $\mathcal{R} = \mathsf{append}$):

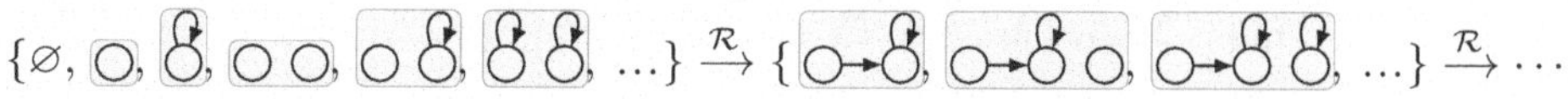

A set-based transition system is *induced* by some condition Init if $X_0 = [\![\mathsf{Init}]\!]$ and Q is the smallest subset of $\mathcal{P}(\mathbf{Arr}_0)$ reachable from X_0. For an induced set-based transition system, the verification question therefore asks whether any reachable set Y intersects with $[\![\mathsf{Bad}]\!]$ for a condition Bad.

Since individual states as well as the set of all states can be infinite, verification on set-based transition systems is in general infeasible. Note, however, that the problem of checking whether a system is *in*correct (a bad state is reachable), is in fact semi-decidable for graphs (rewriting in $\mathbf{ILC}(\mathbf{Graph_{fin}})$): we can enumerate all graphs satisfying Init and while doing this in parallel enumerate the reachable graphs. Once we detect a graph satisfying Bad, we can give the respective answer, i.e., the system is not correct. However, it is well-known that graph transformation systems can encode Turing machines [17] and hence the problem is undecidable. Here we are interested in developing a technique that can (in some cases) definitely show that the system under consideration is in fact correct.

A first step towards a verification method is to use conditions (which have finite size) as a representation for (first-order definable) sets of arrows. This step is justified by the following result, which implies that the set of condition-definable sets of arrows is closed under rule application.

Lemma 8. *For any condition $\mathcal{A}$ and rule $\mathcal{R}$, $[\![\mathrm{sp}(\mathcal{A},\mathcal{R})]\!] = \mathrm{rule}([\![\mathcal{A}]\!],\mathcal{R})$.*

Using the construction given in Sect. 3.5, we can define transitions through strongest postconditions. However, we have to ensure that equivalent but syntactically distinct conditions collapse to the same state. This gives rise to the following definition:

Definition 9 (Condition-based transition system). *Given a reactive system $\mathcal{S}$, a* condition-based transition system *is a tuple $\langle Q, \to, \mathsf{Init}\rangle$ with $\mathsf{Init} \in \mathbf{Cond}_0$, where $Q \subseteq \mathbf{Cond}_0/{\equiv}$ is the set of states, $[\mathsf{Init}]_\equiv \in Q$ is the initial state, and $\to \subseteq Q \times \mathcal{S} \times Q$ is the transition relation, defined by $[\mathcal{A}]_\equiv \xrightarrow{\mathcal{R}} [\mathrm{sp}(\mathcal{A},\mathcal{R})]_\equiv$ for arbitrary $\mathcal{A} \in \mathbf{Cond}_0$ and $\mathcal{R} \in \mathcal{S}$.*

Transitions are well-defined because $\mathcal{A} \equiv \mathcal{B}$ implies $\mathrm{sp}(\mathcal{A},\mathcal{R}) \equiv \mathrm{sp}(\mathcal{B},\mathcal{R})$. Below we will usually omit the explicit construction of $\equiv$-equivalence classes and just talk about conditions, tacitly assuming that they are representatives of the corresponding equivalence classes. Due to Lemma 8, $[\![\_]\!]$ maps any condition-based transition system to an isomorphic set-based one, with initial state $[\![\mathsf{Init}]\!]$.

On condition-based transition systems, the verification problem (is a transition system correct w.r.t. Bad) reduces to checking whether all states reachable from $\mathcal{A}$ entail $\neg\mathsf{Bad}$. The condition-based transition system for our running example has the following initial steps, starting with Init_2 (cf. the set-based transition system above):

$$\mathsf{Init}_2 \xrightarrow{\mathcal{R}} \mathrm{sp}(\mathsf{Init}_2, \mathcal{R}) \xrightarrow{\mathcal{R}} \mathrm{sp}(\mathrm{sp}(\mathsf{Init}_2, \mathcal{R}), \mathcal{R}) \xrightarrow{\mathcal{R}} \cdots$$

The condition in the initial state expresses that there exists a multiset of empty lists. The second state (after a single rule application) allows a single list element in any of the lists. The third state (after two rule applications) allows two list elements in total, either as two one-element lists or as a single two-element list, and otherwise only empty lists. None of these conditions are equivalent.

4.2 Abstract Transition Systems

Condition-based transition systems still do not provide a way to answer the verification question: compared to the set-based transition systems, successors are now representable; however, the definition of transitions relies on entailment, which is undecidable, and the reachable part of the transition system will typically still be infinite and therefore not fully explorable (as in the example above). This is where we introduce predicate abstraction. Instead of conditions of arbitrary complexity, states will be subsets (or conjunctions) of a predetermined set of conditions (the *predicates*), each of which can be either positive, negative or absent (unknown) (e.g. $(\mathcal{P}_1 \wedge \neg\mathcal{P}_3) \in \mathrm{Abs}(\{\mathcal{P}_1, \mathcal{P}_2, \mathcal{P}_3\})$). This guarantees finiteness of the resulting transition system.

Definition 10 (Predicate abstraction). *Let* $P = \{\mathcal{P}_1, \ldots, \mathcal{P}_n\}$ *be a non-empty set of conditions in* $\mathbf{Cond}_0$ *which we will call* predicates. *We define a lattice* $Abs(P)$ *as follows:*

- *The carrier set contains all conjunctions of subsets of* $P \cup \{\neg\mathcal{P}_1, \ldots, \neg\mathcal{P}_n\}$, *quotiented by equivalence* $\equiv$ *(which includes the constants* true *and* false*).*
- *The set is ordered by entailment (*$\models$*).*

For an arbitrary condition $\mathcal{A} \in \mathbf{Cond}_0$, $\overline{\mathcal{A}} := \bigwedge\{\mathcal{Q}' \in Abs(P) \mid \mathcal{A} \models \mathcal{Q}'\}$ *is the strongest element of* $Abs(P)$ *for which* $\mathcal{A} \models \overline{\mathcal{A}}$, *i.e., the best possible approximation of* $\mathcal{A}$ *for the given set of predicates.*

Since $\overline{\mathcal{A}}$ is in general weaker than $\mathcal{A}$, reasoning with $\overline{\mathcal{A}}$ rather than $\mathcal{A}$ results in over-approximation, meaning that our abstract transition system suggests that the reachable sets of arrows are larger than is actually the case. As a result, unsafe states might seemingly be reachable when in reality they are not. Avoiding this requires careful selection of a suitable set of predicates. We will take care of this issue later in Sect. 5.

Definition 11 (Abstract transition system, $\mathrm{sp}^{\#}(\_, \mathcal{R})$). *Given a reactive system* $\mathcal{S}$ *and a set of predicates* P, *an* abstract transition system *is a tuple* $\langle Q, \Rightarrow, \mathsf{Init}\rangle$ *with* $\mathsf{Init} \in P$, *where* $Q \subseteq Abs(P)$ *is the set of states,* $[\mathsf{Init}]_\equiv \in Q$ *is the initial state, and* $\Rightarrow \subseteq Q \times \mathcal{S} \times Q$ *is the transition relation, defined by* $\mathcal{Q} \overset{\mathcal{R}}{\Longrightarrow} \mathrm{sp}^{\#}(\mathcal{Q}, \mathcal{R})$ *where* $\mathrm{sp}^{\#}(\mathcal{Q}, \mathcal{R}) := \overline{\mathrm{sp}(\mathcal{Q}, \mathcal{R})}$.

Hence the abstract transition relation is obtained by computing the strongest postcondition of a condition and then weakening it so that it can be expressed in $\mathrm{Abs}(P)$. The latter requires checking whether $\mathrm{sp}(\mathcal{Q},\mathcal{R}) \models \mathcal{P}_i$ or $\mathrm{sp}(\mathcal{Q},\mathcal{R}) \models \neg\mathcal{P}_i$ for all i and forming a conjunction of those predicates where the check succeeds.

In fact, this approach precisely follows the paradigm of abstract interpretation, based on Galois connections. Let α and γ be mappings from $\mathbf{Cond}_0/\equiv$ to $\mathrm{Abs}(P)$ and back, defined by $\alpha(\mathcal{A}) := \overline{\mathcal{A}}$ and $\gamma(\mathcal{Q}) := \mathcal{Q}$, respectively. We then have the following:

Proposition 12. *Let P be a set of predicates. Then (α,γ) as defined above is a Galois connection between $\mathbf{Cond}/\equiv$ and $Abs(P)$, and $\mathrm{sp}^{\#}$ is the induced over-approximation of sp (i.e., $\mathrm{sp}^{\#}(\_,\mathcal{R}) = \alpha \circ \mathrm{sp}(\_,\mathcal{R}) \circ \gamma$).*

Note that α is (in general) not computable because it involves the entailment problem of first-order logic. A practical solution to its non-computability will be discussed later in the paper in Sect. 5.3.

For our running example, we have used this construction in Sect. 2, first for $P = \{\mathsf{Init}_2, \mathsf{Bad}\}$ and next for $P = \{\mathsf{Init}_2, \mathsf{Bad}, \mathcal{W}_1\}$, to obtain the abstract transition systems induced by Init_2. In the second case, as all states entail $\neg\mathsf{Bad}$, we verified the desired property.

Theorem 13. *Let P be a set of predicates with $\mathsf{Init}, \mathsf{Bad} \in P$. If all reachable states of the abstract transition system with initial state Init entail $\neg\mathsf{Bad}$, the set-based transition system induced by Init is correct w.r.t. Bad.*

5 Counterexample-Guided Abstraction Refinement (CEGAR)

We are now ready to define the full CEGAR loop. In particular, we will explain how to obtain suitable predicates for refinement.

5.1 Obtaining Predicates

In the example from Sect. 2, using only predicates $P = \{\mathsf{Init}_2, \mathsf{Bad}\}$ and initial state Init_2 we found an apparently unsafe abstract state, i.e., one which did not entail $\neg\mathsf{Bad}$, through the trace append, append. However, the condition-based state reached via the same trace was actually safe, and augmenting P with $\mathcal{W}_1$ resulted in a successful proof that, indeed, all reachable states are safe. Hence the general question arises how to refine a set of predicates, given an abstract trace to an unsafe state (i.e., a *counterexample* to correctness) that does not exist on the concrete level (i.e., is *spurious*).

Definition 14 (Spurious counterexample). *Let $\mathcal{S}$ be a reactive system and P be a set of predicates with $\mathsf{Init}, \mathsf{Bad} \in P$, and consider the abstract transition system with initial state Init. A* counterexample *to correctness w.r.t. Bad is a trace $\mathcal{R}_1 \cdots \mathcal{R}_n \in \mathcal{S}^*$ such that $\mathsf{Init} \xrightarrow{\mathcal{R}_1} \mathcal{Q}_1 \xrightarrow{\mathcal{R}_2} \mathcal{Q}_2 \ldots \xrightarrow{\mathcal{R}_n} \mathcal{Q}_n$ where $\mathcal{Q}_n \not\models \neg\mathsf{Bad}$. The counterexample is* spurious *if $\{\mathsf{Init}\}\mathcal{R}_1;\ldots;\mathcal{R}_n\{\neg\mathsf{Bad}\}$.*

Note that checking whether $\{\mathsf{Init}\}\mathcal{R}_1;\ldots;\mathcal{R}_n\{\neg\mathsf{Bad}\}$ is equivalent to each of the following two entailments:

$$\mathrm{sp}(\mathsf{Init},\mathcal{R}_1;\ldots;\mathcal{R}_n) = \mathrm{sp}(\ldots\mathrm{sp}(\mathrm{sp}(\mathsf{Init},\mathcal{R}_1),\mathcal{R}_2)\ldots,\mathcal{R}_n) \models \neg\mathsf{Bad}$$
$$\mathsf{Init} \models \mathrm{wp}(\mathcal{R}_1;\ldots;\mathcal{R}_n,\neg\mathsf{Bad}) = \mathrm{wp}(\mathcal{R}_1,\ldots\mathrm{wp}(\mathcal{R}_{n-1},\mathrm{wp}(\mathcal{R}_n,\neg\mathsf{Bad}))\ldots)$$

Hence we have at least two options for checking spuriousness. In both cases, this involves intermediate predicates $\mathcal{Q}'_1,\ldots,\mathcal{Q}'_{n-1}$ (for strongest postconditions: $\mathcal{Q}'_1 = \mathrm{sp}(\mathsf{Init},\mathcal{R}_1)$ and $\mathcal{Q}'_i = \mathrm{sp}(\mathcal{Q}'_{i-1},\mathcal{R}_i)$ for $1 < i \le n$; for weakest preconditions: $\mathcal{Q}'_{n-1} = \mathrm{wp}(\mathcal{R}_n,\neg\mathsf{Bad})$ and $\mathcal{Q}'_{i-1} = \mathrm{wp}(\mathcal{R}_i,\mathcal{Q}'_i)$ for $1 \le i < n$) such that

$$\{\mathsf{Init}\}\mathcal{R}_1\{\mathcal{Q}'_1\}\mathcal{R}_2\ldots\mathcal{R}_{n-1}\{\mathcal{Q}'_{n-1}\}\mathcal{R}_n\{\neg\mathsf{Bad}\}$$

We then augment P by adding all the $\mathcal{Q}'_i$. In the running example $\mathcal{W}_1$ equals the wp-based $\mathcal{Q}'_1$; adding it to P eliminated the counterexample. This elimination is in fact guaranteed (in a rather obious way) by the underlying theory, as formally stated by the following proposition:

Proposition 15. *Let P (with $\mathsf{Init},\mathsf{Bad} \in P$) be a set of predicates, and (considering* Init *as initial state) let $\mathcal{R}_1,\cdots,\mathcal{R}_n$ be a spurious counterexample to correctness w.r.t.* Bad. *Let $\mathcal{Q}'_1,\ldots,\mathcal{Q}'_{n-1}$ be predicates such that*

$$\{\mathsf{Init}\}\mathcal{R}_1\{\mathcal{Q}'_1\}\mathcal{R}_2\ldots\mathcal{R}_{n-1}\{\mathcal{Q}'_{n-1}\}\mathcal{R}_n\{\neg\mathsf{Bad}\}.$$

Then, in the abstract transition system based on $Abs(P \cup \{\mathcal{Q}'_1,\ldots,\mathcal{Q}'_{n-1}\})$ with initial state Init, *the trace $\mathcal{R}_1,\ldots,\mathcal{R}_n$ leads to a condition entailing* $\neg$Bad; *in other words, it is not a counterexample any more.*

5.2 Idealized Algorithm

The above brings us to the (idealized) CEGAR algorithm already discussed in Sect. 2 and illustrated in Fig. 1. Starting with an initial set of predicates $P = \{\mathsf{Init},\mathsf{Bad}\}$, construct the abstract transition system with initial state Init, adding successor states until either no new states are found or we reach a state $\mathcal{Q}_n$ that does not entail $\neg$Bad. In the former case, the algorithm terminates: verification succeeded, the system is correct w.r.t. Bad. In the latter case, however, the sequence of rules $\mathcal{R}_1,\ldots,\mathcal{R}_n$ from Init to $\mathcal{Q}_n$ is a counterexample; check whether it is spurious by computing either strongest postconditions or weakest preconditions, obtaining additional predicates $\mathcal{Q}'_1,\ldots,\mathcal{Q}'_{n-1}$ as described above.

- If it is spurious, add $\mathcal{Q}'_1,\ldots\mathcal{Q}'_n$ to the current predicate set P to eliminate the spurious counterexample, and restart the analysis.
- If it is not spurious, the algorithm terminates: verification failed, the system is not correct w.r.t. Bad.

Proposition 16. *The idealized algorithm is correct in the sense that the system is correct if it is successful and incorrect if it failed. Moreover, if counterexamples are processed in ascending length (i.e., we always process the* shortest *counterexample), it constitutes a semi-decision procedure.*

5.3 Practical Algorithm

So far we have based our definitions on entailment ($\models$), assuming that it is somehow computable, when in fact the entailment problem is in general undecidable. This affects both the abstraction $\alpha\colon \mathbf{Cond}_0/{\equiv} \to \mathrm{Abs}(P)$ and the check for spuriousness.

A practical implementation of the entailment check can only give approximate answers to this problem and may be unable to prove or disprove some entailments. Hence we can only rely on "provable entailment" $\widehat{\models}$, which is a subrelation of $\models$, not formally characterized but determined by the strength of our proof tools and the available time. Based on this and given $\mathcal{A} \in \mathbf{Cond}_0$, we can define $\widehat{\mathcal{A}} := \bigwedge\{\mathcal{Q}' \in \mathrm{Abs}(P) \mid \mathcal{A} \mathrel{\widehat{\models}} \mathcal{Q}'\}$ (the strongest condition in $\mathrm{Abs}(P)$ for which we can prove that it is implied by $\mathcal{A}$). Since $\widehat{\models}$ is a subrelation of $\models$, we obtain $\overline{\mathcal{A}} \models \widehat{\mathcal{A}}$.

In practice, this is computed by iterating over all predicates $\mathcal{P} \in P$ and checking whether $\mathcal{A} \mathrel{\widehat{\models}} \mathcal{P}$ or $\mathcal{A} \mathrel{\widehat{\models}} \neg\mathcal{P}$. Taking the conjunction of all such predicates yields $\widehat{\mathcal{A}}$. Predicates for which we obtain no result are not included in the conjunction.

Based on that we can define $\widehat{\alpha}\colon \mathbf{Cond}_0/{\equiv} \to \mathrm{Abs}(P)$ via $\widehat{\alpha}(\mathcal{A}) = \widehat{\mathcal{A}}$ as a computable function. Note that $\widehat{\alpha}$ is an over-approximation of α ($\alpha \models \widehat{\alpha}$).

Using such an over-approximation $\widehat{\alpha}$ also affects the abstract transitions. Compared to Definition 11, which used the induced over-approximation $\mathrm{sp}^{\#}(\mathcal{Q}, \mathcal{R}) = \alpha(\mathrm{sp}(\gamma(\mathcal{Q}), \mathcal{R}))$, we now obtain a function $\widehat{\mathrm{sp}}(\mathcal{Q}, \mathcal{R}) := \widehat{\alpha}(\mathrm{sp}(\gamma(\mathcal{Q}), \mathcal{R}))$. This is no longer the induced over-approximation, however, since $\alpha \models \widehat{\alpha}$, we have $\mathrm{sp}^{\#}(\mathcal{Q}, \mathcal{R}) \models \widehat{\mathrm{sp}}(\mathcal{Q}, \mathcal{R})$. Therefore $\widehat{\mathrm{sp}}$ is a safe approximation of sp and a corresponding adaptation of Theorem 13 to such abstract transition systems still holds.

Undecidability of $\models$ also affects detection of spurious counterexamples which involves an entailment check (see Sect. 5.1). A solver might be unable to produce a proof in reasonable time, which means that the algorithm cannot check and eliminate this specific counterexample. Either the method proceeds with another counterexample or stops and reports the found counterexample to the user as a (potential) error.

Hence, while the previous procedure is a semi-decision method (cf. Proposition 16), this is now lost by the additional level of abstraction.

6 Implementation

The CEGAR algorithm described in this paper has been implemented in a prototypical tool[2] [46], instantiated to $\mathbf{ILC}(\mathbf{Graph}_{\mathbf{fin}})$, i.e., graph transformation systems. It is a command-line based tool that verifies a given graph transformation system against given predicates Init, Bad, and automatically derives new predicates from strongest postconditions. The tool is written in Java and makes heavy use of the graph library presented in [5]. For checking satisfiability of

[2] https://git.uni-due.de/sflastol/cegar-prototype.

conditions, the algorithm from [44] is used and has been adapted to cospan conditions.

Previous tests of the satisfiability checker alone have indicated that for many practical examples, the intermediate conditions quickly reach impractical sizes, leading to long runtimes, which might limit the usefulness of the tool. There are several possible optimizations to drastically cut down the size of intermediate results, only some of which have been implemented in the tool so far. This is an avenue for future work.

Nevertheless, there are several examples that can be successfully verified using the CEGAR tool in its current state. One example (`talk_delete2.sgf`) is a system where a rule can delete two nodes at once, Init states that exactly three nodes exist, and Bad states that the graph is empty (for simplicity, edges are assumed to be absent). Another example (`talk_outedge.sgf`) starts from a non-empty graph, adds and deletes edges and their target nodes and asks again whether the empty graph is reachable. The tool can verify these systems to be safe after a single refinement step using strongest postconditions, almost instantaneously. Further successful examples of similar complexity are discussed in [46].

As of now, the tool can immediately spot the error in the unrefined version of our running example (`paper_examples.sgf`), but is unable to verify the refined variant in reasonable time (out of memory after 130 s). We hypothesize that strongest postconditions—as currently implemented—do not always yield the most useful predicates for refinement and using weakest preconditions (as we manually did in the running example) might yield better results. Furthermore, optimizations can be implemented to reduce the size of the conditions both in the CEGAR loop itself and the satisfiability checker. In manual tests, elimination of redundant subconditions has typically resulted in a reduction of more than half of all subconditions at any given step. (A similar question is studied in [33], which reports on the simplification of constraint-guaranteeing conditions.) An additional optimization could be to pair the existing solver with translation to first-order formula and using off-the-shelf first-order logic or SMT solvers.

7 Conclusion and Future Work

Static analysis and verification techniques for graph transformation systems have by now been studied for at least two decades. Graph transformation provides a flexible and powerful modelling technique, but this comes with a trade-off with respect to verification. Due to its inherent complexity, with features such as infinite state spaces and the ability to model dynamic topologies, graph transformation system pose several challenges for verification. We can not give a complete overview over the literature, but we would like to mention that there are contributions based on efficient state space enumeration [38], invariant checking [12,18], over-approximation [26,39], well-structured transition systems [35,45], logic-based approaches [36,37], techniques for termination analysis [6,34], and so on. In many cases, these techniques were adapted from other areas such as string and term rewriting, infinite state verification or program analysis.

A technique that has not yet received much attention in the area of graph transformation is the CEGAR technique adapted in this paper. CEGAR is a well-known and widely studied method for program analysis [8,9,15,19,20,23, 32]. The typical setting is transition systems (Kripke structures) respectively program verification.

To the best of our knowledge, this is the first CEGAR approach to the verification of graph transformation systems that is based on predicate abstraction. In [25] a CEGAR framework based on approximated unfoldings was introduced for graph transformation systems. This however follows a completely different approach – not connected to predicate abstraction – and works only for a restricted class of systems: in particular rules are not allowed to delete nodes and there is no integration of application conditions.

As mentioned above, the prototypical implementation still has scalability issues and a next step would be to introduce several optimization techniques, in particular, for improving efficiency for the entailment problem and covering more entailments. One possibility could be to consider sufficient conditions for entailment that are easier to check, similar to [41].

Another line of research is to improve the automatically generated predicates, in terms of their size and in terms of their usefulness (i.e., whether they help to prove the correctness of the system). This is also related to the scalability question. In CEGAR the typical idea is to use so-called Craig interpolants [19]. Given two formulas φ_1, φ_2 in a program logic (referring to program variables) with $\varphi_1 \models \varphi_2$, the Craig interpolant (of φ_1, φ_2) is a formula ψ that satisfies $\varphi_1 \models \psi \models \varphi_2$ and contains only the variables present in both φ_1, φ_2. The idea is to eliminate spurious counterexamples by computing *both* weakest preconditions and strongest postconditions (where the former entails the latter), but adding the Craig interpolants between both. This typically leads to more compact and better suited predicates. In the setting of conditions it is unclear what the analogue to Craig interpolants actually is, how they can be computed and used. We believe that this is a promising and potentially fruitful line of research.

In this paper we concentrated mainly on applications in the area of graph transformation systems. However, the framework of reactive systems is more general and encompasses also other rewriting systems, such as ground term rewriting based on Lawvere theories. In [44] we showed how to compute shifts in this setting, which enable us to compute pre- and postconditions and instantiate the entire CEGAR framework. It would be worthwhile to further investigate the applicability and scalability of this approach.

Disclosure of Interests. The authors have no competing interests to declare that are relevant to the content of this article.

References

1. Ball, T., Levin, V., Rajamani, S.K.: A decade of software model checking with SLAM. Commun. ACM **54**(7), 68–76 (2011)

2. Beyer, D., Henzinger, T.A., Jhala, R., Majumdar, R.: The software model checker Blast. Int. J. Softw. Tools Technol. Transfer **9**, 505–525 (2007)
3. Bonchi, F., Gadducci, F., Kissinger, A., Sobociński, P., Zanasi, F.: Confluence of graph rewriting with interfaces. In: Yang, H. (ed.) ESOP 2017. LNCS, vol. 10201, pp. 141–169. Springer, Heidelberg (2017). https://doi.org/10.1007/978-3-662-54434-1_6
4. Bruggink, H.J.S., Cauderlier, R., Hülsbusch, M., König, B.: Conditional reactive systems. In: Proceedings of FSTTCS '11. LIPIcs, vol. 13. Schloss Dagstuhl – Leibniz Center for Informatics (2011)
5. Bruggink, H.J.S., König, B., Matjeka, M., Nolte, D., Stoltenow, L.: A flexible and easy-to-use library for the rapid development of graph tools in Java (tool presentation paper). In: Proceedings of ICGT '20. LNCS, vol 12150, pp. 297–306. Springer, Heidelberg (2020). https://doi.org/10.1007/978-3-030-51372-6_17
6. Bruggink, H.J.S., König, B., Zantema, H.: Termination analysis for graph transformation systems. In: Diaz, J., Lanese, I., Sangiorgi, D. (eds.) TCS 2014. LNCS, vol. 8705, pp. 179–194. Springer, Heidelberg (2014). https://doi.org/10.1007/978-3-662-44602-7_15
7. Chaki, S., Clarke, E.M., Groce, A., Jha, S., Veith, H.: Modular verification of software components in C. In: Proceedings of ICSE '03 (25th International Conference on Software Engineering), pp. 385–395. IEEE Computer Society (2003)
8. Clarke, E.M., Grumberg, O., Jha, S., Lu, Y., Veith, H.: Counterexample-guided abstraction refinement. In: Emerson, E.A., Sistla, A.P. (eds.) CAV 2000. LNCS, vol. 1855, pp. 154–169. Springer, Heidelberg (2000). https://doi.org/10.1007/10722167_15
9. Clarke, E.M., Grumberg, O., Jha, S., Lu, Y., Veith, H.: Counterexample-guided abstraction refinement for symbolic model checking. J. ACM **50**(5), 752–794 (2003)
10. Cousot, P.: Abstract interpretation. ACM Comput. Surv. **28**(2) (1996)
11. Cousot, P., Cousot, R.: Abstract interpretation: a unified lattice model for static analysis of programs by construction or approximation of fixpoints. In: Proceedings of POPL '77, Los Angeles, California, pp. 238–252. ACM (1977)
12. Dyck, J., Giese, H.: k-inductive invariant checking for graph transformation systems. In: de Lara, J., Plump, D. (eds.) ICGT 2017. LNCS, vol. 10373, pp. 142–158. Springer, Cham (2017). https://doi.org/10.1007/978-3-319-61470-0_9
13. Ehrig, H., Ehrig, K., Prange, U., Taentzer, G.: Fundamentals of Algebraic Graph Transformation. Monographs in Theoretical Computer Science. Springer, Heidelberg (2006)
14. Ehrig, H., Pfender, M., Schneider, H.J.: Graph grammars: an algebraic approach. In: Proceedings 14th IEEE Symposium on Switching and Automata Theory, pp. 167–180 (1973)
15. Esparza, J., Kiefer, S., Schwoon, S.: Abstraction refinement with craig interpolation and symbolic pushdown systems. In: Hermanns, H., Palsberg, J. (eds.) TACAS 2006. LNCS, vol. 3920, pp. 489–503. Springer, Heidelberg (2006). https://doi.org/10.1007/11691372_35
16. Habel, A., Pennemann, K.H.: Correctness of high-level transformation systems relative to nested conditions. Math. Stuct. Comput. Sci. **19**(2), 245–296 (2009)
17. Habel, A., Plump, D.: Computational completeness of programming languages based on graph transformation. In: Honsell, F., Miculan, M. (eds.) FoSSaCS 2001. LNCS, vol. 2030, pp. 230–245. Springer, Heidelberg (2001). https://doi.org/10.1007/3-540-45315-6_15

18. Heckel, R., Wagner, A.: Ensuring consistency of conditional graph rewriting - a constructive approach. In: Proceedings of the Joint COMPUGRAPH/SEMAGRAPH Workshop on Graph Rewriting and Computation. ENTCS, vol. 2 (1995)
19. Henzinger, T.A., Jhala, R., Majumdar, R., McMillan, K.L.: Abstractions from proofs. In: Proceedings of POPL '04, pp. 232–244. ACM (2004)
20. Henzinger, T.A., Jhala, R., Majumdar, R., Sutre, G.: Lazy abstraction. In: Proceedings of POPL '02, pp. 58–70. ACM (2002)
21. Hülsbusch, M., König, B.: Deriving bisimulation congruences for conditional reactive systems. In: Birkedal, L. (ed.) FoSSaCS 2012. LNCS, vol. 7213, pp. 361–375. Springer, Heidelberg (2012). https://doi.org/10.1007/978-3-642-28729-9_24
22. Hülsbusch, M., König, B., Küpper, S., Stoltenow, L.: Conditional bisimilarity for reactive systems. Logical Methods Comput. Sci. **18**(1) (2022)
23. Jhala, R., McMillan, K.L.: A practical and complete approach to predicate refinement. In: Hermanns, H., Palsberg, J. (eds.) TACAS 2006. LNCS, vol. 3920, pp. 459–473. Springer, Heidelberg (2006). https://doi.org/10.1007/11691372_33
24. König, B.: Hypergraph construction and its application to the static analysis of concurrent systems. Math. Struct. Comput. Sci. **12**(2), 149–175 (2002)
25. König, B., Kozioura, V.: Counterexample-guided abstraction refinement for the analysis of graph transformation systems. In: Hermanns, H., Palsberg, J. (eds.) TACAS 2006. LNCS, vol. 3920, pp. 197–211. Springer, Heidelberg (2006). https://doi.org/10.1007/11691372_13
26. König, B., Kozioura, V.: Augur 2—a new version of a tool for the analysis of graph transformation systems. In: Proceedings of GT-VMT '06 (Workshop on Graph Transformation and Visual Modeling Techniques). ENTCS, vol. 211, pp. 201–210. Elsevier (2008)
27. König, B., Rensink, A., Stoltenow, L., Urrigshardt, F.: Counterexample-guided abstraction refinement for generalized graph transformation systems (full version) (2025). arXiv:2504.08617
28. Lack, S., Sobociński, P.: Adhesive and quasiadhesive categories. RAIRO – Theor. Inf. Appl. **39**(3) (2005)
29. Lambers, L., Orejas, F.: Tableau-based reasoning for graph properties. In: Giese, H., König, B. (eds.) ICGT 2014. LNCS, vol. 8571, pp. 17–32. Springer, Cham (2014). https://doi.org/10.1007/978-3-319-09108-2_2
30. Leifer, J.J.: Operational congruences for reactive systems. Ph.D. thesis, University of Cambridge Computer Laboratory (2001)
31. Leifer, J.J., Milner, R.: Deriving bisimulation congruences for reactive systems. In: Palamidessi, C. (ed.) CONCUR 2000. LNCS, vol. 1877, pp. 243–258. Springer, Heidelberg (2000). https://doi.org/10.1007/3-540-44618-4_19
32. McMillan, K.L.: Interpolation and SAT-based model checking. In: Hunt, W.A., Somenzi, F. (eds.) CAV 2003. LNCS, vol. 2725, pp. 1–13. Springer, Heidelberg (2003). https://doi.org/10.1007/978-3-540-45069-6_1
33. Nassar, N., Kosiol, J., Arendt, T., Taentzer, G.: Constructing optimized constraint-preserving application conditions for model transformation rules. J. Logic Algeb. Program. **114**, 100564 (2020)
34. Overbeek, R., Endrullis, J.: Termination of graph transformation systems using weighted subgraph counting. Logical Methods Comput. Sci. **20**(4) (2024)
35. Özkan, O.: Decidability of resilience for well-structured graph transformation systems. In: Proceedings of ICGT '22. LNCS, vol. 13349, pp. 38–57. Springer, Heidelberg (2022). https://doi.org/10.1007/978-3-031-09843-7_3

36. Poskitt, C.M., Plump, D.: A hoare calculus for graph programs. In: Ehrig, H., Rensink, A., Rozenberg, G., Schürr, A. (eds.) ICGT 2010. LNCS, vol. 6372, pp. 139–154. Springer, Heidelberg (2010). https://doi.org/10.1007/978-3-642-15928-2_10
37. Poskitt, C.M., Plump, D.: Monadic second-order incorrectness logic for GP 2. J. Logic Algeb. Program. **130**, 100825 (2023)
38. Rensink, A.: The GROOVE simulator: a tool for state space generation. In: Pfaltz, J.L., Nagl, M., Böhlen, B. (eds.) AGTIVE 2003. LNCS, vol. 3062, pp. 479–485. Springer, Heidelberg (2004). https://doi.org/10.1007/978-3-540-25959-6_40
39. Rensink, A.: Canonical graph shapes. In: Schmidt, D. (ed.) ESOP 2004. LNCS, vol. 2986, pp. 401–415. Springer, Heidelberg (2004). https://doi.org/10.1007/978-3-540-24725-8_28
40. Rensink, A.: Representing first-order logic using graphs. In: Ehrig, H., Engels, G., Parisi-Presicce, F., Rozenberg, G. (eds.) ICGT 2004. LNCS, vol. 3256, pp. 319–335. Springer, Heidelberg (2004). https://doi.org/10.1007/978-3-540-30203-2_23
41. Rensink, A., Corradini, A.: On categories of nested conditions. In: Principles of Verification: Cycling the Probabilistic Landscape – Essays Dedicated to Joost-Pieter Katoen on the Occasion of His 60th Birthday, Part I. LNCS, vol. 15260, pp. 393–418. Springer, Heidelberg (2025). DOI: https://doi.org/10.1007/978-3-031-75783-9_16
42. Sassone, V., Sobociński, P.: Reactive systems over cospans. In: Proceedings of LICS '05, pp. 311–320. IEEE (2005)
43. Sobociński, P.: Deriving process congruences from reaction rules. Ph.D. thesis, Department of Computer Science, University of Aarhus (2004)
44. Stoltenow, L., König, B., Schneider, S., Corradini, A., Lambers, L., Orejas, F.: Coinductive techniques for checking satisfiability of generalized nested conditions. In: Proceedings of CONCUR '24. LIPIcs, vol. 311, pp. 39:1–39:20. Schloss Dagstuhl – Leibniz Center for Informatics (2024)
45. Stückrath, J., König, B.: A general framework for well-structured graph transformation systems. Inf. Comput. **252**, 79–94 (2017)
46. Urrigshardt, F.: Counterexample-guided Abstraction Refinement for Reactive Systems. Master's thesis, Universität Duisburg-Essen (2024)

Test Case Generation from Graph Transformation Systems Using Deep Reinforcement Learning

Simin Ghasemi[1], Vahid Rafe[2](✉), Mohammad Mehrabi[3], Reiko Heckel[4], and Issam Al-Azzoni[5]

[1] Department of Computer Engineering, Faculty of Engineering, Arak University, 38156-8-8349 Arak, Iran
simin-ghasemi@pnu.ac.ir

[2] Center for Software Reliability, Department of Computer Engineering, City St George's, University of London, London, UK
vahid.rafe@city.ac.uk

[3] Tehran, Iran

[4] School of Computing and Mathematical Sciences, University of Leicester, Leicester, UK
rh122@le.ac.uk

[5] College of Engineering, Al Ain University, Al Ain, UAE
issam.alazzoni@aau.ac.ae

Abstract. Graph transformations can be used to specify and analyse software systems by modelling operations as rules and generating the labelled transition system (LTS) as a representation of system behaviour. Model-based testing (MBT) often uses model checking over LTS to discover paths that satisfy certain test requirements. Significant challenges include the size of the state space and the complexity of model checking. Meta-heuristic search-based approaches try to cope with this problem by exploring only a small portion of the LTS to produce paths that cover maximal test objectives. Despite acceptable results in small case studies, these approaches also do not scale well.

MBT approaches using graph transformation face the same limitations as LTS-based MBT in general, exacerbated by the more complex nature of graph-based LTS. However, apart from the behavioural information of the LTS, here we are able to use the graph structure of states and rules to define and evaluate test objectives. This enables a new method based on deep reinforcement learning to generate test suites for systems specified through graph transformations.

We use the reward/penalty mechanism of reinforcement learning to optimise the selection of moves within the state space, enabling the generation of test cases based on prior decisions. Our goal is to achieve greater coverage of test objectives while minimising the size of the test cases. The method has been implemented in GROOVE, an open-source toolset for designing and model checking graph transformation systems. Experimental results in well-known case studies demonstrate that our

M. Mehrabi—Independent Researcher

J. Endrullis and M. Tichy (Eds.): ICGT 2025, LNCS 15720, pp. 178–201, 2025.
https://doi.org/10.1007/978-3-031-94706-3_9

approach generates test cases with improved coverage scores while reducing the cost of testing compared to existing works.

Keywords: Model Checking-based Testing · Test Suite Generation · Graph Transformation Systems · Deep Reinforcement Learning · Neural Network

1 Introduction

Software testing is an essential but expensive part of software quality assurance. Automated test generation can reduce the cost of testing and improve the effectiveness of tests to discover errors. Model-Based Testing (MBT) [35] is an approach to generate and execute test cases from behavioural models of the system under test. Earlier fault detection reduces the cost of testing and increases the quality of the system under test. Various strategies have been used in MBT to generate more effective tests at a lower cost.

Model checking [25] has been applied to software testing for state-based models. Model checkers [5] were originally designed as formal verification tools capable of producing paths representing counterexamples/witnesses for the violation/satisfaction of a desired property. In Model Checking-based Testing (MCT), a model-checker is employed to take the test model as input and use test objectives as reachability properties to derive paths representing test scenarios. Although MCT ensures high coverage of test objectives for small models, its scalability is limited by state-space explosion, the complexity of model checking, and the generation of redundant test cases. One solution to this problem is to use search and optimisation methods to heuristically explore only a relevant portion of the state space. Genetic Algorithms (GA) [18], Particle Swarm Optimisation (PSO) [9] and Ant Colony Optimisation (ACO) [7] have been used, but there is still room for improvement in terms of eliminating redundant test cases and enhancing accuracy and efficiency. One challenge with these techniques is that they generate test cases in multiple iterations, making it difficult to identify and remove redundant test cases generated at different steps. Additionally, the heuristic functions used for test case selection can sometimes behave randomly, leading to the generation of incorrect or inaccurate test cases (such as a random crossover of test cases). Lastly, additional steps are often required to assess test case quality and eliminate redundancy or inaccuracies, introducing extra post-processing efforts and reducing overall efficiency. Graphs can represent, visualise, and analyse complex software models. Graph Transformation Systems (GTS) [10] use rules to represent pre- and postconditions of operations and model the behaviour of a system as a graph transition system, i.e., labelled transition systems (LTS) with graphs as states and graph transformations as transitions. Established algorithms and tools exist to simulate transformations, generate a graph-based LTS and analyse it through model checking.

However, both graph-based MCT and testing based on heuristic search and optimisation suffer from the same limitations as the generic approaches described above, in particular if they treat the graph transition system as a generic LTS without exploiting the graphical structure.

In this paper, we propose a testing approach for GTS that uses Deep Reinforcement Learning (DRL) to explore state space, finding test cases with the best coverage. The search begins at the initial state and selects actions randomly at first, then explores the state space based on the reward/penalty received for previous actions, ultimately generating semi-optimal test cases with high coverage and low cost. DRL is also able to improve over time by learning from the environment. The novel contributions of this paper are as follows:

- We develop a DRL-based approach to generate test cases for systems specified through GTS that cover more test objectives.
- The proposed approach is capable of storing network weights, allowing them to be reloaded in subsequent executions to reduce computational costs.
- The approach performs efficiently and is more scalable in terms of the time taken to generate test cases.
- The generated test suite can be executed at lower cost because it is smaller and contains shorter test cases.

The remainder of this paper is organised as follows. Section 2 describes the basic concepts of testing with GTS and DRL. Section 3 introduces the state of the art in MBT and reviews the existing work on GTS-based testing. Section 4 describes our approach in detail. The experimental results and the discussions are presented in Sect. 5. Finally, Sect. 6 concludes the paper and suggests future work.

2 Background

This section reviews the concepts of GTS testing and deep reinforcement learning.

2.1 Testing Graph Transformation Systems

Software testing is crucial today due to the increasing complexity of software systems and the need for reliability in critical applications. Automation testing is a good alternative to manual testing, but the problems of cost and uncertainty in the environment still remain [35]. Thus, to fully address these issues, MBT is used. MBT uses models to describe the environment and its behaviours, which can be easily extended when a feature is added or changed in the system. This approach leads to more efficient testing practices.

Model-checking [5] is a powerful procedure that can be utilised to automatically generate test cases from the model by exploring the state space to identify faults. In the MCT [1] procedure, a model of the system under test and a set of test objectives, expressed as reachability properties, are given to the model-checker. The behaviour of a system modelled as a GTS can be represented by an LTS providing a starting point for MCT. A test case is a finite path in the state space that starts at the start graph and satisfies certain test objectives expressed by reachability properties or more complex temporal formulas. The aim is to generate a test suite as a set of test cases that together satisfy the test objectives [20]. One of the most mature and effective tools for model checking graph transformation systems is GROOVE [28]. To revisit the concepts of GTS using GROOVE, interested readers are referred to the appendix.

2.2 Coverage Criteria

In MBT approaches that use GTS, criteria such as state coverage, rule-dependency coverage, transition coverage, path coverage, and rule coverage serve as test objectives to direct the testing process. As the rule-dependency coverage [19] gives an estimate of the def-use relations (as used in code-based testing) and could lead to the detection of potential bugs, many studies use it as a coverage criterion.

In GTS, the rules have some data sharing and affect each other. For example, according to Fig. 6, the application of the BookRoom rule appoints a vacant room to the guest by creating an edge labelled 'bookingInfo', which could be read later by the production rule CheckOut. In other words, the application of the rule CheckOut depends on the successful use of BookRoom to the system state. This is a type of rule dependency, where one rule creates/updates an entity (node/edge) and another rule uses (i.e. read/delete) it. The following is the definition of rule dependency: (For a more accurate definition, refer to [19])

Definition 1 ***(Rule-Dependency).*** *Given two production rules R_i and R_j, we say R_j is dependent on R_i if there are transformation steps $t_i = (S \overset{R_i}{\Longrightarrow} S_i)$ and $t_j = (S_i \overset{R_j}{\Longrightarrow} S_j)$ such that t_i enables t_j and S is a state graph.*

Data-flow relations between rules are considered where one rule can enable another, e.g., when it creates a node or edge or updates an attribute and another rule uses, reads, or deletes it, or when the first rule deletes entities that would violate the negative application conditions of the second. Hence, possible relationships between rules include *Create_Read (C_R), Create_Delete (C_D), Create_Update (C_U), Delete_Nac (D_N), Update_Read (U_R) and Update_Update (U_U).* In C_R a rule creates an entity that another rule reads, in C_D a rule creates an entity that another rule deletes, etc.

2.3 Deep Reinforcement Learning

Reinforcement learning is a method that aims at learning to control a set of actions based on feedback from a given environment. Formally, this can be described as a Markov Decision Process. In this framework, machines that learn and make decisions are referred to as agents, and everything external interacting with the agents represents the environment. This interaction forms the basis for decision making and learning, where the agent aims to maximise cumulative rewards. At first, the agent's knowledge is zero, but through interaction with the environment, it acquires information and learns how to achieve the defined goals [33]. At each step of reinforcement learning, the agent receives a state from the environment and performs an available action to transition to the next state. Based on the agent's action, it will receive a reward or penalty from the environment.

Deep reinforcement learning (DRL) uses particular neural networks, such as the Double Deep Q-Network (DDQN) [17]. In DDQNs, one neural network chooses the best action while the other one calculates the value of the selected action used to update the network's weights.

Some important terminology: A *step* denotes the execution of an action by an agent and the corresponding transition to a new state. An *episode* is a sequence of steps from the start to a final state. The *experience replay memory* is used to store the agent's past experiences. An *ϵ-greedy algorithm* is an exploration policy where the agent chooses the current best option most of the time and occasionally chooses a random option with low probability ϵ.

3 Related Work

Based on different techniques that existing approaches have used, we organise them in the following:

3.1 MBT in General

Model-based testing [35] is an effective method for validating the relationship between a model and its implementation where test generation is based on the model rather than the code of the system under test [15].

Most approaches to MBT rely on the mainstream Unified Modelling Language (UML). For example, [26] proposes an MBT approach for web applications using UML and Web diagrams, and [22] uses finite-state machines to reduce redundancy and improve efficiency. Some tools, such as Alloy [21], generate exhaustive test cases in bounded search spaces using SAT solvers. Alloy is a formal modelling language that allows the user to produce models of software systems by abstracting their key characteristics. Although our approach uses MBT techniques, it differs from these methods because it utilises GTS and optimisation, making it a search-based technique.

3.2 Model-Checking and Search-Based Techniques

Others have approached the expansion of MBT in different ways. For example, [36] provides a framework based on bounded model checking with coverage-guided fuzzing to generate test cases. They combined model checking with fuzzing for test case generation to generate initial seeds for fuzzers in order to give more interesting and effective seeds than manually created ones. The authors in [32] suggest a method called ISR-MCF to generate test cases in MBT. This method uses three search techniques: NSGA-II-LLH, SPEA2-LLH, and PSO-LLH. This method uses reinforcement learning to check how each operator performs and then picks the best one to create the test case. The study shows that ISR-MCF, when used with NSGA-II-LLH, performs better than the other methods. In [13], the authors study various methods of model checking-based testing and review the state of the art. They also investigated the MCT issues in [12] such as the quality and usability of the generated tests and proposed some techniques to overcome the problems. Our approach also employs model checking and optimisation-based search techniques; however, it is primarily based on GTS.

3.3 GTS and Optimisation-Based Techniques

Some approaches utilise GTS for MBT. In [16], the authors present a methodology for testing components described in a high-level language using GTS. They employed model checking to find adequate test sequences for a given requirement.

Another challenge for MCT is state-space explosion. Some research uses search-based optimisation and AI methods to overcome this, such as Bee Colony, Genetic Algorithms (GA), Particle Swarm Optimisation (PSO), Bat Algorithms and Gravitational Search Algorithms [2,3,14,23,27]. These solutions partially explore the state space to extract paths with high coverage. [23] proposes a search-based hybrid GA and PSO algorithm (HGAPSO) to generate test cases from models specified through GTS. They defined test generation as an optimisation problem that aims to maximise the static data dependency of test paths. They used Groove to implement HGAPSO, and the experimental results show that the approach covers most of the test objectives, but there are challenges when defining the test objectives statically and HGAPSO cannot be applied in complex systems. Due to its limitations, [4] extend HGAPSO to a memetic Algorithm (MA) to perform robustness testing. Their main goals were to maximise coverage and minimise testing costs. The effectiveness of the strategy is evaluated through a type of mutation analysis at the model level. The experimental results demonstrate that the approach outperforms existing ones in fault detection. They also propose another approach in [3] in order to test software models specified through GTS. In this work, a GA has been used to maximise coverage of rule, state, transition, and data flow criteria. The results show that data-dependency coverage works better than other methods in complicated systems with many possible states and related rules. This work has a few drawbacks, like the computation complexity of the approach and also the length of the generated tests. In [14], the authors use Ant Colony Optimisation (ACO) to generate entire test suites instead of individual test cases. The results show that it creates better test suites in terms of size and coverage, but its main disadvantage is the cost of generating the state space and finding the test paths. In a recent work on MCT, the authors in [2] propose an approach that uses two data mining algorithms, Apriori and FP-Growth, to avoid exhaustive state-space exploration. They first explore a portion of the state space using a BFS strategy and then apply data mining algorithms to extract frequent patterns guiding the exploration process. However, this method has two main drawbacks: its reliance on static dependencies to calculate the test case fitness and the need for an additional reduction phase to minimise the test suite, which significantly increases computational complexity.

Among the related approaches, we have selected HGAPSO [23], GA [3], and WholeAcoT [14] for comparison in different case studies. These approaches were chosen because they share the same objectives as our study, and their code is accessible.

4 Test Generation Through Deep Reinforcement Learning

The main challenge of MBT is the quality of the generated test cases and test suites in terms of coverage and test size. The key objectives are to achieve high coverage, minimise test size, and develop an executable algorithm that avoids state-space explosion and excessive testing costs. Several approaches in the literature utilise AI methods to explore state space heuristically and choose promising paths. To the best of our knowledge, although most of these methods have improved MBT and provided useful results, the desired criteria are not satisfactorily covered. An effective method for exploring the state space is one that actively acquires knowledge during traversal and utilises this knowledge to guide the exploration of subsequent states. One possible way to design this method is a mechanism that calculates reward and penalty values to guide it toward better paths, a principle that lies at the core of reinforcement learning algorithms. In this paper, we propose MoTDeReL as an approach to create and optimise test cases for GTS, based on DRL. MoTDeRel stands for MOdel-based testing through DEep REinforcement Learning. Our aim is to cover more test objectives at each step based on accumulated experience while also reducing the costs associated with testing and test generation.

Figure 1 illustrates our approach using two flowcharts, which we divided to simplify the explanation. According to Fig. 1(a), MoTDeReL initially receives the input model along with the appropriate parameter values. The process begins with the start graph of the input model as the initial state (later referred to as the current state). To facilitate a move (action, rule) within the state space, the approach employs the steps outlined in the second flowchart depicted in Fig. 1(b). In this flowchart (as the second flowchart), action selection is performed using the ϵ-greedy algorithm. Initially, a random number is generated; If this number is less than ϵ (an input parameter), the action is selected randomly; otherwise, it is chosen based on the recommendations of the neural network. The neural network can either be a pre-trained model or one with newly loaded weights. After applying the chosen action, its reward is calculated using the fitness function, and the action is stored in the experience replay memory. If the memory size reaches a predefined limit, a random batch from memory will be selected for training the network. Upon completion of the steps in the second flowchart (Fig. 1(b)), the approach resumes with the steps in the first flowchart (Fig. 1(a)). After the movement is executed, the length of the traversed path is checked. If the length of test cases reaches the maximum allowed for a test case or attains a final state, the path terminates. Additionally, the maximum length of the test suite restricts the number of generated test cases. If the fitness of a generated test case is repetitive up to a maximum number of allowed repeated rewards, that test case is disregarded. Finally, the generated test cases are printed as part of the final test suite. In the following subsections, we will explain each part of the proposed approach in more detail.

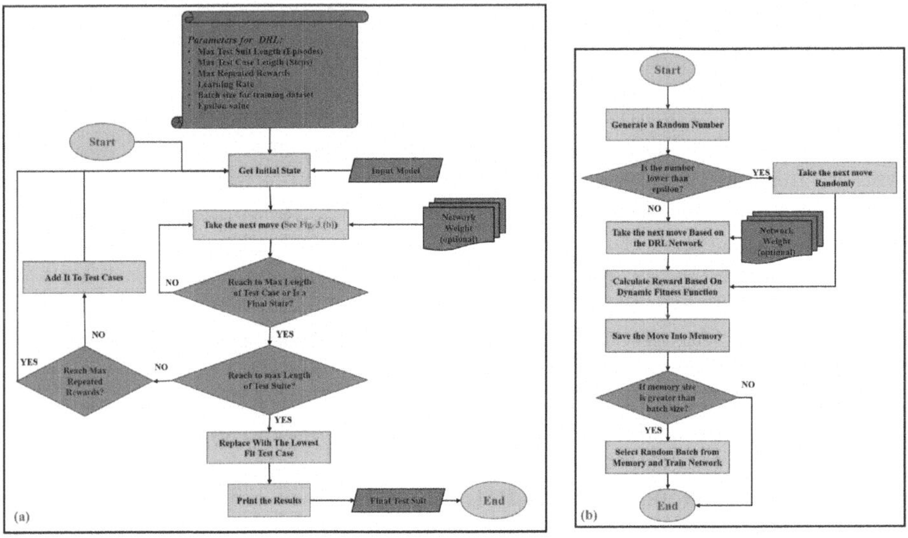

Fig. 1. The general flowcharts of the proposed approach (a) Main flowchart (b) Second flowchart

4.1 Neural Network Configuration

The core components of DRL are its neural networks. Neural networks consist of many simple processing nodes that are arranged in layers to learn weights between connections through several iterations of training data. In MoTDeReL we use a Double Deep Q-Network (DDQN) employing a multi-layer perceptron with three layers: an input layer, a hidden layer, and an output layer. The activation functions for the hidden layer and the output layer are the rectified linear unit and IDENTITY, respectively. The number of neurons in the hidden layer is set to two-thirds of the number of neurons in the input layer, as this configuration has shown optimal results in many case studies [24]. The number of neurons in the output layer is a variable parameter and will match the maximum number of applicable rules of the running model.

4.2 Parameter Analysis

There are different parameters in the DRL that should be analysed and initialised to determine the proper values for them. In this subsection, the most important parameters are explained:

The parameter ϵ is used for the ϵ-greedy algorithm. Its value is a number between 0 and 1. The higher its value, the closer the agent's behaviour is to random selection of actions. A low value for ϵ causes the agent to rely more on the recommendations of the neural network (exploitation) rather than exploring random actions. This means that the agent behaves more purposefully according to its learned policy.

In DRL, the value of ϵ is decreased in each step to balance exploration and exploitation. As ϵ decreases, the agent explores less and takes more advantages of learned experiences. Decreasing ϵ at each step ensures sufficient exploration early on, while gradually shifting to exploitation later in the process. There is another parameter that determines the rate at which ϵ is reduced over time. This parameter controls the reduction in ϵ in each step, gradually changing the agent's behaviour from exploration to exploitation. The larger this parameter, the earlier ϵ reaches zero. Small values are usually used for this parameter.

The learning rate directly affects the speed of convergence. Values of 0.1, 0.001 and 0.00001 have been obtained and validated in various experiments. With high learning rates, the algorithm may converge to a local minimum, while low learning rates can result in slow convergence. To achieve optimal results requires the use of an adaptive optimiser. One commonly used parameter for this purpose is Root Mean Square Propagation (RMSProp).

The number of actions (rules) is another important parameter to consider, as it defines the maximum number of output transitions a state can have in determining the number of neurons in the output layer of the neural network. It should be defined according to the problem setting. In many large case studies, more than 200 actions may be required.

When an agent relies solely on its previous experiences to learn, it can lead to an unstable system where the agent forgets its past experiences. To prevent this, the agent's experiences should be stored in its experience-replay memory. Its size must be carefully chosen: A larger size increases storage usage but allows the agent to retain more past experiences, while a smaller size may cause the agent to forget older experiences and rely more heavily on recent ones.

The batch size of the training dataset determines how many samples are processed in each neural network training iteration. A larger batch size can enhance prediction accuracy, but requires more time and storage, while a smaller batch size may lead to poor future predictions. Therefore, balancing speed and accuracy is crucial.

4.3 Reward Engineering

Rewards management is crucial in reinforcement learning framework, as they guide the agent's decision-making process by providing feedback on the consequences of its actions. To design a more accurate reward function, we should assign rewards to each action (rule) based on the relative worthiness or impact of that action in achieving the desired goal.

In our MBT, the goal is to generate a test suite containing test cases with the highest coverage and shortest length while minimising time and effort. For an accurate reward function, these parameters must be considered; however, the nature of DRL allows us to eliminate some of them. For the coverage score of generated test cases, in each step we count the data dependencies between the rules and return this number as a reward. The more data dependencies a test case covers, the higher its reward. Based on this reward function, the DRL agent aims to select rules that maximise dependency coverage.

For the length of the generated test cases, we initialise the discount factor (γ) value accordingly. The discount factor is an important parameter in reinforcement learning that influences the agent's valuation of future rewards, the balance between exploration and exploitation, and the stability of the learning process. It determines the degree to which agents prioritise long-term rewards over short-term ones. If $\gamma = 0$, the agent will be completely myopic and learn about actions that produce an immediate reward. And if $\gamma = 1$, the agent will evaluate each of its actions based on the total sum of all future rewards. If we choose a value near 1 for γ, the behaviour of the neural network will lead to a longer test case, and if the value was near zero, the length of the test case will be very short. According to these explanations, we choose the value for γ as 0.95 to prevent the agent from generating very short test cases with low coverage scores.

Algorithm 1. The Reward Function algorithm

1: **Input**: Current state of the model = sequence of selected rules in the state space
2: **Output**: R = reward of the current state
3: **Initialise** *dependencies* = {} // The pair of dependent rules as C_R, C_D, C_U, D_N, U_R, and U_U
4: **Extract** dependencies of rules
5: **Delete** redundant pairs from dependencies
6: **Set** R as size of dependencies
7: **Return** R

For the time effort involved in the test-case generation, there is no need to parametrise the reward function, as the nature of DRL enables the agent to quickly learn the test case generation process and produce suitable test cases with minimal time effort. Although most rule selections in the early stages of DRL are random, after several steps, the neural network is appropriately weighted, and rule selection becomes more deliberate.

According to the above explanation, we set the reward function to calculate the data-dependency values of the rules. The reward function used for MoT-DeReL is presented in Algorithm 1. The input of the algorithm is the sequence of selected rules in the state space as current state, and the output is the reward value for the current state of the selections. In line 3 of the algorithm, the set of dependencies is initialised. Then, in line 4, we extract the data-dependency relation of selected rules. Dependencies is a set of pairs of rules which have one type of data-dependency relation as C_R, C_D, C_U, D_N, U_R and U_U. Then in line 5, we delete repetitive pairs and in line 6 count the number of pairs in dependencies and set it as the value for the reward R. In other words, this reward value is the fitness function of the current state of the agent in state space.

4.4 Feature Selection

Feature selection is the process of designing a model that is easier to understand and interpret by humans [8]. In large state spaces with numerous features, identifying the most relevant features serves to reduce the dimensionality of the feature space and improve model efficiency.

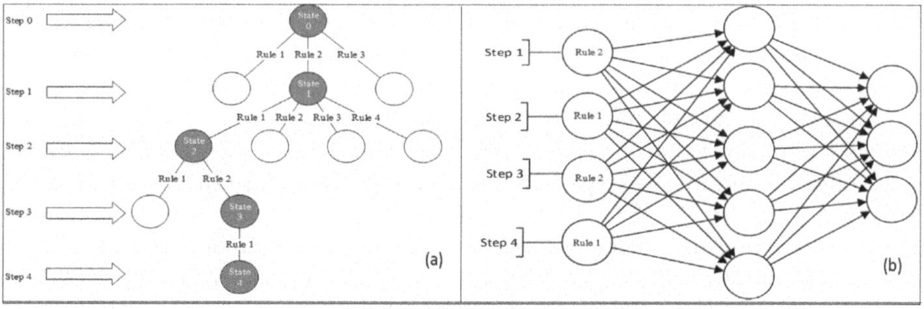

Fig. 2. An example for feature selection in MoTDeReL

In MoTDeReL, the inputs to the neural network correspond to the sequence of rules applied, from the initial state to the current state. The number of inputs is also equal to the maximum number of steps the agent can take in the environment. For example, if the agent is allowed a maximum of 4 steps per episode, the neural network will have 4 inputs, all initialised to 0 at the beginning. The maximum number of steps allowed for the agent is determined by the maximum length of the test case, which is specified in the input parameters by the user. The user usually specifies it according to the input model and the maximum length to cover more dependencies.

An example is shown in Fig. 2 where the agent is allowed to take 4 steps, resulting in 4 inputs for the neural network. Suppose that the agent selects the rules in the following sequence: 2, 1, 2, and 1 (Fig. 2(a)). Consequently, the first input to the neural network becomes 2 in the first step. In the second step, the second input is updated to 1. The third input changes to 2 in the following step. In the final step, the fourth input changes to 1, after which the agent cannot proceed further. Therefore, the final inputs to the neural network are 2, 1, 2, and 1, which match the order of selection of the rules shown in Fig. 2(b). As explained previously, the number of neurons in the output layer is determined by the user to cover all applicable rules of the model. In this example, the output layer has 3 neurons, which means that the network can estimate one of three rules for the next movement.

4.5 Handling Invalid Actions

In MoTDeReL the agent in each step selects one rule according to the recommendation of the neural network. Sometimes, this rule is not applicable to the current state. These types of actions are referred to as invalid actions. To address this issue, a large output layer is created. After filtering the outputs using applicable rules, the highest Q-value is selected.

4.6 Double Deep Q-Network Implementation

The idea of MoTDeReL is illustrated in Algorithm 2 and Algorithm 3. In Algorithm 2, the next move execution on state space is presented. At first, in line 1, the *epsilon* parameter, the maximum steps for updating the target network (*MSUTN*), the maximum number of repetitions in a test case (*MFR*), the current state of the model and *allRewardsUntilNow* array are received. The user specifies the initial value for *epsilon* parameter of DRL. The *MSUTN* parameter specifies the number of steps required to regularly update the target network. The *MFR* is used to finalize the test case if the fitness value does not improve after a series of identical fitness evaluations. The *allRewardsUntilNow* array is an array which contains all gathered rewards until now in prior episodes. It will be updated in this algorithm. The algorithm finally will return the updated current state i.e., the next traversed state. In line 3, the experience replay memory $\mathcal{D}$ with capacity $\mathcal{n}$, the main neural network Q_θ, the target neural network $Q_{\theta'}$ with weight 0 and batch size β are initialised. The algorithm starts with the extracting the applicable rules of the current state in line 4 denoted as *rules*. Then, according to the *ϵ-greedy* algorithm, a random number between 0 and 1 is generated as *rand*. Based on the value of *rand* and its comparison to the *epsilon* an action (rule) is selected. If the rand value is lower than the *epsilon*, the action (rule) will be selected randomly (line 7). Otherwise, it is predicted using the main neural network (line 9). In line 11, the selected action is performed, the next state is explored. The reward of applied action on the current state will be calculated through Algorithm 1 in line 12. In line 13, the agent stores the last step as an experience in the experience replay memory $\mathcal{D}$. In line 14, if the size of the $\mathcal{D}$ is equal or greater than the batch size β, a random batch of experiences is retrieved from $\mathcal{D}$ and used to train the main neural network. As stated in lines 17 to 21, if the next state is a terminal state, the target value will be equal to the reward amount. Otherwise, the main neural network selects the best action for the next state, and the target neural network is used to estimate its Q-value, which is then added to the reward value. Here, the target value refers to the value used to update the Q-value of a state-action pair during training. For each value of the loop, the algorithm performs a gradient descent with respect to the network parameters. In line 24, if the steps value equals *MSUTN*, the weights of the two neural networks (main and target) are synchronized. In lines 28 and 30, the agent takes one step and moves to the next state and reduces the value of *epsilon*. In line 33, it counts the number of repeated rewards (e.g., *fitness*) in the path (e.g., *allRewardsUntilNow*) and if this number is greater than the MFR, the loop is broken. This is done to return the test case if the fitness does not improve after a number of repetitions. Finally, the Algorithm returns the updated current state at line 36.

Algorithm 3 presents the algorithm of MoTDeReL approach. In this algorithm, first, the GTS model of the system under test is provided as the input model. In addition, the number of algorithm episodes (M), the number of steps of the agent per episode (N), the maximum number of selected test cases (MST) as input parameters is received from the user. For each episode, the algorithm generates a test case and, for each step of the episode, it traverses a transition of the state space or performs an action. The MST determines the total number of test cases that are ultimately generated. In line 2, the output of the algorithm is determined which is the finally generated test cases. In line 3, the selected test case array is initialised. The first loop of the algorithm in line 4 is used to run multiple episodes. For each episode, the initial state of the input model is taken as s_t and *allRewardsUntilNow* is initialised. *allRewardsUntilNow* array will be used to store all the rewards (fitness) achieved for different test cases. In line 7 the *for* loop is used to run multiple steps. At each step, a movement on the state space will occur, representing an application of a rule which changes the current state to the next one. This is done using Algorithm 2 (line 8). The execution of Algorithm 2 leads to a sequence of states as a TC. After finishing the episodes or breaking out of the loop in line 10, the proposed approach checks if the optimal number of test cases has been reached. If so, the test case with the lowest fitness in the *selectedTestCases* array is replaced; otherwise, the new test case is added to *selectedTestCases*. Finally, in line 16, after termination of all episodes, the *selectedTestCases* is returned as the output.

5 Evaluation[1]

In this paper, we propose an approach based on deep reinforcement learning to test the software models specified using GTS. The MoTDeReL approach has two main objectives: first, to generate a test suite that maximises the coverage of the test objectives, and second, to develop a cost-aware test generation methodology that addresses the state space explosion problem in complex systems. To evaluate our solution against these objectives, we implemented MoTDeReL using the Groove toolset and assessed its coverage score and cost-effectiveness. In the following, we consider three research questions and answer them based on our findings.

[1] The Groove Toolset Version Designed for This Research Along with the Necessary Case Studies Are Available at https://github.com/SGhasemi1987/MoTDeReL.

Algorithm 2. The Test Case Generation Algorithm

Input: ϵ as epsilon, $MSUTN$ as Maximum steps for update target network, MFR as Maximum number of fitness repetition for a test case, s_t as current state, $allRewardsUntilNow$ array
Output: updated s_t
Initialise experience replay memory $\mathcal{D}$ to capacity n, primary network $\mathbf{Q}_\theta$ and target network $\mathbf{Q}_{\theta'}$ with weights 0, batch size β
Extract rules as the list of applicable rules for s_t
Generate a random number as $rand$
if $rand$ is lower than ϵ **then**
    Select a random action a_t from 0 to count($rules$)
else
    Select $a_t = \arg\max Q_\theta(s_t)$ to count of rules
end if
Execute a_t, explore the next state s_{t+1}
Set reward $r_t = R(s_t, a_t)$ // Algorithm 1
Store (s_t, a_t, r_t, s_{t+1}) in experience replay memory $\mathcal{D}$
if size($\mathcal{D}$) $\geq \beta$ **then**
    Select batch = Sample random batch from $\mathcal{D}$
    for each (s_j, a_j, r_j, s_{j+1}) in batch **do**
        if s_{t+1} is a Terminal State **then**
            $Target = r_j$
        else
            $Target = r_j + \gamma Q_{\theta'}(s_{j+1}, \arg\max_a Q_\theta(s_{j+1}, a))$
        end if
        Perform a gradient descent step on $(Target - Q_\theta(s_{j+1}, a))^2$
    end for
    if $steps$ equals $MSUTN$ **then**
        Reset $Q_{\theta'}$ weights = Q_θ
    end if
end if
Set $s_t = s_{t+1}$
if $\epsilon > threshold$ **then**
    Reduce ϵ
end if
Store fitness in allRewardsUntilNow
if count(r_t in $allRewardsUntilNow$) $> MFR$ **then**
    Break
end if
Return s_t

There are several approaches in the literature that use MBT for graph transformations. The most relevant are HGAPSO [23], GA [3], and WholeAcoT [14]. Hence, we compared our proposed approach with these approaches.

We first describe the experimental setup. Then we evaluate the performance and cost of MoTDeReL in comparison with similar approaches and provide answers to the research questions.

5.1 Experimental Setup

We implemented MoTDeReL using Groove. Although primarily a model checker designed for verifying graph transformation systems, we adapted Groove to generate effective test cases and test suites. The experiments were carried out on a PC with an Intel Core (TM) i7-8565U CPU at 1.80 GHz and 16 GB of memory. In this experiment, we used data dependencies including C_R, C_D, C_U, D_N, U_R and U_U as our coverage criteria. As stated in Subsect. 2.2, data-dependency gives a suitable estimation of the def-use relations among rules, and this is the motivation to reveal the bugs of the system under test. In order to specify suitable values for these parameters, we ran the experiment multiple times and chose the optimal values. The MST and step values that determine the maximal number and length of test cases, respectively, are variable for each case study.

The parameter ϵ is initially set to 0.2, allowing the agent to take some random actions. To gradually reduce randomness over time, we apply a decay factor of 0.995, ensuring that the epsilon decreases at each step. This approach progressively shifts the agent from exploration to exploitation.

The learning rate, a key parameter in neural network training, is set to 0.001, a commonly used value that facilitates a fast convergence to a global optimum. Additionally, we chose RMSProp as the optimiser to effectively regulate the learning rate.

After evaluating various case studies, we determined that a maximum action output parameter of 400 is optimal. This value is sufficiently large to accommodate most models while maintaining efficiency.

According to the explanation in Subsect. 4.3, we set the discount factor to 0.95. The agent utilises this value as part of its decision-making process. Addi-

Algorithm 3. The MoTDeReL Approach

Input: Model as A GTS model of the system under test, N = Steps, MST as Maximum number of selected test case
Output: SelectedTestCases
Initialise SelectedTestCases array
for *episodes* = 1 to M **do**
  s_t = initial state of the Model state space
  Initialise allRewardsUntilNow array
  for *steps* = 1 to N **do**
    s_t = **NextMoveExecution**(s_t) // Algorithm 2
  end for
  if count($selectedTestCases$) > MST **then**
    Replace s_t with the lowest fitness testcase in selectedTestCases
  else
    Add it to SelectedTestCases
  end if
end for
Return SelectedTestCases

tionally, we set the memory size for the experience replay to 1000 to achieve a balanced trade-off between memory usage and learning stability. In addition, the batch size is set to 8 to achieve a balance between speed and accuracy. A very high batch size increases computational time and storage requirements, so we chose this value to effectively manage these trade-offs. In addition, we set the *MSUTN* value to 5 to update the target network as often as possible. The summary of the parameter setting for the experiments is listed in Table 1.

Table 1. Parameters of MoTDeReL

Parameter	Value
episode	10
step	variable
MST	variable
MFR	5
epsilon	0.2
decrease value of epsilon	0.995
learning rate	0.001
adaptive optimiser	RMSProp
Max action output	400
Max steps for update target network	5
discount factor	0.95
experience replay memory size	1000
batch size	8

Table 2. Rule Numbers, Step and MST for each case study

Case Study	Rule Numbers	MST	Step
DPs	6	1	20
HMS	7	1	20
OSS	20	4	30
BTS	34	7	15
TAS	43	7	15
SCRP	63	10	30

For the purpose of evaluation, some well-known case studies are selected from the literature, including the Dining Philosophers problem (DPs) [31], the Hotel Management System (HMS) [19], the Bug Tracker System (BTS) [29], the Online Shopping System (OSS) [11], the Travel Agency System (TAS) [34] and the Scanflow Cash Register Protocol (SCRP) [6]. The case studies have different sizes but all of them can encounter state explosion due to the use of a fairly large start graph. Table 2 presents the number of rules for each case study, serving as a metric of their relative sizes. Although the DPs and HMS case studies are relatively small, they are fundamental examples commonly used in the literature to compare different strategies. To incorporate more realistic scenarios, we included BTS, TAS, OSS, and SCRP as real-world case studies. Also, We have determined different values for step and MST parameters with respect to each case study. These values were chosen to generate applicable test suites. In Table 2, the step and MST values are listed for each case study.

5.2 Experimental Results

To compare the performance of MoTDeReL against other strategies [3,14,23] we executed each method under similar conditions ten times and then calculated the average results. The experimental results are shown in Table 3. The first column lists the case studies, while the second, third, and fourth columns show the results for each approach, including the coverage score and test size. The coverage score is the total number of rule dependencies covered in the generated test cases, while the test size is the sum of the length of each test case in the resulting test suite.

Based on the values in Table 3, in the rest of this section, we analyse the data and address the research questions.

Q1: How much has MoTDeReL improved the coverage score compared to the state-of-the-art? The blue numbers in Table 3 are the results of the coverage values. The highlighted values in the table indicate the best results for each case study. A comparison of these values shows that the MoTDeReL approach achieves a higher coverage of the test objectives compared to GA, HGAPSO, and WholeAcoT. As the size of the case study increases, the difference in coverage scores achieved by the MoTDeReL approach becomes more significant. The HGAPSO approach produces the lowest coverage scores in most cases, whereas the GA approach achieves better coverage than HGAPSO, although still lower than both WholeAcoT and MoTDeReL. Although the coverage score of the WholeAcoT approach is higher than that of GA and HGAPSO, MoTDeReL outperforms all other approaches in terms of coverage.

Q2: How much has MoTDeReL reduced testing costs compared to the state-of-the-art? In software testing, cost-effective tests are essential to ensure an efficient testing process and developing high-quality software. When a smaller test case reveals more defects, its cost becomes significantly lower compared to larger tests with the same output. We aim for tests to achieve high objective coverage while remaining compact in size.

Table 3. Comparison of average results for coverage and test size for different approaches

Approaches → **Case Studies ↓**		**HGAPSO**[23]	**GA**[3]	**MoTDeReL**	**WholeAcoT**[14]
DPS	**Coverage**	2.0	17.1	18.2	17.5
	Test Size	19.3	29.3	19.9	17.3
HMS	**Coverage**	8.0	33.9	54.3	34
	Test Size	18.7	32.8	19.5	18
BTS	**Coverage**	49.2	74.5	95.5	85.9
	Test Size	105.1	79.0	88.4	29.2
OSS	**Coverage**	46.0	122.9	223.2	207.8
	Test Size	103.6	74.3	119.0	113.6
TAS	**Coverage**	55.7	203.0	370.7	286.7
	Test Size	103.2	57.5	83.0	99.9
SCRP	**Coverage**	108.0	390.8	981.6	457.1
	Test Size	292.3	193.0	256.5	207.5

One commonly used metric for evaluating approaches is the length of the generated test cases. The black numbers in Table 3 indicate the size of the generated test cases for each approach, with the smallest test size in each case study is highlighted. For test size, we calculated the total size of the final generated test suite, which includes multiple test cases. Based on the black values in Table 3, the test cases generated by the MoTDeReL approach are not consistently shorter in most cases. However, upon closer examination, tests with shorter lengths cover significantly fewer test objectives, making their reduced size less beneficial for these approaches. To illustrate this point, we define the concept of testing cost as an efficiency metric [30] that describes the capacity of the specified test to fulfill the test objectives along with its size which affects the testing process:

Definition 2 ***testing cost:*** *Given a test suite TS with length of L and coverage value of CV, the testing cost is defined as* $\frac{L}{CV}$.

Figure 3 represents the testing cost of generated tests associated with HGAPSO, GA, WholeAcoT and MoTDeReL. The goal is to have smaller test suites with higher coverage values, resulting in lower testing costs. According to this figure, the proposed approach generates test suites with appropriate costs in most case studies, making it more suitable for complex problems. Although in DPs and BTS, the cost of WholeAcoT is lower than that of MoTDeReL, its low coverage level in these case studies outweighs this advantage.

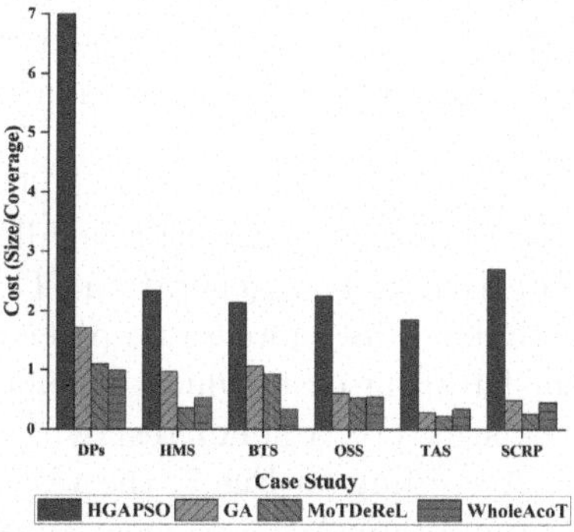

Fig. 3. Comparison of testing Cost for MoTDeReL, HGAPSO [23], GA [3] and WholeAcoT [14] in different case studies

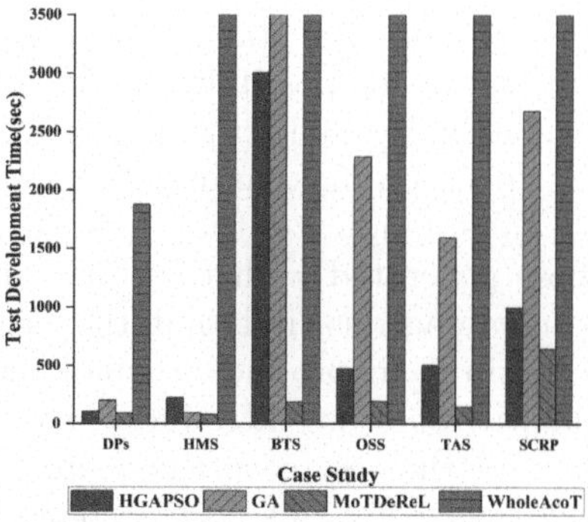

Fig. 4. Comparison of test development time for MoTDeReL, HGAPSO [23], GA [3] and WholeAcoT [14] in different case studies

Q3: How does MoTDeReL works in terms of time compared to state-of-the-art methods? A key criterion for assessing the efficiency of a new strategy is the time required to complete the test design and development process. This time represents the cost of preparing the test, and it is desirable to generate tests in the shortest possible time.

Figure 4 shows the time required for test generation by each approach for all case studies. As shown in the figure, the MoTDeReL approach is the most time-efficient compared to all other approaches in every case study. Although the time difference between the MoTDeReL approach and both GA and HGAPSO is not very significant for small test cases, it becomes particularly notable for large and complex models. The time complexity of WholeAcoT is excessively high, making it an inefficient approach for complex case studies.

5.3 Discussion

In this section, we discuss the experimental results to validate the objectives of our test suite generation approach and address some potential weaknesses. Table 3 shows that MoTDeReL generates test cases and test suites that cover more test objectives and can detect more defects in the models. This performance is attributed to the well-trained neural network within the DRL structure. In MoTDeReL, the agent traverses the state space and receives rewards or penalties based on the coverage score of its actions. The neural network leverages this feedback to distinguish between effective and ineffective actions, thereby refining its decision-making process for future moves. This iterative learning mechanism allows the network to progressively improve its performance by prioritising actions that produce favourable outcomes while avoiding those that do not. Since these moves are guided by neural network recommendations, the exploration of the state space is conducted heuristically, enabling the discovery of paths with extensive coverage. This approach ensures that the system efficiently navigates through potential states, prioritising areas of interest while maintaining a broad exploration of possibilities. As a result, the neural network can identify and exploit high-value trajectories, balancing exploration and exploitation to optimise decision making over time.

Table 3 indicates that the tests generated by MoTDeReL are not smaller than those generated with the other approaches. Although generating smaller tests is advantageous, it is not the primary objective of test generation. The ideal scenario is to produce a small test suite that covers more test objectives. To quantify this, we introduced a metric called testing cost, which calculates the ratio of the test coverage score to the test size. For a small test with a high coverage score, this cost becomes smaller. According to Fig. 3, the test cost of MoTDeReL is lower than HGAPSO, GA, and WholeAcoT in most cases. This suggests that MoTDeReL generates an efficient test suite, improving the test objective coverage while maintaining a shorter length.

The data in Fig. 4 show that test generation in MoTDeReL is faster than other approaches. Since HGAPSO, GA and WholeAcoT utilise multiple generations to produce semi-random test cases and test suites, they require more time to produce the final generation. In contrast, MoTDeReL uses a previously trained neural network to select actions in new iterations, leading to improved test cases in terms of coverage score.

From the experimental results and discussions presented, we can conclude several advantages of the MoTDeReL approach:

- Generates acceptable test suites with high coverage compared to related approaches.
- Produces effective test suites with low testing cost relative to other methods.
- The runtime of MoTDeReL is significantly lower than that of other approaches.
- MoTDeReL improves with each iteration as it learns, progressively improving its performance until it achieves full coverage.
- Due to the nature of DRL, the results of each learning cycle can be stored and applied to larger or smaller host graphs, eliminating the cold start issue.

However, there are some weaknesses that should be addressed in future work.

- The length of generated tests could be shorter in certain models or case studies.
- In limited-time systems, a zero-knowledge agent may initially produce suboptimal tests due to the time required for DRL to acquire sufficient knowledge.
- Given MoTDeReL's reliance on rewards and penalties, even minor changes to the reward function can have significant effects.

6 Conclusion

We proposed an approach that employs DRL to generate test cases and test suites for systems specified by graph transformation. This approach traverses the state space to explore desired test paths that cover a large number of test objectives. It utilises the penalty/reward mechanism of DRL to assess the effectiveness of selected actions and adjusts the neural network's weights based on this feedback. The next action is then selected using the neural network's recommendations.

We implemented the approach alongside other related methods in the Groove toolset and evaluated them on the basis of coverage score, testing cost, and test generation time. The experimental results indicate that our approach improves the testing process by generating test cases with a high coverage score at a lower cost and in significantly less time. Despite these improvements, there are areas for future work:

- Instead of randomly retrieving experiences, we can employ more intelligent techniques, such as prioritised experience replay.
- Although double DQN demonstrate strong overall performance in DRL, alternative methods such as the Actor-Critic method could also be explored.

Reward functions are crucial in every reinforcement learning algorithm, so optimising them could yield further enhancements.

7 Appendix: Graph Transformation in Groove

In this section, we present a Hotel Management System (HMS) originally introduced in [19] revised in [3] for MBT to recall concepts of GTS using the Groove tool set. The revised HMS includes multiple rooms and registered guests within a hotel. Each guest has the option to reserve at most one available room, leading to the generation of the relevant bill. Before proceeding to checkout, the guest must settle the bill. Figure 5 shows a simple start graph of the HMS in Groove. Figure 6 shows two sample rules of the HMS, which include BookRoom and CheckOut. In this figure, a solid black line denotes readers, blue dashed lines symbolise erasers, green solid lines represent creators, and forbidden elements, constituting negative application conditions, are indicated by red dashed lines. The complete description of this example and its rules is given in [4].

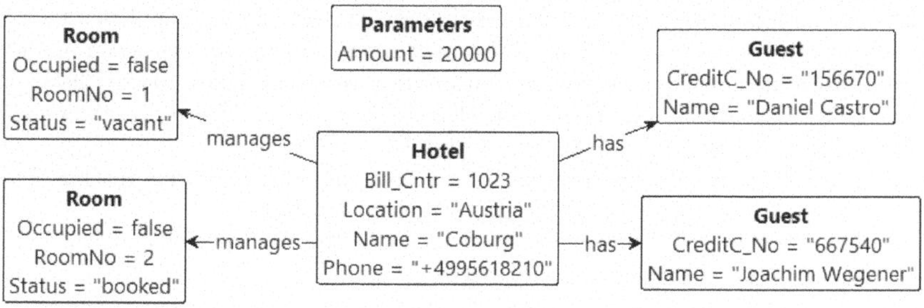

Fig. 5. A start graph for the HMS [4]

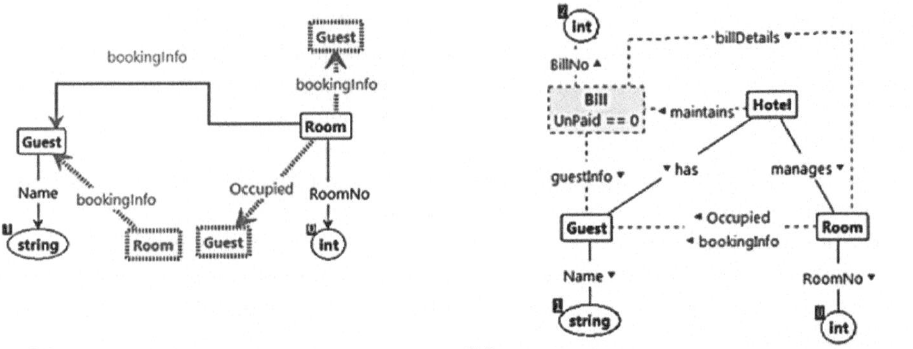

(a) *BookRoom(int RoomNo, String Name)* (b) *CheckOut(int RoomNo, String Name, int BillNo)*

Fig. 6. Two sample rules of the HMS [4]

Disclosure of Interests. The authors have no competing interests to declare that are relevant to the content of this article.

References

1. Anand, S.: An orchestrated survey of methodologies for automated software test case generation. J. Syst. Softw. **86** (1978)
2. Asgari Araghi, M., Rafe, V., Khendek, F.: Using data mining techniques to generate test cases from graph transformation systems specifications. Autom. Softw. Eng. **31**(1), 17 (2024)
3. Bahrampour, A., Rafe, V.: Using search-based techniques for testing executable software models specified through graph transformations. Int. J. Mach. Learn. Cybern. **11**(12), 2743–2770 (2020). https://doi.org/10.1007/s13042-020-01149-9
4. Bahrampour, A., Rafe, V.: Using memetic algorithm for robustness testing of contract-based software models. Artif. Intell. Rev. **54**, 877–915 (2021)
5. Baier, C., Katoen, J.P.: Principles of model checking (2008)
6. Bruijn, V.: Model-based testing with graph grammers. Ph.D. thesis, University of Twente (2013)
7. Dorigo, M., Birattari, M., Stutzle, T.: Ant colony optimization. IEEE Comput. Intell. Mag. **1**, 28–39 (2006)
8. Duboue, P.: The art of feature engineering: essentials for machine learning (2020)
9. Eberhart, R., Kennedy, J.: Particle swarm optimization. In: Proceedings of the IEEE International Conference on Neural Networks, pp. 1942–1948 (1995)
10. Ehrig, H., Rozenberg, G., Kreowski, R.H.J.: Handbook of graph grammars and computing by graph transformation **3** (1999)
11. Engels, G., Güldali, B., Lohmann, M.: Towards model-driven unit testing. Models Softw. Eng. Workshops Symp. MoDELS **9**, 182–192 (2006)
12. Fraser, G., Wotawa, F., Ammann, P.: Issues in using model checkers for test case generation. J. Syst. Softw. **82**, 1403–1418 (2009)
13. Fraser, G., Wotawa, F., Ammann, P.: Testing with model checkers: a survey. Softw. Test. Verification Reliab. **19**, 215–261 (2009)

14. Ghasemi, S., Rafe, V., Bahrampour, A., Heckel, R.: Whole test suite generation from graph transformation specifications using ant colony optimization. Softw. Qual. J. **33**, 1–37 (2025)
15. González, C., Varmazyar, M., Nejati, S., Briand, L., Isasi, Y.: Enabling model testing of cyber-physical systems. In: Proceedings of the 21th ACM/IEEE International Conference on Model Driven Engineering Languages and Systems, pp. 176–186 (2018)
16. Gönczy, L., Heckel, R., Varró, D.: Model-based testing of service infrastructure components. Testing of Software and Communicating Systems: 19th IFIP TC6/WG6. 1 International Conference, pp. 155–170 (2007)
17. Hasselt, H.V., Guez, A., Silver, D.: Deep reinforcement learning with double qlearning. In: Proceedings of the AAAI Conference on Artificial Intelligence (2016)
18. Haupt, R., Haupt, S.: Practical genetic algorithms (2004)
19. Heckel, R., Khan, T., Machado, R.: Towards test coverage criteria for visual contracts. Electron. Commun. EASST **41** (2011)
20. Hong, H.S., Lee, I., Sokolsky, O., Ural, H.: A temporal logic based theory of test coverage and generation. In: Katoen, J.-P., Stevens, P. (eds.) TACAS 2002. LNCS, vol. 2280, pp. 327–341. Springer, Heidelberg (2002). https://doi.org/10.1007/3-540-46002-0_23
21. Jackson, D.: Software Abstractions: logic, language, and analysis. MIT Press (2012)
22. Jadhav, P., Lanke, G., Shrivastava, A., Patil, V., Gupta, S.: Efficient test case generation using model-based testing, and model paradigm approach. In: International Conference on Recent Trends in Computing, pp. 817–828 (2023)
23. Kalaee, A., Rafe, V.: Model-based test suite generation for graph transformation system using model simulation and search-based techniques. Inf. Softw. Technol. **108**, 1–29 (2019)
24. Mehrabi, M., Rafe, V.: Using deep reinforcement learning to search reachability properties in systems specified through graph transformation. Soft. Comput. **26**, 9635–9663 (2022)
25. Mohalik, S., Gadkari, A., Yeolekar, A., Shashidhar, K., Ramesh, S.: Automatic test case generation from simulink/stateflow models using model checking. Softw. Test. Verification Reliab. **24**, 155–180 (2014)
26. Nabuco, M., Paiva, A.: Model-based test case generation for web applications. In: International Conference on Computational Science and Its Applications, pp. 248–262 (2014)
27. Rafe, V., Mohammady, S., Cuevas, E.: Using bayesian optimization algorithm for model-based integration testing. Soft. Comput. **26**, 3503–3525 (2022)
28. Rensink, A.: The groove simulator: a tool for state space generation. Appl. Graph Transformations Ind. Relevance: Second Int. Workshop **2**, 479–485 (2003)
29. Runge, O., Khan, T., Heckel, R.: Test case generation using visual contracts. Electron. Commun. EASST **58** (2013)
30. Schieferdecker, I., Hoffmann, A.: Model-based testing. IEEE Softw. **29**(1), 14–18 (2012)
31. Schmidt, A., Varró, D.: CheckVML: a tool for model checking visual modeling languages. Int. Conf. Unified Model. Lang. 92–95 (2003)
32. Sulaiman, R., Jawawi, D., Halim, S.: Cost-effective test case generation with the hyper-heuristic for software product line testing. Adv. Eng. Softw. **175**, 103335 (2023)
33. Sutton, R.: Reinforcement learning: An introduction (2018)

34. Thöne, S.: Dynamic Software Architectures: A Style-Based Modeling and Refinement Technique with Graph Transformations. Ph.D. thesis, University of Paderborn (2005)
35. Utting, M., et al.: Recent advances in model-based testing. Adv. Comput. **101**, 53–120 (2016)
36. Yang, Y.: Improve model testing by integrating bounded model checking and coverage guided fuzzing. Electronics (Basel) **12**, 1573 (2023)

Applications of Graph Transformation for Modeling

Graph-Transformational Threat Modeling

Lars Friederichs(✉) and Aaron Lye(✉)

German Aerospace Center DLR, Institute for the Protection of Maritime Infrastructures, Bremerhaven, Germany
{lars.friederichs,aaron.lye}@dlr.de

Abstract. Many well-established threat modeling approaches use graphs as underlying structures. However, the formalism of graph-transformation is not applied. In this paper, we propose a graph-transformational approach to threat modeling. We take a more detailed look at a transformation from attack-defense trees to stochastic games. This leads us to further research questions that we believe are suitable to be answered in the language of graph transformations.

1 Introduction

In order to develop secure systems by finding and mitigating vulnerabilities, threat-modeling methods are used to create an abstraction of the system, profiles of potential attackers, including their goals and methods, as well as a set of potential threats which may arise. Over the years many threat-modeling methods have been developed (cf., e.g. [1–4] for an overview). Many well-established threat modeling approaches utilize graphs as underlying structures. However, the formalism of graph-transformation is not applied.

In this paper we start with the investigation of benefits of using graph transformations for security research. This paper focuses on threat modeling and more specifically, on attack defense trees (ADTrees) and their transformations into stochastic security games. This model transformation is not new but based on prior work. Our main contribution is to formalize the algorithms presented in [5] with graph transformations, which we believe to be a more appropriate language to represent this since ADTrees as well as stochastic games use graphs as underlying structures. Furthermore, our formalization unlike the presentation in [5] takes the compositional nature of trees into account and makes the effect a single operation has on the resulting game more clear.

The rewriting formalism which we use throughout this paper is the well-established double-pushout approach as introduced by Ehrig, Pfender and Schneider in [6]. For the notion of graph transformation confer, e.g., Ehrig et al. [7,8]. Our graph transformations are actually over attributed typed graphs. However, we omit some of the details for simplicity. Moreover, in the following we use the concept of graph-transformation units (confer Kreowski et al. [9]) to specify initial and terminal graph classes and control conditions by extended regular expressions to restrict rule applications. The established notions of graph

J. Endrullis and M. Tichy (Eds.): ICGT 2025, LNCS 15720, pp. 205–217, 2025.
https://doi.org/10.1007/978-3-031-94706-3_10

transformation provide a framework with a formal semantics, a strong notion of concurrency wrt rule applications, non-determinism of matches, the ability to analyze complexity and runtime of algorithm by means of finding matches in the host graph and the length of derivations.

The model transformation from ADTrees into a stochastic games is just one example of various which can be found in the literature of threat modeling. Often graphs are used as underlying structures which are evaluated or transformed using a set of rules. However, the formalisms used are not uniform and to the best of our knowledge never use established approaches of graph transformation such as the double-pushout approach, single-pushout approach, hyperedge replacement, node replacement, etc. (see e.g. [10]). Therefore, the approaches and transformations are difficult to relate to each other and not straightforwardly compositional. Security research is of high interest and applications of graph transformation in this direction can contribute to this area. Threat modeling in particular could be a relevant avenue for such efforts, and the contribution of this paper could be seen as a first step in bridging threat modeling and graph transformation research.

The paper is organized as follows. In Sect. 2 we discuss some preliminaries on attack trees and dynamic variations, in particular, attack-defense trees as well as stochastic two player games. In Sect. 3 we discuss the model transformation from attack-defense trees to stochastic games. In Sect. 4, we show an attempt at formalizing a size reduction for graphs resulting from the transformation. This leads us to further research questions that we believe are suitable to be answered in the language of graph transformations. Further research directions are discussed in Sect. 5.

2 Attack Trees, Attack-Defense Trees and Stochastic Two Player Games

In this section we recall attack trees, some dynamic variations of attack trees as well as the basic notion of attack-defense trees and stochastic two player games.

2.1 Attack Trees and Dynamic Variations of Attack Trees

Attack trees (in several variants) are the usual approach for expressing possible ways of an attack exploiting a vulnerability of a system (cf. e.g. [11] for an overview). Attack trees are usually atomic propositions, i.e. Boolean-valued expressions formed from variables, constants and predicate symbols. The propositions are represented as a tree in order to represent the refinement of the statement. The non-terminal vertices represent predicate symbols and the leafs represent (atomic) actions of the attacker. The edges relate the statements. Ingoing edges of a set of child vertices are usually considered as alternatives ($\vee$) or joint condition ($\wedge$). The first formal model of attack trees introduced by Schneier in [12] aimed at describing a possible attack over a system by refining the main attack goal into sub-goals using either an operator $\vee$ or an operator $\wedge$ to coordinate those refinements. Today usually probabilistic trees are considered, where actions are assigned probabilities and possibly costs. Since their introduction attack trees have been an active research area and several variations have been proposed. The attack tree model is also a subject of increasing attention in the community of formal methods with a lot of different formal approaches (see [13] for a survey).

The analysis conducted over attack trees is "static" in the sense that the attacked system does not evolve during the attack. As such, there is no concept of a goal happening before or after another. To represent this different approaches exist in the literature. In [14] a path semantics for attack trees over a transition system is proposed. Temporally ordered variants of the logical operators were introduced following [15] denoted by ($\overrightarrow{\wedge}$) for the sequential conjunction and in later sources ($\overrightarrow{\vee}$) for the sequential disjunction. Here an order is imposed on the outgoing edges with the implied meaning of such an operation usually being that it only makes sense to attempt to satisfy the later subtrees if the previous subtrees succeeded or failed respectively. In order to model more realistic scenarios where the attacker is able to adapt actions according to the behavior of the environment - typically, a defender who tries to protect the system - in [16], inspired by the path semantics from [14] a new two-player strategy semantics is introduced without adding additional syntactic elements. Another approach to solving the lack of interaction between attacks on the system and possible defenses put in place is presented in [17] where attack-defense trees (ADTrees) are formalized. This formalization is largely similar to [12] with the addition of an additional counter operator (here denoted $\sim$) that basically switches the roles of attacker and defender for the subtree below it. Finally, in [5] this formalization is further extended by the sequential operators.

In the following subsections we provide a formal definition of ADTrees and stochastic two-player-games to then reframe the transformation from ADTrees to stochastic games developed in [5] in the context of graph transformations.

2.2 Formal Syntax of ADTrees

The author of [5] takes all sequential compositions to happen hierarchically above any non-sequential compositions and the counter-measure operator $\sim$. The following type system mirrors that of [5] with slight changes and corrections. Let $Actions_A$ and $Actions_D$ be disjoint action sets for attacker and defender, respectively.

$$\begin{aligned} t &::= nt \mid \overrightarrow{\wedge}(t,t) \mid \overrightarrow{\vee}(t,t) \\ nt &::= \alpha \mid \wedge(nt,nt) \mid \vee(nt,nt) \mid {\sim} nt \\ \tau &::= A \mid D \end{aligned}$$

$$\vdash \alpha\colon A \text{ if } \alpha \in Actions_A \qquad \vdash \alpha\colon D \text{ if } \alpha \in Actions_D \qquad \vdash t\colon A$$

$$\frac{\vdash t_1\colon A \quad \vdash t_2\colon A}{\vdash \overrightarrow{\wedge}(t_1, t_2) : A} \qquad \frac{\vdash t_1\colon A \quad \vdash t_2\colon A}{\vdash \overrightarrow{\vee}(t_1, t_2) : A}$$

$$\frac{\vdash nt_1 : \tau \quad \vdash nt_2 : \tau}{\vdash \wedge(nt_1, nt_2) : \tau} \qquad \frac{\vdash nt_1 : \tau \quad \vdash nt_2 : \tau}{\vdash \vee(nt_1, nt_2) : \tau}$$

$$\frac{\vdash nt\colon A}{\vdash \sim nt\colon D} \qquad \frac{\vdash nt\colon D}{\vdash \sim nt\colon A}$$

ADTrees then are the parse-trees belonging to the language t of sequential trees. Additionally two functions $c\colon Actions_A \cup Actions_D \to \mathbb{R}_{\geq 0}$ and $p\colon Actions_A \cup Actions_D \to \mathbb{R}_{\geq 0}$ representing the cost and success probability of actions are given.

Given a Boolean assignment signifying success or failure of the basic actions the overall success of the ADTree t can then be calculated as the classical evaluation of a Boolean formula, where the sequential operators $\overrightarrow{\vee}$ and $\overrightarrow{\wedge}$ are simply evaluated like their non-sequential counter-parts and $\sim$ is taken as a Boolean negation operator $\neg$.

2.3 Stochastic Two Player Games

A stochastic two player game is directed attributed typed graph $G = (V, E, l)$ with a labeling function l on the edges and a designated initial vertex $s_0 \in V$. Here $V := S_A \cup S_D \cup S_P \cup S_{final}$ represents a set of states partitioned into sets S_A and S_D of non-deterministic states belonging to the two players A and D respectively, a set S_P of probabilistic states and a set S_{final} of terminating states without outgoing edges. Additionally, we have edge labels/attributes partially defined by $l_A\colon \bigcup_{v \in S_A} \delta^+(v) \to L_A \times \mathbb{R}_{\geq 0}$, $l_D\colon \bigcup_{v \in S_D} \delta^+(v) \to L_D \times \mathbb{R}_{\geq 0}$, $l_P\colon \bigcup_{v \in S_P} \delta^+(v) \to [0, 1]$ representing the label and cost of possible actions for non-deterministic states in S_A and S_D and the probability of transitions for the outgoing edges of states in S_P. Furthermore, we require $\sum_{e \in \delta^-(v)} l_P(e) = 1$ for all $v \in S_P$. In our case L_D and L_A are powersets of all basic actions. Furthermore we choose to label the terminating states with attached loops with boolean attributes to signify success and failure.

This definition of two-player games is slightly different from the standard definition of stochastic games where all players have the possibility to choose an action at every state and the choices made determine a probability distribution on the outgoing edges of a state. But it can be easily seen that our definition of turn-based action as found in [5] can be recovered as a special construction in the standard definition of taking simultaneous action.

For a small example of what a concrete ADTree and stochastic two-player game might look like, refer to Fig. 1a and Fig. 1b.

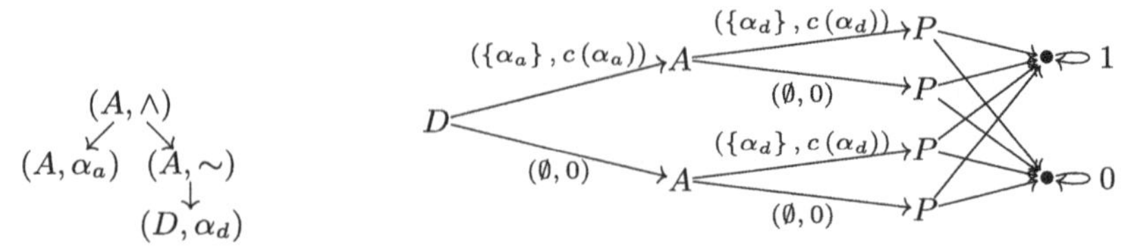

(a) ADTree corresponding to the formula $\wedge(\alpha_a, \sim(\alpha_d)) : A$.

(b) Stochastic two player game corresponding to the ADTree in Fig. 1a under the transformation in Sect. 3.1.

Fig. 1. Example ADTree and corresponding two player stochastic game. Note: the probabilities on the outgoing edges from probabilistic nodes where left out for readability.

3 Model Transformation: Probabilistic ADTree → Stochastic Two Player Game

In [5] an attack-defense tree as described above is transformed into a stochastic two player game. The goal behind this transformations is to give additional semantics to attack defense trees by utilizing the well established theory of stochastic games. The authors focus is on the use of different temporal logics to analyze the resulting games while the transformation itself is paid no further attention and only presented in an awkward procedural manner. In this section we adapt this model transformation by means of graphs transformation. We believe this better shows the effects different operators in the tree have on the resulting game. We do this transformation in two steps first we present the transformation of a non-sequential attack-defense-tree into a game. Then we discuss the sequential composition of games.

3.1 Transformation of a Non-sequential Attack-Defense Tree into a Game

A non-sequential tree nt is transformed into a game in Algorithm 1 in [5] as follows. For $X \in \{A, D\}$ let $Actions_{X|t} = \{a \in Actions_X \mid a \text{ appears in } t\}$ denote the set of actions of the attacker/defender respectively that appear in the tree. First, an initial state s_0 is added from which an edge leads to a separate state s_{B_D} for each $B_D \subseteq Actions_{D|t}$ this represents the actions of the defender. Then from each of these states get an edge to a new state $s_{B_D}^{B_A}$ for each $B_A \subseteq Actions_{A|t}$ this represents the possible actions of the attacker. With this order it is assumed that the attacker has knowledge about which defensive actions the defender has attempted to perform. Finally, two states with loops labeled 1 and 0 respectively, representing the success or failure of nt, are added. From the previously created attacker states an edge is created to these final states if there exists a non-zero probability that with the attempted actions nt could evaluate to 1 (resp. 0) and this probability is added as a label to the edge. The edges in the first two steps are each labeled with the cost of performing the selected subset of action.

Graph Transformation Unit 1 (ADTree → Game)

initial *Non-sequential attack-defense-tree*
rules

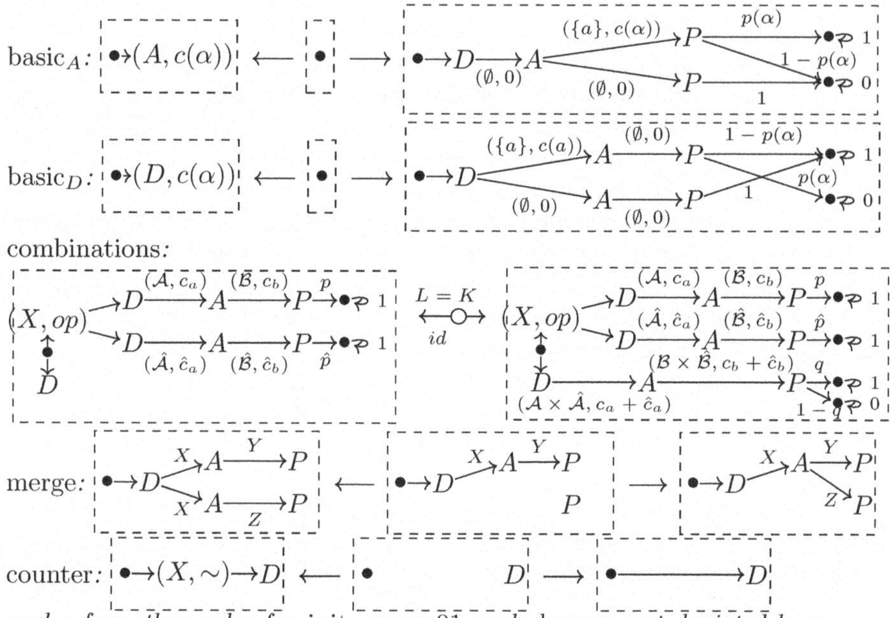

and a few other rules for init, merge01 *and* cleanup *not depicted here.*
control *1;2 where*

1. $(basic_A|basic_D)!; init!$
2. $((combinations; try\ merge; merge01)!; try\ counter)!; cleanup\ rules$

In the rules $basic_A$ and $basic_D$ the bottom of an ADTree, the basic actions of the attacker and defender, are matched and transformed into a sort of normal form game, where there is still a turn for the opposite player but no available actions to take. Furthermore the games success and failure are always measured from the perspective of the attacker, so inaction of the defender leads to guaranteed success (cf. $bacis_D$). The probabilities are given by the additional functions p and c defined in Sect. 2.2. The rule *init* attaches a vertex labeled D hanging from each parent of an operator vertex $\wedge$ or $\vee$. This rule has a negative application condition such that it can only be applied once.

Building on this base case the recursive combination of two games G_1 and G_2 using $op \in \{\wedge, \vee\}$ can then be seen as kind of product, where all possible combinations of paths in G_1 and G_2 from the initial vertex to the vertex with loop labeled 1 are explored. This is done in the rule *combinations*. The final success probability of a combination depends on *op* and of which player type the vertex in the tree is. It is calculated under the assumption that the outcomes of the action sets of the two games are independent. The probability q in the right-hand side of the rule *combinations* is given by relations $q\,(A, \wedge) = p \cdot \hat{p}$, $q\,(A, \vee) = 1 - (1 - p) \cdot (1 - \hat{p})$, $q\,(D, \wedge) = q\,(A, \vee)$ and $q\,(D, \vee) = q\,(A, \wedge)$. We use negative application conditions for this rule-schema such that each rule is applied exactly once for each matching.

After each application of the *combination* rule, if there the different paths with the same choice for the defender should be merged and furthermore the different newly generated final vertices should also be merged. This is done with the rule *merge*, where X, Y, Z are pairs consisting of sets of actions and the respective costs, addresses the first of these merging operations. The rule *merge01* for merging the final states with loops labeled 0 and 1 is be designed similarly.

Finally we are left to consider how to deal with the counter ($\sim$) operator. Since we used the type information attached to the tree with the (correct) perspective of the attacker in $basic_A$ and $basic_D$ and when calculating probabilities for *combinations* our games do not need to be inverted, hence, such vertices can simply be skipped as done in the rule *counter*. In a sense in the previous steps the ADTree was treated as if all counter operators had been propagated to the leafs using deMorgans rules and then only worrying about the impact of the role-reversal on the most basic games.

During the transformations old games are not deleted, so a few *cleanup* rules ares necessary to remove all intermediate games and the tree structure to only yield the final game.

For the control condition we use extended regular expressions where | denotes alternative, ; denotes sequential composition, *try* denotes the application if possible, and ! denotes application as long as possible.

3.2 Sequential Composition of Games

After generating the individual non-sequential trees Algorithm 2 in [5] is used to recursively sequentially compose two games G_1 and G_2 into the final game using $op \in \left\{\overrightarrow{\wedge}, \overrightarrow{\vee}\right\}$. First a disjoint union of G_1 and m copies of G_2 is made, where m is the number of final states of G_1. Then the key steps of this algorithm in the form of graph transformations are the following. The rules in the schema *connect* connect copies of G_2 to final states of G_1. Afterwards using rules of the schema *eval* we match on paths leading from a final state of G_1 to a final state of G_2 and change the Boolean attribute at the matched final state of G_2 by evaluating op on the old attribute values ($bool_1$ and $bool_2$).

Graph Transformation Unit 2 (Sequential composition of games)

initial *Disjoint union of a game G_1 and m copies of a game G_2 where m is the number of vertices in G_1 with a loop carrying a Boolean attribute. For each initial vertex v of one of the games exists a vertex w_v with an edge (w_v, v) indicating if v is initial for G_1 or one of the copies of G_2.*

rules

connect$_{k_1,k_2,bool}$ *for* $k_1, k_2 \in \mathbb{N}_{>0}, bool \in \{0, 1\}$*:*

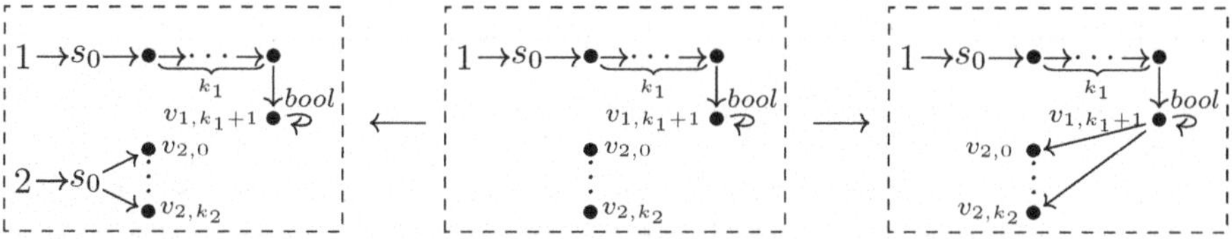

$\text{eval}_{k_1,bool_1,bool_2}$ *for* $k_1 \in \mathbb{N}_{>0}, bool_1, bool_2 \in \{0, 1\}$*:*

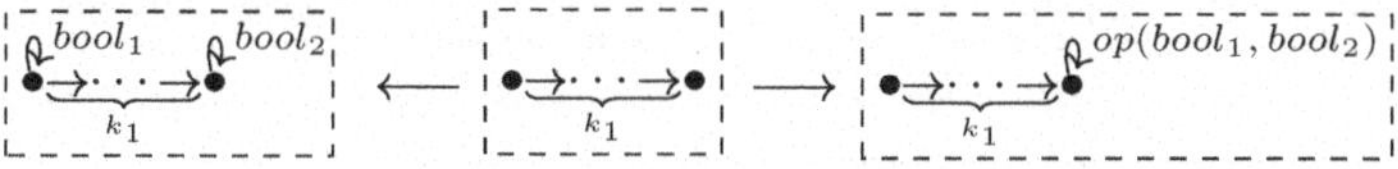

We take note of the fact that this transformation produces games whose size grows in $\mathcal{O}(2^n)$ where n is the number of maximal non-sequential subtrees.

4 Recognizing Possible Optimizations to Games

For the informal description of this algorithm in [5] the author remarks that if the complete tree has only conjunction as a sequential operator appearing, then in the last step of Algorithm 2 all final vertices of the resulting game with matching attached loop could be merged. The resulting game will then not grow exponentially depending on the number of phases. Intuitively this sounds reasonable, but should similarly hold for games where the final tree has only disjunctions in its sequential part. In the following we present a sketch of how graph-transformations can play a role in showing this formally.

To start we will make a slight change to how the game is generated in that at the end we create two vertices with attached loops labeled 1 and 0 respectively and connect them to the final states of our complete game if they have a matching labeled loop. This does not change the game, but graphically represents the fact that for the outcome of the complete game we consider all final states representing success (similarly failure) the same.

Now let us define two simple graph properties on acyclic graphs and graph transformation units to test for them.

Definition 1. *Let G be an acyclic graph, let V_G denote the vertex set of G, and let $v, w \in V_G$. We say w is* inevitable descendant *of v if w lies on every inclusion maximal path starting in v.*

The intuition behind the name being that no matter which path we take from v we are guaranteed to eventually end up at w. As an example consider a game whose sequential part only consists of conjunctions ($\overrightarrow{\wedge}$) any vertex v not on a path consisting of only successes of the non-sequential games can then only lead to failure. Therefore this vertex and the final vertex w labeled with loop labeled 0 conform to the definition above. The recognition of such a pair of vertices can be encoded in a graph transformation-unit as described below.

Graph Transformation Unit 3 (Inevitable descendent test)

initial *Acyclic graph with two loops labeled* s *and* e.
rules

start:
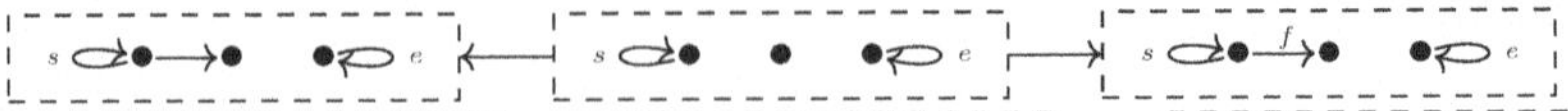

not-end:
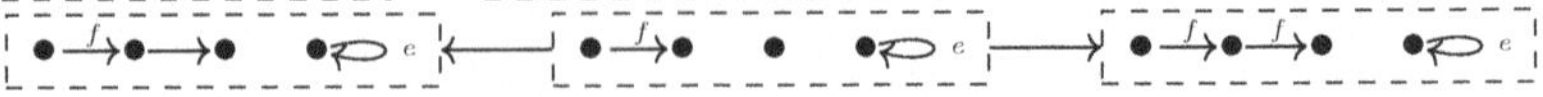

at-end:

terminal *Exists no vertex v with incoming edge marked* f (found) *that does not have outgoing edges unless v has a loop marked E attached to it.*

For the second property let us first introduce the following notation. Let G be an acyclic graph and $v \in V(G)$ a vertex in G. We define G_v to be the induced subgraph of G formed by v and all its descendents in G.

Definition 2. *Let G be an acyclic graph and $v \in V_G$. Then we define $G_v := \{w \in V_G \mid \exists$ path from v to $w\}$.*

Observation 1. *If for three vertices $v, w, t \in V_G$, t is an* inevitable descendant *of v and w then strategy optimization on the subgraph $G_v \setminus G_t$ (and $G_w \setminus G_t$) becomes single objective (only strategy cost can still be optimized) since the probability distribution of paths to final vertices from v (and w) is independent of strategic choices made in $G_v \setminus G_t$ (and $G_w \setminus G_t$). If now $G_v \cong G_w$, we could reduce the size of our game graph by changing all incoming edges to w to point to v instead and removing all vertices not reachable from the initial state, since an optimal strategy on $G_v \setminus G_t$ is surely optimal on $G_w \setminus G_t$ as well.*

This optimization is encoded in the following graph transformation unit, after first marking the appropriate vertices.

Graph Transformation Unit 4 (Game reduction)

initial *Acyclic graph with unique vertex s_0 with no incoming non-loop edges; a loop labeled s_0 is attached at s_0; two loops labeled* v *and* w *not attached to s_0.*
rules *for $k \in \mathbb{N}_{>0}$*

reroute:
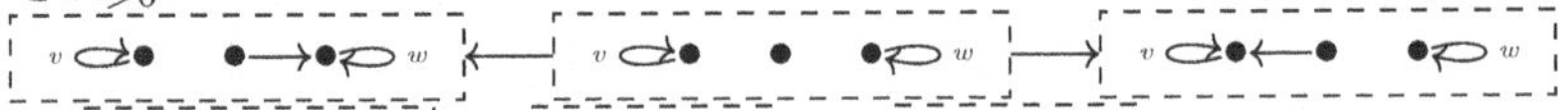

remove-B_k
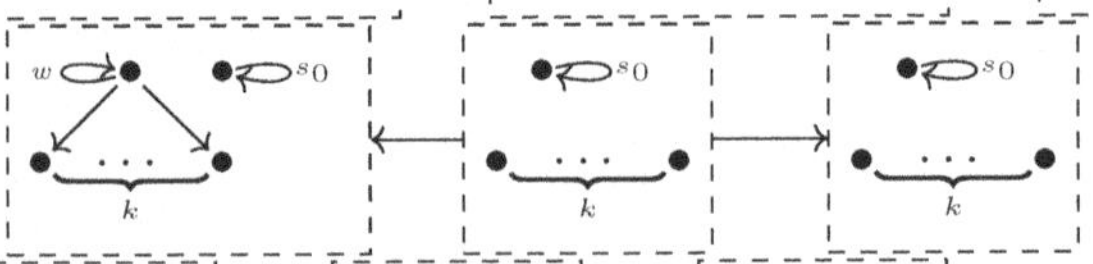

remove$_k$:
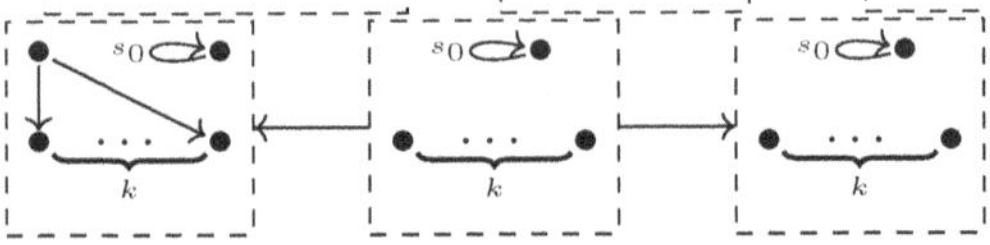

The initial state s_0 of our game also has no incoming edges, so needs to be marked to not be deleted. For all other vertices not having incoming edges means not being reachable in the process of the game, so they could be deleted.

Furthermore, checking for two acyclic graphs G_1 and G_2 with a single vertex without incoming edges each if $G_1 \cong G_2$ can also be expressed as a simple graph transformational unit. To apply this to the case of v and w in a game-graph G as above we assume to have some operation to copy G_v and G_w out of G.

Graph Transformation Unit 5 (Game equivalence)

initial *Disjoint union of two acyclic graphs with unique vertices without incoming non-loop edges; every vertex has a loop with label* 0.

rules *for* $k_1, k_2, k, i, j \in \mathbb{N}_{>0}$

init_{k_1,k_2}:

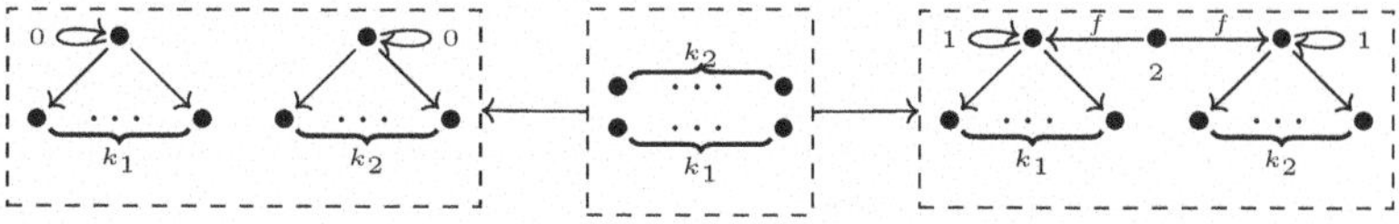

$\text{expand}_{i,k}$:

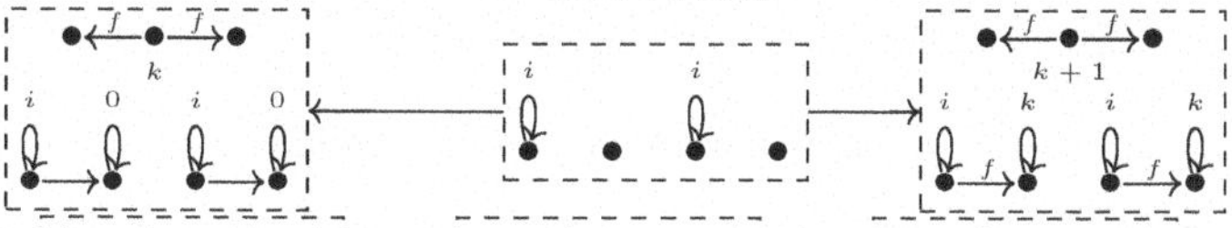

$\text{find-edge}_{i,j}$:

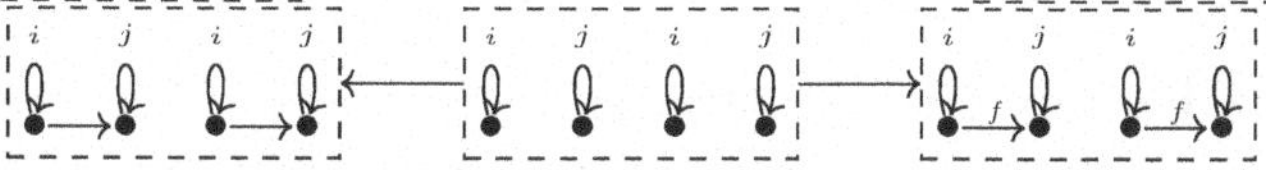

terminal *All edges are labeled with* f *(*found*) except for loops.*

Optimizing a game can now be done by repeatedly searching for pairs of vertices v and w with properties as detailed above and applying the modification transformation. Additionally appropriate clean up of auxiliary labels, edges and vertices would need to be done in between the steps.

Finally, using this method on the example game from [5] as shown in Fig. 2a we arrive at a reduced version of the game whose topology can be seen in Fig. 2b. This corresponds to the actual graphic found in [5] and the informal description therein. The advantage of our approach would be that the same size optimization is analogous for a game with only sequential disjunctions as one would intuitively expect and could also be applied to other games not necessarily generated with their approach.

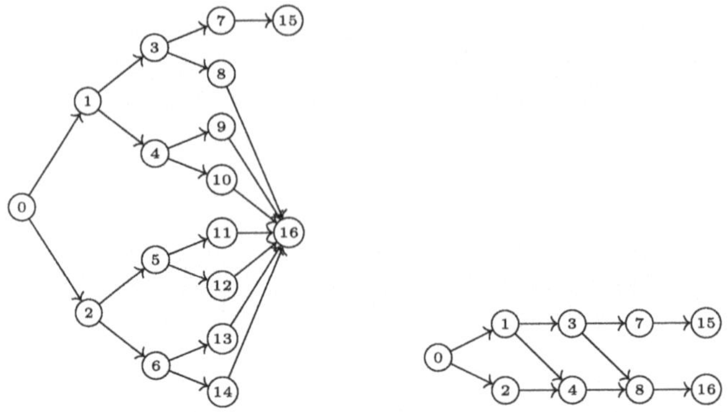

(a) Topology of a game with two sequential conjunctions.

(b) Topology of a game with three sequential conjunctions after reduction.

Fig. 2. Reduction of an example game

5 Open Research Questions

There are various research directions where graph transformation may be a useful framework in the context of threat modeling.

1. An adjacent approach to attack trees are attack graphs (cf. [18]) which represent all possible states a system may hold and the transition between those states. Moreover, transformations from attack trees to attack graphs as in [19] and vice versa like in [20] have been proposed. Formalizing and extending such approaches with graph-transformation yields the ability to relate the approaches, compose constructions straightforwardly and, following the lines of [21] to prove correctness of reductions. Ultimately we aim to develop a more unified language that is able to faithfully capture a range of different current approaches in threat modeling and transformations between them and we believe graphs and graph-transformations to be a promising setting for this.
2. There are also various threat modeling approaches based on data-flow diagrams (cf. e.g. [22]). A usual approach consists of the following steps: Define data flow diagrams modeling the systems data flows, data stores, processes and external entities; Systematically iterate over all model elements and map predefined threats to diagram elements; Identify threat scenarios, i.e., consider each identified threat entry point against all known threat categories; Prioritize threats; Elaborate mitigation strategies; Perform mitigations. Elaborating this approach with graph-transformation may be worth.

3. Another approach for designing safe and secure systems uses explicit system models described in system description languages such as UML or SysML. This enables to express different views of complex systems, e.g. hierarchical decomposition of system components, system dynamics (finite state machines, sequence diagrams), etc. A system model can be seen as a collection of typed attributed nested graphs with appropriate graph transformations describing its dynamics. In order to relate vulnerabilities and system components adequately, the technical/abstraction level of vulnerability and system description must match. This may be an interesting application for hierarchical graph-transformation and SysML-Sec [23] is a good starting point.
4. Graph transformation units are just one possible approach as a structuring mechanism for graph transformation algorithms. We belief a promising direction is to investigate alternative structuring mechanisms to extend the expressivity of graph transformation as a modeling language.

Acknowledgements. We are grateful to the anonymous reviewers for their valuable comments that led to various improvements.

Disclosure of Interests. The authors have no competing interests to declare that are relevant to the content of this article.

References

1. Shostack, A.: Threat Modeling: Designing for Security. Wiley (2014)
2. Bodeau, D.J., McCollum, C.D., Fox, D.B.: Cyber threat modeling: survey, assessment, and representative framework. Technical report, Homeland Security Systems Engineering and Development Institute (HSSEDI)/MITRE Corporation (2018)
3. Xiong, W., Lagerström, R.: Threat modeling - a systematic literature review. Comput. Secur. **84**, 53–69 (2019). https://doi.org/10.1016/j.cose.2019.03.010
4. Tarandach, I., Coles, M.J.: Threat Modeling. O'Reilly Media, Inc. (2020)
5. Aslanyan, Z.: Formal analysis of graphical security models. English. Ph.D. thesis. Technical University of Denmark (2016)
6. Ehrig, H., Pfender, M., Schneider, H.J.: Graph-grammars: an algebraic approach. In: 14th Annual Symposium on Switching and Automata Theory, Iowa City, Iowa, USA, 15–17 October 1973, Proceedings, pp. 167–180. IEEE Computer Society (1973). https://doi.org/10.1109/SWAT.1973.11
7. Ehrig, H., et al.: Fundamentals of Algebraic Graph Transformation. Monographs in Theoretical Computer Science. An EATCS Series. Springer (2006)
8. Ehrig, H., et al.: Graph and Model Transformation - General Framework and Applications. Monographs in Theoretical Computer Science. An EATCS Series. Springer (2015). https://doi.org/10.1007/978-3-662-47980-3
9. Kreowski, H.-J., Kuske, S., Rozenberg, G.: Graph transformation units - an overview. In: Degano, P., De Nicola, R., Meseguer, J. (ed.) Concurrency, Graphs and Models. LNCS, vol. 5065, pp. 57–75. Springer (2008)
10. Rozenberg, G. (ed.) Handbook of Graph Grammars and Computing by Graph Transformation, Vol. 1: Foundations. World Scientific, Singapore (1997)

11. Kordy, B., Piètre-Cambacédès, L., Schweitzer, P.: DAGBased Attack and Defense Modeling: Don't Miss the Forest for the Attack Trees (2013). arXiv: 1303.7397 [cs.CR]
12. Schneier, B.: Attack trees. Dr. Dobb's J. (1999)
13. Widel, W., et al.: Beyond 2014: formal methods for attack treebased security modeling. ACM Comput. Surv. **52**(4), 75:1–75:36 (2019). https://doi.org/10.1145/3331524
14. Audinot, M., Pinchinat, S., Kordy, B.: Is my attack tree correct? In: Foley, S.N., Gollmann, D., Snekkenes, E. (eds.) ESORICS 2017. LNCS, vol. 10492, pp. 83–102. Springer, Cham (2017). https://doi.org/10.1007/978-3-319-66402-6_7
15. Jhawar, R., et al.: Attack trees with sequential conjunction. CoRR abs/1503.02261 (2015). arXiv: 1503.02261
16. Brihaye, T., Pinchinat, S., Terefenko, A.: Adversarial formal semantics of attack trees and related problems. In: Ganty, P., Monica, D.D. (eds.) Proceedings of the 13th International Symposium on Games, Automata, Logics and Formal Verification, GandALF 2022, Madrid, Spain, 21–23 September 2022, vol. 370, pp. 162–177. EPTCS (2022). https://doi.org/10.4204/EPTCS.370.11
17. Kordy, B., Mauw, S., Radomirović, S., Schweitzer, P.: Foundations of attack–defense trees. In: Degano, P., Etalle, S., Guttman, J. (eds.) FAST 2010. LNCS, vol. 6561, pp. 80–95. Springer, Heidelberg (2011). https://doi.org/10.1007/978-3-642-19751-2_6
18. Sheyner, O., et al.: Automated generation and analysis of attack graphs. In: 2002 IEEE Symposium on Security and Privacy, Berkeley, California, USA, 12–15 May 2002, pp. 273–284. IEEE Computer Society (2002). https://doi.org/10.1109/SECPRI.2002.1004377
19. Schiele, N.D., Gadyatskaya, O.: A novel approach for attack tree to attack graph transformation. In: Luo, B., Mosbah, M., Cuppens, F., Ben Othmane, L., Cuppens, N., Kallel, S. (eds.) CRiSIS 2021. LNCS, vol. 13204, pp. 74–90. Springer, Cham (2022). https://doi.org/10.1007/978-3-031-02067-4_5
20. Pinchinat, S., Acher, M., Vojtisek, D.: ATSyRa: an integrated environment for synthesizing attack trees. In: Mauw, S., Kordy, B., Jajodia, S. (eds.) GraMSec 2015. LNCS, vol. 9390, pp. 97–101. Springer, Cham (2016). https://doi.org/10.1007/978-3-319-29968-6_7
21. Kreowski, H.-J., et al.: A graph-transformational approach for proving the correctness of reductions between NP-problems. In: Heckel, R., Poskitt, C.M. (eds.) Proceedings of the Thirteenth International Workshop on Graph Computation Models, GCM@STAF 2022, Nantes, France, 6th July 2022. EPTCS, vol. 374, pp. 76–93 (2022). https://doi.org/10.4204/EPTCS.374.7
22. Hernan, S., et al.: Uncover security design flaws using the STRIDE approach. MSDN Mag. (2006). https://learn.microsoft.com/en-us/archive/msdn-magazine/2006/november/uncover-securitydesign-flaws-using-the-stride-approach
23. Roudier, Y., Apvrille, L.: SysML-Sec: a model driven approach for designing safe and secure systems. In: 2015 3rd International Conference on Model-Driven Engineering and Software Development (MODELSWARD), pp. 655–664 (2015)

Graph Rewriting for User State-Based Dialogue Adaptation in Real-Time: An Application in Personalized Interview Training

Deeksha Adiani[1(✉)], Timothy J. Vogus[1], Nilanjan Sarkar[1], and Medha Sarkar[2]

[1] Vanderbilt University, Nashville, TN 37235, USA
deeksha.m.adiani@vanderbilt.edu
[2] Middle Tennessee State University, Murfreesboro, TN 37132, USA

Abstract. Personalized learning is most effective as individuals have different learning styles and support needs. Many e-learning systems rely on static dialogue management, which offer limited flexibility to adapt to user behaviors and open discourse. This paper proposes a graph rewriting framework to adapt dialogue flow to user states in real-time. We propose its application in virtual job interviews—a key step toward employment—where personalized simulation training can be particularly valuable for individuals with diverse needs, including autistic users who may face challenges in social communication. Our framework leverages real-time multimodal data—speech, eye gaze, and physiology (e.g., heart rate)—to infer user states. We identified three key states in interviews: engagement, stress, and stuckness. Dialogue scripts are modeled as attributed, directed graphs, incorporating constraints to ensure consistency. Each interview question is represented as a node, with edges reflecting the flow of dialogue. The interview structure (host graph) was designed in collaboration with domain experts. Using graph rewriting, the framework dynamically modifies dialogue flows to enable contextually appropriate adaptations such as redirecting focus when the user goes off-topic, providing hints when stuck, or rephrasing questions for clarity—actions typically performed by coaches during mock interviews. Functionality of the grammar is demonstrated using interview scripts developed with experts, showcasing feasibility for virtual job interviews with simulated user states. This approach offers a personalized simulation experience that enhances realism, supports learning, and promotes the need for inclusive interviews.

Keywords: Graph Grammar · Dynamic Dialogue Management · Graph Representations · Job Interviews

J. Endrullis and M. Tichy (Eds.): ICGT 2025, LNCS 15720, pp. 218–245, 2025.
https://doi.org/10.1007/978-3-031-94706-3_11

1 Introduction

Virtual social skills training (SST) platforms are designed to help individuals develop effective communication [65,69,87], emotional regulation [63,67,108], and collaborative interaction and conversational skills [6,9,26,62,68,74,106]. These platforms often simulate scenarios [47], such as job interviews [7,13,59,90–92,94,101], group interactions [81], customer service [43,53,96], negotiations [22, 31,49,78], or workplace interactions [24,42,83]. They provide a controlled environment, especially for individuals with special needs such as neurodiverse individuals, including autistic users and those with attention-deficit/hyperactivity disorder (ADHD), to practice complex interpersonal scenarios to help overcome social anxiety while building confidence [5]. These platforms rely on dialogue management systems to simulate realistic conversations. Realistic simulations, however, require understanding the user's behaviors, emotional states or affect[1], or learning progress [32,34]. Without these features, interactions can feel rigid or unrealistic [76]. These systems must adapt to user behaviors and emotional states, potentially providing realistic and supportive training environments.

Affective dialogue systems aim to infer and respond to users' emotional and cognitive states using multimodal data signals such as speech, eye gaze, and posture. For instance, AutoTutor [48] employed a Finite State Machine (FSM) and rule-based logic to adjust dialogue flow based on learning progress, while Ayedoun et al. [11] used detected silence and uncertainty to inform system responses. Similarly, Pitterman and Pitterman [85] integrated emotion recognition using Hidden Markov Models (HMMs) and speech analysis tools like PRAAT [17]. Although these approaches successfully leverage multimodal data, they rely heavily on episodic, rule-based dialogue management. FSMs and similar architectures require predefined mappings between all possible user responses and system actions, often resulting in rigid dialogue flows and limited scalability. As the number of state variables increases, FSMs are susceptible to the state explosion problem, significantly increasing system complexity [107].

More recent end-to-end models like Heimerl et al.'s Generative Adversarial Network (GAN)-based approach [57], use multimodal inputs (e.g., gesture, posture, eye gaze) to provide feedback and drive interaction in virtual job interviews. While promising, these models require large annotated datasets linking real multimodal user data to appropriate system actions (e.g., providing feedback, choosing next dialogue). Virtual social skills platforms provide valuable support for autistic individuals, people with schizophrenia, and those facing mental health challenges; however, collecting and labeling the necessary training data for these systems remains a time-intensive and resource-demanding process. In settings where data is limited, proven user-state inference methods offer practical alternatives [11,13,14,20,50,93], while domain-general dialogue adaption the systems need, can be tackled via a different approach.

[1] Affective Computing focuses on recognizing and processing human emotions or affects [84,98].

Graph representations model dialogue flow with nodes as states and edges as transitions, enabling structured interaction [7,23,51]; however, in prior work, these graphs are typically fixed once authored. To address this limitation, we propose a graph rewriting (GR)-based dialogue management framework that allows real-time, modular adaptation of dialogue flow based on inferred user states from multimodal data (e.g., stress, engagement, or being stuck). Building upon our own virtual job interview training system [2,4], we define graph rewrite rules informed by domain experts and literature. While applicable to a range of social scenarios, our approach supports overlapping user states, where the user can be in more than one state at a given moment (e.g., a user can be stressed and stuck or engaged and stressed). These rules allow dialogues to evolve dynamically through—contextual redirection, simplification and clarification of content, or re-engagement of disengaged users through interactive prompts—potentially providing more personalized and enhanced adaptive interactions in virtual SST and human-agent interactions.

2 Related Work

Graph Rewriting (GR) has shown to be a versatile framework for modeling and transforming complex systems across domains such as biochemical sciences, engineering, and software development. In biology, GR is used to model protein structures, molecular interactions, and biochemical networks [44,54,55,61]. In engineering, it supports tasks like generating building layouts, infrastructure modeling, and interior planning [21,73,102]. In software development, it supports graph database queries, schema evolution, database sanitization, program optimization, network analysis, and visual tool development [12,41,72,100,105]

These diverse applications are enabled by core fundamental methods in GR such as Double Pushout (DPO) and Single Pushout (SPO) methods [27,38], grounded in category theory [10,15,80]. GR grammars—collections of rewrite rules—can operate on typed, attributed, or labeled graphs, enabling context-sensitive transformations by associating nodes and edges with data types such as strings, integers, and Booleans [37]. Additional approaches, such as hyper-edge replacement grammars [36], node replacement grammars [40], and term rewriting [86], support representation of more complex relationships or hierarchical structures. As GR systems are typically tailored to specific domains and tools, there is no universal mechanism for rule design [16,102]. Nevertheless, tools like GREW, GReAT, GROOVE, and ATOM [8], have been developed to support GR rule development and execution.

Graph representations are widely used for dialogue management systems such as Google DialogFlow[2], PyDial[3], and OpenDial[4]. Virtual SST, virtual job interviews, and conversational agents often model dialogue as labeled, directed graphs, with nodes representing dialogues and edges defining transitions based

[2] https://cloud.google.com/dialogflow.

[3] https://pydial.cs.hhu.de/.

[4] https://sites.google.com/site/opendialtoolkit.

on user responses or system logic [23]. Gaffney and Dagger [29] reviewed tools for soft skills training that use graph-based visualizations, highlighting those for teaching psychiatric interviewing, while AutoTutor [33,48] dynamically guides conversations in e-learning using graph-based structures. Zapata Rivera et al. [109] extended this approach for virtual agents in conversation-based assessments, using graph visualizations to structure agent dialogues and interactions.

In task-oriented scenarios, Jonell et al. [66] modelled dialogue data as graphs built dynamically during interactions, while Anderson et al. [7] used the TARDIS frame-work to represent job interview scripts as graphs with nodes for questions and edges for transition conditions. ConvGraph [51] represents task-oriented dialogues (e.g., of a chatbot for booking tickets) as sequences of states between the system and the user, and new dialogue paths can be traversed to create new instances to improve the system's decisions on what dialogue to choose next. Similarly, Walker et al. [103] also modeled dialogue as graphs to track user entities and their dynamic interactions over time, further showcasing the suitability of graphs for managing complex dialogue systems. Amongst the above-mentioned works, Walker et al. [103] demonstrated the use of graph rewriting for dialogue management using a graph query language to create transformation rules, demonstrating that graphs are suited to modelling dialogue states in complex open-ended domains.

Building on prior work, we represent conversational flows as labeled, typed, attributed, and directed graphs. Some attributes (e.g., *is_stuck*, *is_engaged*, *is_stressed*, *is_offtopic*) dynamically update based on user states, which are inferred from multimodal user data. While existing approaches have used attributed graphs, they primarily focus on static attribute updates through pre-defined rules (e.g., updating a counter $n := n - 1$) [16,18,39,56,70]. These methods have limited flexibility to adapt dialogue structures dynamically in response to real-time user states.

A key limitation in existing dialogue modeling is the reliance on FSMs, which offer a clear structure and allow for data-driven dialogue management, but require predefining all transitions between dialogue states. While effective in constrained scenarios, FSMs become difficult to scale when handling multiple, overlapping user states (e.g., a user being both *stressed* and *stuck*), often resulting in large and unwieldy static graphs.

Our proposed framework addresses this by applying typed, attributed graph rewriting to support real-time adaptability. This enables the dialogue structure to evolve dynamically based on inferred user states while maintaining semantic constraints through context-sensitive grammar. It has potential to be a more flexible and expressive approach to adaptive dialogue modeling, improving responsiveness in human-agent interactions.

3 Method

3.1 Expert Involvement

To ensure our system met real world needs we consulted with three job coaches and a professor of business and management (also a co-author), who specializes in social skills training (e.g., workplace interaction, negotiation) and inclusive interview practices. These experts work with clients who have diverse needs, including autistic individuals, neurodiverse individuals, and those with attention deficiency hyperactivity disorder (ADHD), for whom a data-driven adaptive social skills training or interview training system can be a useful aid.

The job coaches brought valuable, hands-on experience in preparing candidates for mock interviews. Their feedback highlighted difficulties candidates have often faced such as stress or anxiety, lack of confidence, candidates going off-topic, and challenges with open-ended questions such as "Tell me about a time...". Interviewees sometimes stray off-topic, mentioning unrelated points with phrases like "Oh! I just remembered...", making it important to gently redirect their attention to the original question.

We asked job coaches about candidates requesting to skip questions. While uncommon in actual interviews, coaches saw value in this practice during mock sessions, as it allows candidates to regain composure or confidence before returning to difficult questions. If the candidate is unable to respond, we can prompt them with a hint on what the interviewer is looking for, and if they are still struggling, then the question and its follow-up questions can be skipped altogether. They recommended revisiting skipped questions systematically before moving to new topics, ensuring interview coherence and flow. When we probed for more details, they emphasized that skipped questions should be revisited before moving to a new topic, rather than at the end or randomly, to maintain focus and interview flow.

Additionally, we consulted an expert in business and management (a co-author) specializing in interviewing techniques. Their insights helped refine stress management strategies and the structuring of complex, behavior-based questions within our system, aligning it with professional best practices. For example, a challenging question like "Tell me about a time you disagreed with a colleague and how you handled it" can be broken into manageable parts: a) "What was the disagreement about?" and b) "How did you resolve it?". Visual-verbal prompts (e.g., clarifying when happened, who was involved, what actions occurred) were recommended to aid recall and detailed responses, especially beneficial for individuals with diverse cognitive needs [77]. Although simplified compared to real-life interactions, these adaptations establish a foundation for dynamic, adaptive virtual conversations.

Based on these insights, the proposed framework system must: 1) rephrase questions if a user struggles to respond, 2) skip questions when a user remains stuck after receiving hints, and reintroduce them systematically, 3) provide prompts (e.g., "thank you for sharing, but can you please tell me... repeated question)," to redirect conversations back on topic when the candidates stray, 4) offer simpler, alternative questions if the user appears stressed or faces difficulty, 5) break down complex questions to reduce cognitive load, and 6) allow users to request clarifications as needed.

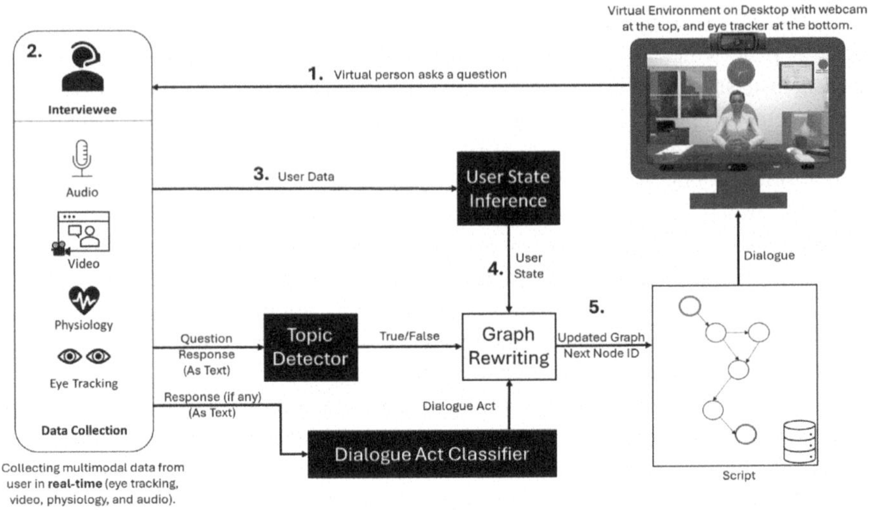

Fig. 1. Framework of adaptive dialogue management starting from the virtual environment with the virtual person speaking, followed by the user's turn during which *User Data* collected is sent to the *User State Inference* system which returns a *User State*. Meanwhile, the *Dialogue Act* is predicted via the *Dialogue Act Classifier* and the *Topic Detector* checks whether the user response is on or off topic (True/False). The *User State* is input to the *Graph Rewriting* module program which outputs the *Updated Graph* and the *Next Node ID*. The *Dialogue* in that node is output to the virtual platform and the cycle continues. When the dialogue does not require a response from the user, steps 2, 3, and 4 are not used in the decision-making process.

3.2 Groundwork and User State Inference

Social cues, both verbal and nonverbal, play a crucial role in human interactions that convey valuable information about affective states such as engagement, frustration, confusion, anxiety, and disengagement [19,30,32,33,75,97]. Affective computing enables recognition and response to these states [84]. Using multimodal data, including physiological signals, eye gaze, speech, facial expressions, and body posture, we can understand and respond to user states [45,46,89]. In our prior work [2,4], we explored the potential of these affective states in job interview contexts, highlighting the importance of task-specific user states such as engagement, stress or anxiety, and stuckness. By fusing multimodal data [35,71], these states can be inferred to inform dynamic decision-making in real-time [28,104]. Speech data, which is a critical modality, is captured to recognize the dialogue acts (e.g., statements, user questions, yes and no responses, acknowledgments, etc. [25,95]) which our prior work demonstrated can be identified with high accuracy using a fine-tuned machine learned transformer model [3]. Given an established reliable method exists for detecting user states through multimodal data (explored in [15,24,64,67]), as well as for detection of responses on or off topic (explored in [14,19,58,60]), GR rules determine the adaptive decisions based on these inferences. Figure 1 illustrates the adaptive dialogue management framework in which the GR module is crucial, and shows where user state inference fits in this framework.

3.3 Graph Definition and Constraints

Since our framework focuses on dialogue, the graph must follow semantic constraints inherent to conversational dynamics that address various dialogue structures, including yes/no questions, open-ended questions, user response dependencies, and follow-up dialogue. Yes/no questions expect binary responses, typically "yes" or "no" with some additional lines for explanation [95]. Open-ended or behavioral questions (e.g., "Tell me about a time when..."), on the other hand often require further clarifications, rephrasing, or follow-ups. Questions may be abandoned or skipped, altered or rephrased, or tried again. In addition, users may become disengaged or go off topic during the conversation, requiring a prompt or redirection to help guide the user back to the main topic [52]. We follow the six requirements suggested by experts as discussed in Sect. 3.1 to design the interview graph and the patterns and rules for the GR grammar.

Graph Definition. The host graph G is defined as $G = (V, E, A)$ and consists of a set of nodes V, directed edges $E \subseteq V \times V$, and an attribute mapping function A that assigns attributes to nodes and edges. Each edge $e \in E$ connects a source and target node in V. The graph includes a start node v_{start}, identified by $\mathbf{id}(v_{start}) = 'start'$ with no incoming edges, and an end node v_{end}, identified by $\mathbf{id}(v_{end}) = 'end'$ with no outgoing edges. Each node $v \in V$ has attributes defined as $A(v) = \{k_i \rightarrow v_i\}$, where k_i is a key and v_i is a value (both strings). The value domain supports Booleans (true or false), lists, and nested maps, enabling flexible and hierarchical attribute modeling for example, `"key": true`, `"key": [true, false, false, true]`, or `"key": {"key": [value1, value2, value3]}`.

A mandatory set of attributes provide contextual information about the state of dialogue. The attribute **id** (string) uniquely identifies the node, while section (string) specifies the conversation topic (e.g., technical skills, behavioral). The attribute dialogue (string) holds the dialogue text (e.g., "What are your interests?"), and **user_response** (string) stores any response from the user. Attribute **response_required** (Boolean) indicates whether a user response is necessary to proceed. State flags **is_stuck**, **is_engaged**, and **is_stressed** (Boolean) allow multiple states and support seven combinations (*engaged*, *stuck*, *stressed*, *engaged-stuck*, *stuck-stressed*, *stressed-engaged*, and *engaged-stuck- stressed*). The **type** attribute categorizes nodes as 'normal' (default), 'follow-up' (sequential question), 'prompt' (redirect), 'copy' (repeat or rephrase), 'hint' (assist), 'broken-down' (simplified in parts), 'system-response' (answer to any user query). Additional flags include **is_offtopic** (Boolean), indicating off-topic responses, and **dialogue_act** (string) representing the meaning or speech act of the user response such as 'yes-answer', 'no-answer', 'other-answer' (e.g., "I don't know"), or 'user-query' (e.g., "When will I here back about the job?"), as predicted by the Dialogue Act Classifier in [2] (see Fig. 1). Traversal attributes include traversed, skipped, and abandoned (all Boolean), denoting the node's traversal. Each node $v \in V$ has the required attributes defined by $A(v) = \{$ **id**, **section**, **dialogue**, **user_response**, **response_required**, **is_stuck**, **is_engaged**, **is_stressed**, **type**, **is_offtopic**, **dialogue_act**, **traversed**, **skipped**, **abandoned** $\}$ and $extra\_attributes(v) \in \{\emptyset, \texttt{map(string, value)}, \texttt{null}\}$ allows nodes to store additional information, while relevant, may not be directly used by production rules for rewriting.

Each edge $e \in E$ has intrinsic attributes represented by $A(e)$. These attributes are key-value pairs defined as $A(e) = k_i \rightarrow v_i$, same as $A(v)$, with the required attributes label (string) indicating the transition type '' (empty), 'yes_path' (affirmative response), 'no_path' (negative response), 'alternate' (indicating alternate path of questions), 'alt_path' (follow-ups from the first question in the alternate path), 'follow-up' (follow-up to any normal node), 'broken-down' (indicating the series of simplified questions one after another), and **section_change** (Boolean) which marks the edge where the section changes from current to the next (e.g., from greetings to technical skills), which is crucial for the retrial of skipped questions stored during the current section. Edges can also include $extra\_attributes(v)$ to hold supplementary information beyond what is necessary for rule-based operations.

Patterns and Rules. A pattern P is defined as $P = (V_p, E_p, A_p)$ consisting of a set of nodes V_p, edges E_p, and node and edge attributes $A_p(v)$ and $A_p(e)$, respectively. Patterns capture key interaction scenarios—e.g., repeated stress, stuck states, user clarifications, or skipped content—and are prioritized for matching.

A rewrite rule consists of a left-hand side (LHS), defining the pattern P_{LHS} to match in the host graph, and a right-hand side (RHS), defining the transformed structure. To adapt the dialogue flow based on user interactions, we define a set of rewrite rules that govern how the conversation evolves. A successful pattern match triggers a transformation rule with a left-hand side (LHS) pattern and a right-hand side (RHS) replacement structure. Figure 2 shows a rule where a user's "yes" response triggers deletion of the 'no_path' (and its follow-ups if any).

A detailed account of the patterns and rules have been provided as supplementary material[5].

Handling Retries. To manage retries of skipped questions, a supplementary skip graph $G_{skip} = (V_{skip}, E_{skip}, A_{skip})$ is introduced. Skipped questions are stored in this graph until the interviewer (the system) decides to revisit them. Experts suggested that the retries should occur before moving on to another topic. The skip graph mirrors the structure and attribute constraints of the

[5] https://github.com/VU-RASL/adaptive-dialogue-graph-grammar.

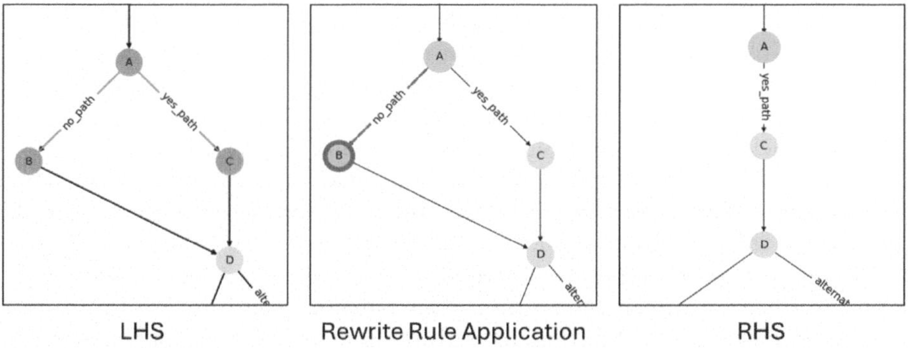

Fig. 2. Rule-Yes-Answer where LHS shows the matched instance (in orange) for the pattern identifying yes/no questions to which the user responded with a yes-answer. The corresponding rule is applied where the 'no_path' is removed. On deleting Node B, ReGraph' s rewrite method deletes Node B and its outgoing edges. RHS shows the modified subgraph within the host graph. (Color figure online)

host graph G. The node set V_{skip}={`init_skip`} ∪ $\{v|A(v)$[**skipped**]=True}, where $V_{skipped}$ contains nodes flagged with $A(v)$ [**skipped**]=True, holds skipped dialogue nodes. Directed edges $E_{skip} \subseteq V_{skip} \times V_{skip}$ retain their attributes, including **label** and *extra_attributes*. Nodes are moved from G to G_{skip} when the rule marks them as skipped. This design preserves the structure and relationships of the skipped subgraph, enabling seamless reintegration through GR when needed.

3.4 Implementation

Our framework uses ReGraph[6], a Python library for Sesqui Pushout (SqPO) graph rewriting, while also supporting Single and Double Pushout rewriting. Graphs are stored as Java Script Object Notation (JSON) with nodes and edges represented as key-value attributes (e.g., **id**, **section**, **dialogue**), enabling flexible, domain-general dialogue representations. ReGraph was built on the NetworkX[7] Python library for graph representations, where the JSON graph is an object with a list of nodes and a list of edges, i.e., nodes: [ ], edges: [ ], within which each node is a dialogue object with the key-value pairs defined above, and each edge is an object with a *source* and *target*, and additional attributes are presented as the *extra_ attributes* defined above. Thus G and G_{skip} are both graphs in the format described above—an object with a list of nodes and a list of edges.

During development, we found ReGraph's built-in matching to be inefficient for graphs exceeding roughly 10 nodes, due to its exhaustive search strategy. Although we did not formally analyze worst-case complexity, performance

[6] https://github.com/Kappa-Dev/ReGraph.
[7] https://networkx.org/.

notably degraded with increasing graph size. The matcher is based on NetworkX's DiGraphMatcher enhanced with custom attribute-matching functions. To improve efficiency, we enable early termination upon the first match and restrict the search space to a local neighborhood around the current node, inspired by neighborhood-controlled graph rewriting [64]. ReGraph allows specifying a subset of nodes for subgraph matching, enabling search within a portion of the graph rather than the entire structure. We extracted node ids for the search space or the neighborhood, at a certain depth. Figure 3 shows an example of a depth-1 neighborhood of the "current" Node 5, which can be extended to the end node if needed. Given the nature of our patterns, successors of the current node (or dialogue) are included along with the previous and pre-previous nodes, which are also relevant. The implementation of the pattern matching and rule application process is detailed in Algorithm 1 in Appendix 1. As the number of follow-ups is unknown, we use Depth-First Search (DFS) to locate nodes for removal based on attributes (**type**='follow-up') and edge labels (e.g., 'alt_path', 'yes_path', 'no_path', or 'broken_down'). In the interview graph, nodes with multiple outgoing paths (e.g., 'yes_path' and 'no_path') are designed to eventually converge at a common node, which marks the start of the next question or dialogue sequence. To trace this convergence point, we use a depth-first search (DFS) to find the path from the current node to the meeting node. This process is illustrated in Sect. 3.5.

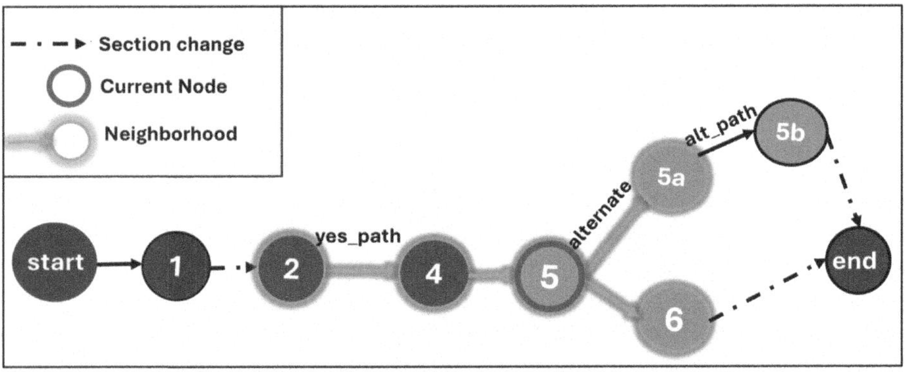

Fig. 3. The interview graph represents the current node 5 and its neighborhood consisting of previous two nodes, 2 and 4, and the successors of 5 itself, 5a and 6. Pattern matching is conducted on a subgraph as shown, which optimizes the search process.

3.5 Dialogue Management

While the grammar includes patterns and rules address various scenarios, this section focuses on select examples specifically relevant to the interview context for which it was developed. We applied the rules to a sample interview graph with example responses to illustrate dialogue management in practice.

Figure 4 shows the dialogue between an interviewee and interviewer in a data entry clerk interview, represented both as a conversation and as a dialogue graph. In addition, we use a supplementary skip graph (G_{skip} labeled "SKIP GRAPH") which is also a NetworkX graph that temporarily stores questions that the user is unable to answer for later reintroduction. The graph also tracks abandoned nodes (a list labeled "ABANDONED"), which represent discarded follow-up questions that have not been traversed; if a question is traversed, it is marked abandoned but not added to the ABANDONED list. Initially all the nodes are **type**='normal.' These visuals aid the discussion below on various interview scenarios. Initially at Node `1` (see Fig. 4), the user's state is identified as engaged and there are no instances of any pattern matched. Moving to Node `2`, assuming a Yes/No question is encountered (e.g., "Do you have any prior work experience?"), the user replies "Yes...", leading to an engaged-stressed state. Pattern P_{yes_answer} is matched, triggering **Rule-Yes-Answer** (see Fig. 2), which deletes the 'no_path' and its follow-ups, abandoning Node `3` and proceeding to Node `4`. If the user had responded negatively or unclearly (e.g., "I don't know"), the system would match P_{no_answer} or P_{other_answer} and apply the corresponding rule. Below we discuss the special cases the GR handles.

Handling Stuck States: Figure 5 illustrates how the framework handles moments when the user gets stuck. On the first instance (P_{stuck_once}), **Rule-Stuck-Repeat** rephrases or repeats the question. If the user remains stuck (P_{stuck_twice}), the **Rule-Stuck-Repeat Again** is triggered, inserts another repetition or rephrasing. After a third failed attempt (P_{stuck_thrice}), **Rule-Stuck-Hint** inserts a helpful hint. If the user still cannot respond ($P_{stuck\_after\_hint}$), **Rule-Stuck-Hint-Skip**, removes the question and follow ups from host graph, moving them to the skip graph (G_{skip}) for later. Before removing the follow up nodes, we retrieve the successor of the last follow-up for the current question. After removal, we reconnect the current node to that successor. In Fig. 4, the hint node `5_1_1_hint` is excluded from G_{skip}. Instead, the original question Node `5` (and follow-ups, if any) are moved and labeled as skipped $A(5)$[**skipped**]=True, highlighted in yellow). Job coaches appreciated this adaptive handling, as it gives users time, appropriate prompts, and ensures important questions are revisited.

Handling Uncertain Responses: In the event the user says "I don't know" or "I'm not sure" (P_{other_answer}) after being stuck once, twice or even after receiving a hint, the question is marked as abandoned and traversed using **Rule-Abandon-Other-Answer** ($A(5)$[**abandoned, traversed**]=True), and the follow-ups are added to the ABANDONED list. nodes list. In Fig. 4 at Node `6`, the user responds with "I don't know", triggering the rule via $A(6)$ [**dialogue_act**]='other-answer.' Node `6` is marked as abandoned ($A(6)$[**abandoned**]=True), but since it has already been traversed, it is not added to the ABANDONED list. Instead, any unvisited follow-ups from this

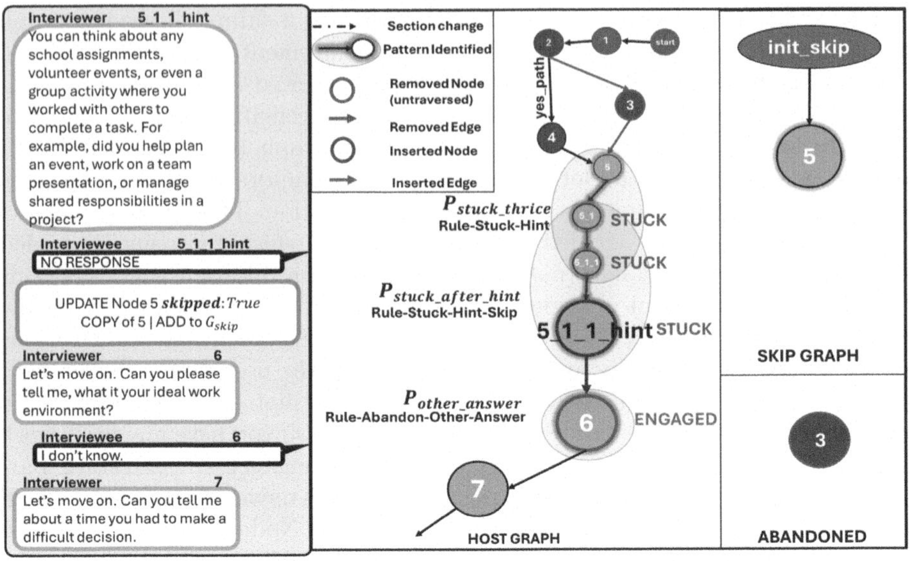

Fig. 4. Interview dialogue depicted as a textual conversation (left) and a dialogue graph (center) with a supplementary skip graph (right). The skip graph temporarily stores unanswered questions for later retrial, while the ABANDONED list includes discarded follow-up questions. Initially, all nodes are of type 'normal'. After a 'yes-answer' at Node 2, leading to an engaged state at Node 4, the user becomes stuck at Node 5. A rephrased version is presented at Node `5_1`, followed by a hint at `Node 5_1_1_hint` upon subsequent stuck states. When the user remains stuck for the third time, **Rule-Stuck-Hint-Skip** moves the original Node `5` to the skip graph. At Node `6`, the user's "don't know" response results in a traversed but abandoned node.

node are marked as abandoned and then removed from the host graph and added to this list.

Handling Stressed States: Managing stress during interviews is a key aspect of social skills training (SST). If a predefined "alternate" path exists in the script (G), the system can reduce user stress by switching to easier questions or topics. In Fig. 5, at Node `8`, the user is asked an abstract question-commonly used to assess problem-solving or creativity-but responds, "I don't know. What does that have to do with the job?" This real example, shared by a job coach, reflects how users may struggle with unexpected questions. Here the user is again in an *engaged-stressed* state consecutively for the third time. Although our Dialogue Act Classifier (which processes the input text as a whole instead of separate sentences) would typically match $P_{other\_answer\_alt}$ based on "I don't know," we prioritize the user's state. Therefore $P_{stressed\_thrice\_alt}$ is matched and the **Rule-Stressed-Alternate** is applied. This removes the original edge to Node `9`. If Node `9` has no other incoming edges, the rule deletes it and any related follow-ups ($A(v)$[**type**]='follow-up') from the HOST GRAPH. If an "alternate" path is not available, $P_{stress\_all\_combo}$ (*stressed-stuck*, *stressed-engaged*, or *stressed-*

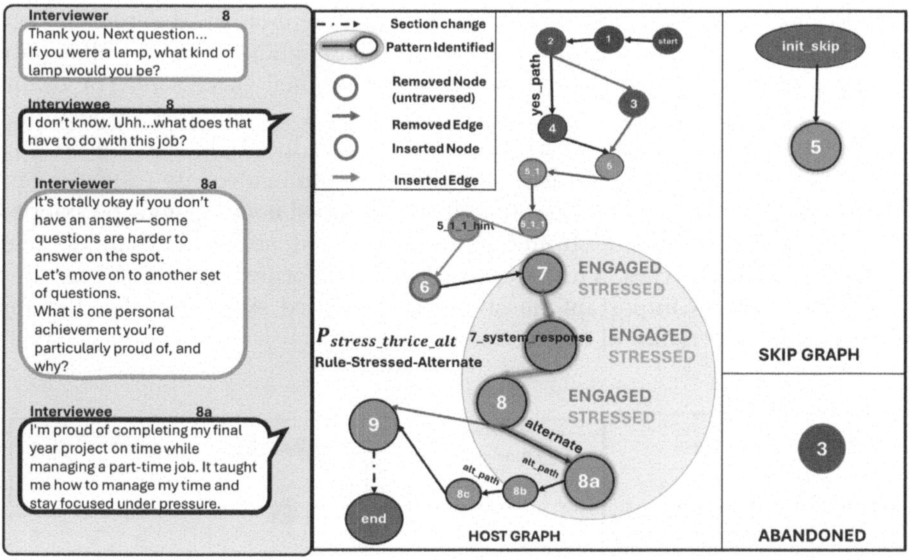

Fig. 5. Illustration of an interaction where the user was stressed three times in a row. At Node 7, the interviewee is in ENGAGED-STRESSED state. They responds with a clarification question, e.g., "What do you mean?" (not shown) which is evident by the next Node `7_system_response`. This implies that the interviewer responded to the clarification. Here the interviewee is STRESS-ENGAGED again, but they were able to answer the question this time. We then move to Node 8 where the user was ENGAGED-STRESSED again. Since, Node 8 has an 'alternate' direction, the pattern $P_{stressed\_thrice\_alt}$ has a match and we delete the path from Node 8 to Node 9 and take the alternate direction where the next node is Node 8a.

stuck-engaged) is matched and **Rule-Stressed-Break-Down**, introducing a simplified subgraph of broken-down questions. If the user remains stuck, **Rule-Stuck-Abandon-Broken-Down** is applied (not shown), abandoning follow-ups to avoid unnecessary cognitive load-an approach supported by job coaches.

Handling Off-Topic Responses: Visually, handling off-topic responses works the same way as shown in Fig. 5. On first occurrence a prompt node is inserted (**Rule-OffTopic-Once-Prompt**). A second off-topic response triggers another prompt (**Rule-OffTopic-Twice-Insert-Prompt**). If the user goes off-topic three times, and an alternate path exists, the system switches to it and removes the original path (**Rule-OffTopic-Thrice-Alternate**). If no alternate path is available, the node and its follow-ups are moved to G_{skip} for later reintroduction.

Revisiting Skipped Nodes: In Fig. 6, after traversing Nodes 8a-8c, the system reaches Node 9, where the user gives an engaged response. This triggers $P_{section\_change}$ is matched, signaling a transition to a new section ("technical") and applying **Rule-Insert-Skip**. As shown in Fig. 6, previously skipped questions are reinserted into the dialogue, with the first skip node relabeled as **init_skip_technical** to inform the user of the revisitation. Skipped nodes (marked **skipped**=True and highlighted with a yellow)-are reintroduced between the current node and its original successors. For example, the previously skipped technical question, "Tell me about a project that you worked on in a team with others," is reintroduced. If the user replies with $A(v)$[**dialogue_act**]='other-answer' or gives an inadequate response, any follow-ups linked to that node are abandoned. Skipped nodes cannot be skipped again. Once all skipped nodes from G_{skip} are reinserted, no further changes occur upon subsequent matches. This adaptive feature, informed by discussions with job coaches ensures important questions are revisited systematically, thereby enhancing the experience of the interview training.

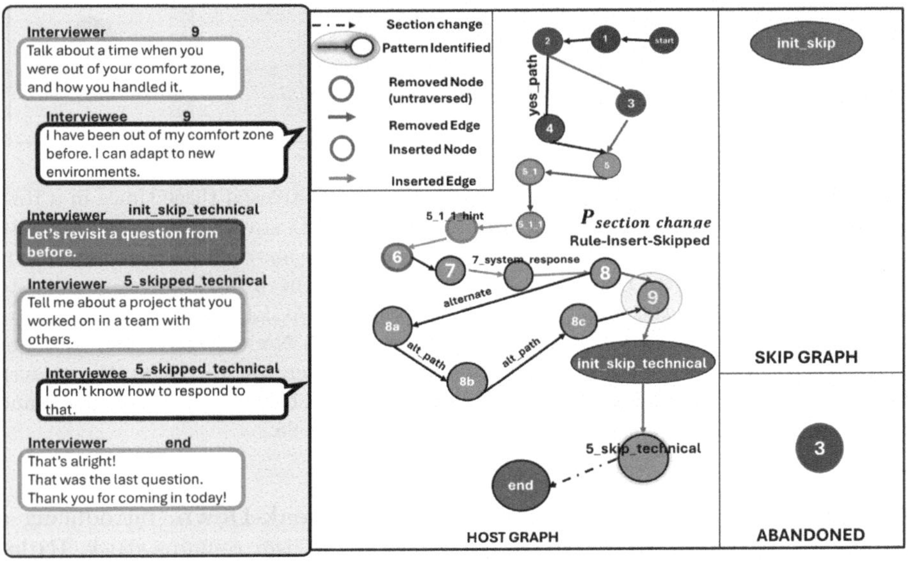

Fig. 6. Illustration of how skipped nodes are reintroduced again. Node 9 is the last node in the current section which is indicated by the dashed arrow between 9 and Node **end**. Since the SKIP GRAPH in Fig. 5 had nodes after the **init_skip** node, we remove the skipped nodes in the preserved pattern and relabel the nodes to avoid conflict. The source of the edge (9 → **end**) where **section_change** edge label is True. This edge's source is now the terminal node in the skip graph (Node **5_skip_technical**). We add an edge from 9 to the **init_skip_technical** which on the left chat view, provides dialogue to inform the user that a previous question is being revisited.

3.6 Verfication of Graph Properties

The interview graph, structurally, must satisfy key properties. While multiple properties apply—such as connectivity, reachability, acyclicity, termination, and degree constraints. Since *acyclicity* and *reachability* implies *termination*, and *reachability* implies *connectivity* [88], we need to verify only three properties—*reachability*, *acyclicity*, and *degree constraints*.

We utilized NetworkX functions on the converted Regraph NXGraph host graph. *Reachability* was checked using **has_path**(G, *source*="start", *target*="end") and *acyclicity* was tested using **is_directed_acyclic_graph**(G), and *degree constraints* were determined by counting paths with **all_simple_paths**(G, *source* = "start", *target* = "end"), ensuring only one unique path remained after transformations. These checks, returning True/False, were applied before and after rule execution on test cases.

Test cases were generated that included input conditions suggested to us by the job coaches: 1) the is unable to respond to several questions in within a section of the interview, 2) the user is struggling with open-ended questions, and 3) the user doesn't understand a question and continuously asks for clarifications. Although edge test cases reflect the extreme conditions which only a minority of users may encounter, they have the potential to reveal critical issues. Edge cases included conditions such as 1) the user is engaged throughout the interview (baseline), 2) the user is unable to respond to any question, 3) the user is stressed throughout, 4) the user is stuck throughout the interview, and 5) the user repeatedly asks questions. We believe these edge cases can cover the conditions suggested by the experts.

We used interview scripts from our prior work that had been designed with stakeholders and domain experts from interview coaching centers. These scripts were in a format compatible with our interview simulator, and we converted them to the format that is accepted by ReGraph and NetworkX. We traversed the script using Algorithm 1 (see Appendix 1) which we modified to include saving of subgraphs on which the rule was applied. Since no tool deemed compatible with our work, we developed our own graphics demonstrating counterexamples from when a property was violated. The results are discussed in the next section.

4 Results

We discuss the results from conditions 1–5 below. The interview script used was from our interview simulator given to us by job coaching centers, also stakeholders of this work during the last study (see prior work [2]). It consisted of 53 nodes including the `start` and `end` nodes.

4.1 Testing for Engagement Edge Case

The *engagement* state edge case examines a scenario where the user is *engaged* throughout the interview, ensuring that *baseline* graph properties are met when no negative user state or condition is encountered. An issue arose with the yes/no pattern at Node `3` (See Fig. 7a). Suppose Node `3` asks, "Do you have any previous work experience?" a "yes" response leads to a follow-up in Node `4` (e.g., "Great! Please tell me more!"), then proceeds to Node `5` which is another question ("e.g., Do you have experience with Tableau?"). If they say "no" at Node `3`, then Node `5`

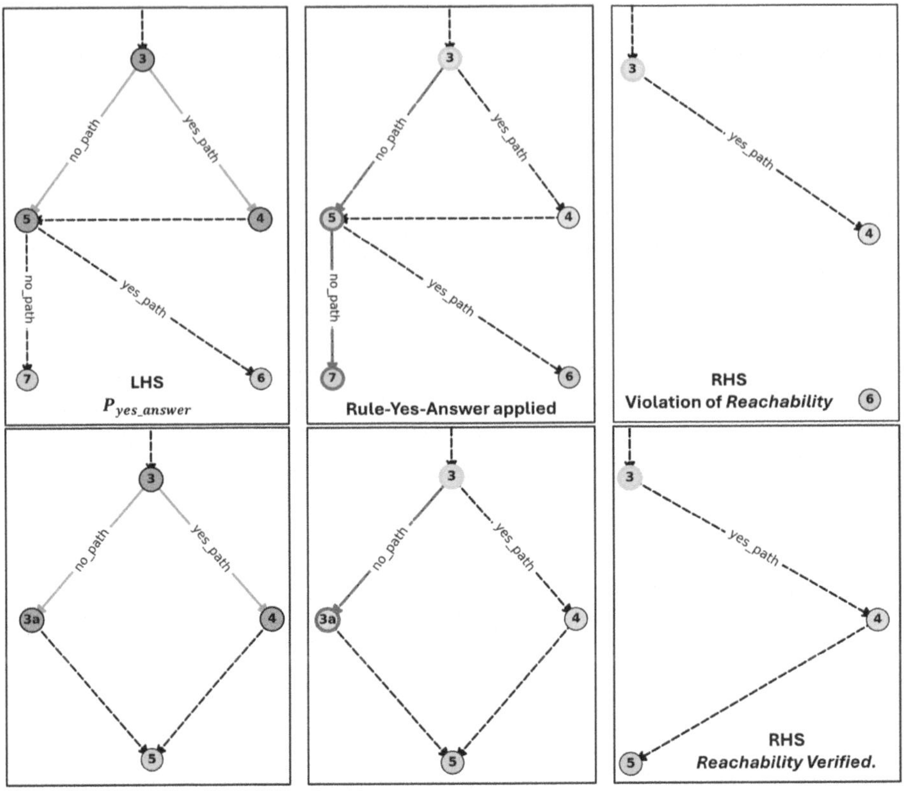

Fig. 7. Figures showing a rule application P_{yes_answer} (center) on an instance of a matched pattern (LHS), with the output after transformation (RHS), a) Illustration of a reachability violation: when **dialogue_act**='no-answer', the 'no_path' and its follow-ups are deleted, causing the graph to become disjoint after Node 4. This occurred because the 'yes_path' and 'no_path' both led to Node 5. b) The fix required inserting a follow-up node (e.g., Node 3a) after each yes/no question(bottom-left), ensuring the 'yes_path' doesn't have an incoming edge to the direct successor of Node 3. Dashed arrows indicate a section or topic change. With this structure, reachability is preserved (bottom-right).

is the next question, which is a direct successor of Node 3 along the 'no_path'. Figure 7a shows the issue where on applying **Rule-Yes-Answer**, Node 5 is removed which automatically removes the outgoing edges from Node 5 that are considered follow-ups along the 'no_path'. To prevent any dangling edges, the rule specifies deletion of edge 3 $\rightarrow$ 5 as well. This leads to a disjointed graph and thus, violating *reachability*. Further, if Node 5 is of type 'follow-up' even though it is a new question to a new section, the 'no_path' along 5 erroneously gets deleted. To address this, we require that the direct successor of a yes/no question along the 'no_path' must be a follow-up node that does not require a response (e.g., the new Node 3a) as shown Fig. 7b. This renders a direct edge from Node 3 to Node 5 invalid; instead, the edge should go from Node 3a to

Node 5. Additionally, the first meeting node on the 'yes_path' must not coincide with the direct successor of the no path, and vice versa.

Another minor issue arose when the **dialogue_act** was a statement, instead of yes-answer in a yes/no pattern. Semantically, 'statement' (e.g., "I have worked at {company} before") is still a positive response. For this test, we treated both a 'statement' and a 'yes-answer' as affirmative. Ideally, future work could replace 'yes-answer' with 'statement' and use sentiment analysis to distinguish a positive statement from negative one that might imply an affirmative response. This would make the rule less dependent on dialogue act labels and more on sentiment, i.e., either positive, negative, or neutral. All issues noted above were resolved before proceeding to the next test case.

4.2 Testing for Stuck Edge Case

We tested the edge case where the user remains stuck and fails to respond to any questions. One key design criterion was to ensure that reintroduced (previously skipped) questions cannot be skipped again. This is enforced through the pattern matching logic, which includes a **skipped**=False condition. Once a node is skipped, it is marked with **skipped**=True, preventing the skip pattern from matching it again upon reintroduction. Figure 8 in Appendix 2 shows a snippet of the graph generated after the test results. No graph properties were violated.

4.3 Testing for Stressed Edge Case

In the "all stressed" edge case, we encountered an infinite loop. At Node 6, due to the user being stressed and the absence of an alternate path, the node split into `8_split_0` and `8_split_1` to ease the response. However, `8_split_0` also triggered the same pattern due to repeated stress, causing continuous splits and preventing termination. We capped traversal at 100 steps. The issue arose because there was no variable breaking the repeated stress pattern. To resolve this, we modified pattern $P_{stress\_all\_combo}$ to only match when the *stressed* state is first encountered on a node with **type**='normal'. This fix ensured the graph terminated as expected. Note: a 'broken-down' question can only be repeated once; if the user is still stuck or stressed, the remaining nodes are abandoned and the flow moves on. Figure 9 in Appendix 2 shows an example of inserted 'broken-down' questions on consecutive stress states.

4.4 Testing Edge Case User Clarifications

The last criteria was to demonstrate what the system would do if the user repeatedly asked for clarifications throughout section of the interview with open-ended questions. For this we used the engagement edge case values from Sect. 4.1, but set the **dialogue_act** to 'user-query'. Figure 10 in Appendix 2 shows a portion of the graph on repeated clarifications. We can see that the system clarifies only once and then accepts whatever response the user provides and moves on to the next node. The resulting graph satisfied all required properties.

5 Discussion and Conclusions

Adaptive interview training systems address the limitations of rigid, scripted interactions by dynamically adjusting dialogue based on user responses and cognitive states. Autistic job seekers, in particular, face challenges with social interaction, response times, and open-ended questions, requiring structured redirection and simplified alternatives [77,82,99]. While 1-on-1 mock interviews offer tailored support, they are costly and not as accessible, making virtual training a scalable alternative. However, existing simulators lack real-time adaptability and do not integrate implicit social cues from multimodal user data. This work introduces graph rewriting (GR) with user-state inference to dynamically adjust dialogue flow based on engagement, stress, and stuckness. Key features include: 1) hints or rephrased questions for stuck users, 2) follow-up questions tailored to cognitive and social needs.

Although it is not claimed here that GR rules are the most efficient way of implementing adaptivity, they offer a flexible, rule-driven approach without requiring large amounts of training data to predict dialogue flow. They act as an alternative method to avoid the state explosion problem that FSMs may posses on incorporating user behavior states. Since interview scripts are naturally structured as graphs with attributes, GR provides a promising foundation for real-time dialogue adaptation.

Despite the potential, there are limitations in this work. Dynamic follow-up question generation is not yet fully implemented but could be improved using Generative AI (e.g., ChatGPT [1]) for context-aware responses. Topic detection also needs refinement, as responses must be classified as on-topic or off-topic—an area where large language models could enhance accuracy by learning semantic patterns. Additionally, current transformation rules have only been tested in one interview context, requiring broader evaluation across varied interview structures. While the current work demonstrates the functionality of the proposed framework and its technical aspects, we acknowledge a limitation in the evaluation of its effectiveness in terms of generating useful and meaningful conversations. The current evaluation primarily focuses on the correctness of rule applications within test scenarios, ensuring that the graph properties are maintained. However, assessing the real-world utility of the framework in generating engaging and effective dialogues is a critical next step. To address this, we plan to conduct a survey-based evaluation with job coaches in the near future. They will review interview sessions generated by the framework, providing feedback on whether they would have taken the same actions in response to user inputs and whether the conversation generated is meaningful.

Future work will expand pattern recognition as real-user interactions introduce new dialogue patterns, refining transformation rules for better adaptability. The system will evolve toward procedural content generation using generative graph grammar, similar to dynamic video game level generation [79]. Future iterations will feature customizable interviewer profiles and extend to social skills training (e.g., negotiation, customer interactions). We will integrate this grammar into our interview simulator to further test feasibility in mock interview simulations.

Acknowledgments. We would like to thank all the experts for participating in discussions regarding the semantics of this project. We would also like to acknowledge our funding source from NSF NRT grant 19-22697 (K. Stassun, PI).

Disclosure of Interests. The authors have no competing interests to declare that are relevant to the content of this article.

Appendix 1

Algorithm 1: Graph Traversal with Pattern Matching

```
   Input: G, G_skip, P
   Output: G'
1  current_node ← "start";
2  while current_node ≠ "end" do
3      Display current_node[dialogue];
4      multimodal_data ← get_user_data();
5      utterance ← multimodal_data["user_response"];
6      dialogue_act ← get_user_data(utterance);
7      is_offtopic ← topic_detector(utterance);
8      user_state ← multimodal_data["user_state"];
9      Update A(current_node) with {utterance, dialogue_act, user_state,
       is_offtopic};
10     (pattern, instance) ← match_pattern(G, current_node);
11     if pattern ≠ ∅ then
12         apply_transformation_rule(G, pattern, instance);
13         Update G_skip;
14     current_node["traversed"] ← True;
15     (pattern, instance) ← match_pattern(G, current_node);
16     if pattern = P_section_change then
17         apply_transformation_rule(G, pattern, instance);
18         Update G_skip;
19     current_node ← get_next_node(G, current_node);
```

Algorithm 1 takes as input, the host graph G, the skip graph G_skip, and a priority-ordered list of patterns P, then begins traversal at the "start" node and repeats until the "end" node is reached. In each iteration, it first retrieves the user's multimodal data (line 4), extracts the utterance to determine the dialogue act and off-topic status (line 5 and 6), and infers the binary state variables **is_stuck, is_engaged**, and **is_stressed** (line 8). It then scans the patterns in P by priority, applies the matching rule to update G, and incorporates any skipped nodes into G_{skip}. Next, it examines the outgoing edge for a **section_change**=True flag (line 16) and, if set and if the skip graph has nodes and edges for reinsertion, inserts them before proceeding. Finally, it advances along the appropriate edge to the next node and repeats the process.

Appendix 2

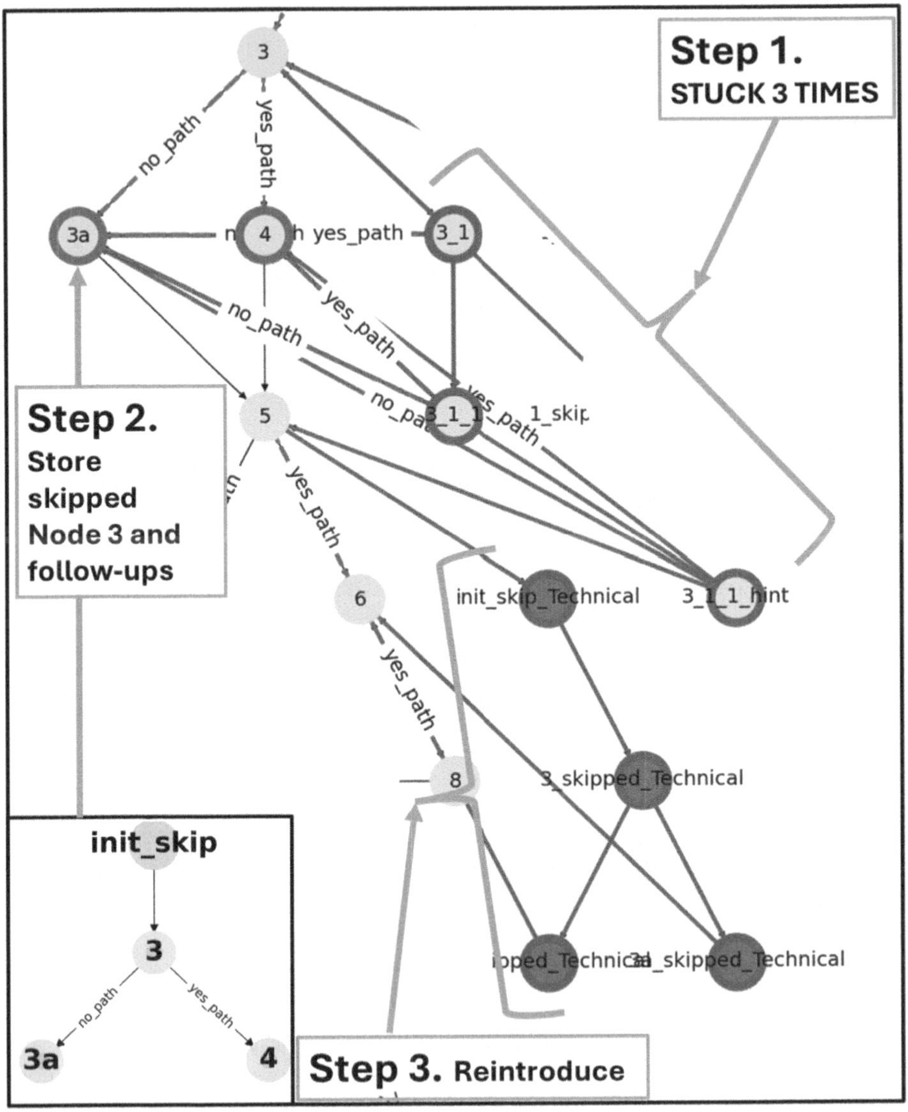

Fig. 8. Snippet of the results from the stuck state edge case test. Step 1, the user is stuck 3 times in a row at Node 3. Step 2, Node 3 and it's followups are inserted into the skip graph. Step 3, the skip graph nodes are relabled and inserted on section change.

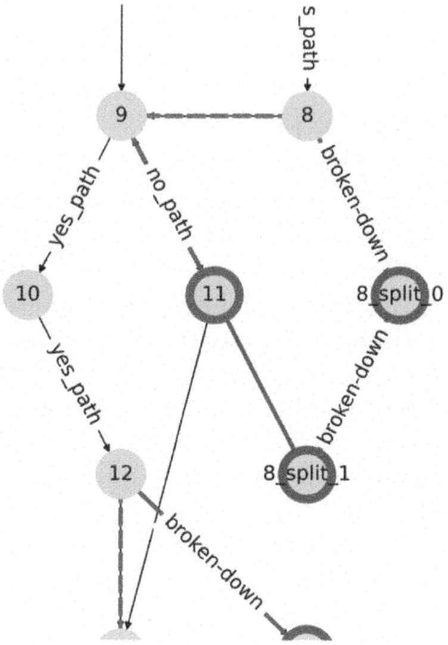

Fig. 9. When the user is stressed for the last three states, and there is no alternate path provided rule **Rule-Stressed-Break-Down** inserts nodes implying broken-down, simpler questions (here 8_split_0) instead of the question at Node 8.

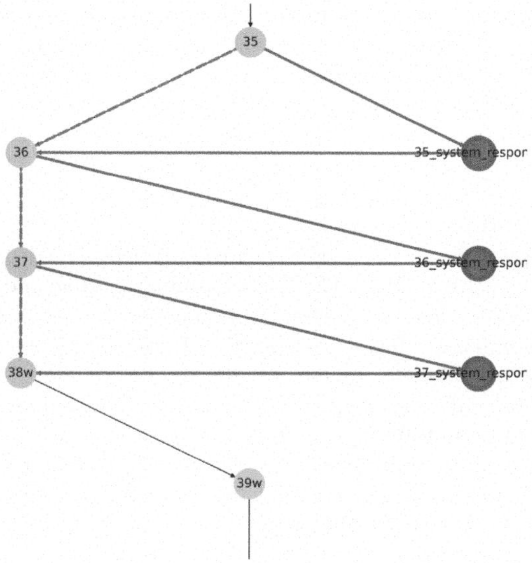

Fig. 10. A snippet of the interview graph showing responses from the system inserted after each question (in green to the right with the id followed by '_system_response'). This illustrates handling of user questions. (Color figure online)

References

1. Chatgpt (2022). https://openai.com/blog/chatgpt
2. Adiani, D., et al.: Multimodal job interview simulator for training of autistic individuals. Assist. Technol., 1–18 (2023)
3. Adiani, D., Colopietro, K., Wade, J., Migovich, M., Vogus, T.J., Sarkar, N.: Dialogue act classification via transfer learning for automated labeling of interviewee responses in virtual reality job interview training platforms for autistic individuals. Signals **4**, 359–380 (2023). https://doi.org/10.3390/signals4020019
4. Adiani, D., et al.: Career interview readiness in virtual reality (CIRVR): a platform for simulated interview training for autistic individuals and their employers. ACM Trans. Access. Comput. **15** (2022). https://doi.org/10.1145/3505560
5. Ali, M.R., Hoque, E.: Social skills training with virtual assistant and real-time feedback. In: Proceedings of the 2017 ACM International Joint Conference on Pervasive and Ubiquitous Computing and Proceedings of the 2017 ACM International Symposium on Wearable Computers, UbiComp 2017, pp. 325–329. Association for Computing Machinery, New York (2017). https://doi.org/10.1145/3123024.3123196
6. Amat, A., et al.: Design of a desktop virtual reality-based collaborative activities simulator (VIRCAS) to support teamwork in workplace settings for autistic adults. ieeexplore.ieee.org (2023)
7. Anderson, K., et al.: The TARDIS framework: intelligent virtual agents for social coaching in job interviews. In: Reidsma, D., Katayose, H., Nijholt, A. (eds.) ACE 2013. LNCS, vol. 8253, pp. 476–491. Springer, Cham (2013). https://doi.org/10.1007/978-3-319-03161-3_35
8. Andrei, O., Kirchner, H., Kirchner, H.E.: Graph rewriting and strategies for modeling biochemical networks. In: International Workshop on Natural Computing and Applications-NCA (2007). https://inria.hal.science/inria-00146362v3
9. Arnold, M., et al.: The use of virtual reality roleplay to improve communicative and digital skills in a transnational collaborative learning environment: Implications for teaching and learning in social work management education. In: Handbook of Applied Teaching and Learning in Social Work Management Education: Theories, Methods, and Practices in Higher Education, pp. 377–413 (2023). https://doi.org/10.1007/978-3-031-18038-5_19
10. Awodey, S.: Category Theory. Oxford University Press (2010)
11. Ayedoun, E., Hayashi, Y., Seta, K.: Adding communicative and affective strategies to an embodied conversational agent to enhance second language learners' willingness to communicate. Int. J. Artif. Intell. Educ. **29**(1), 29–57 (2018). https://doi.org/10.1007/s40593-018-0171-6
12. Bardohl, R.: Application of graph transformation to visual languages-state of the art and further ideas (1997)
13. Baur, T., Damian, I., Lingenfelser, F., Wagner, J., André, E.: NovA: automated analysis of nonverbal signals in social interactions. In: Salah, A.A., Hung, H., Aran, O., Gunes, H. (eds.) HBU 2013. LNCS, vol. 8212, pp. 160–171. Springer, Cham (2013). https://doi.org/10.1007/978-3-319-02714-2_14
14. Behoora, I., Tucker, C.S.: Machine learning classification of design team members' body language patterns for real time emotional state detection. Des. Stud. **39**, 100–127 (2015). https://doi.org/10.1016/j.destud.2015.04.003
15. Bell, J.L.: Types, Sets, and Categories, vol. 6, pp. 633–687. North-Holland (2012). https://doi.org/10.1016/B978-0-444-51621-3.50009-8

16. Blostein, D., Fahmy, H., Grbavec, A.: Issues in the practical use of graph rewriting. In: Graph Grammars and Their Application to Computer Science: 5th International Workshop Williamsburg, VA, USA, 13–18 November 1994 Selected Papers, vol. 5, pp. 38–55 (1996)
17. Boersma, P., Heuven, V.V.: Speak and unSpeak with PRAAT. Glot Int. **5**, 341–347 (2001)
18. Boisvert, B., Féraud, L., Soloviev, S.: Graph transformations, proofs, and grammars
19. Bosch, N., et al.: Automatic detection of learning-centered affective states in the wild. In: Proceedings of the 20th International Conference on Intelligent User Interfaces, pp. 379–388 (2015)
20. Boyer, K.E., Grafsgaard, J.F., Ha, E.Y., Phillips, R., Lester, J.C.: An affect-enriched dialogue act classification model for task-oriented dialogue. In: Proceedings of the 49th Annual Meeting of the Association for Computational Linguistics: Human Language Technologies - Volume 1, pp. 1190–1199. Association for Computational Linguistics (2011)
21. Breunig, M., et al.: Collaborative multi-scale 3D city and infrastructure modeling and simulation. In: The International Archives of the Photogrammetry, Remote Sensing and Spatial Information Sciences **XLII-4-W4**, pp. 341–352 (2017). https://doi.org/10.5194/ISPRS-ARCHIVES-XLII-4-W4-341-2017
22. Broekens, J., Harbers, M., Brinkman, W.-P., Jonker, C.M., Van den Bosch, K., Meyer, J.-J.: Virtual reality negotiation training increases negotiation knowledge and skill. In: Nakano, Y., Neff, M., Paiva, A., Walker, M. (eds.) IVA 2012. LNCS (LNAI), vol. 7502, pp. 218–230. Springer, Heidelberg (2012). https://doi.org/10.1007/978-3-642-33197-8_23
23. Cai, Z., Graesser, A.C., Hu, X.: ASAT: Autotutor script authoring tool. In: Design Recommendations for Intelligent Tutoring Systems: Authoring Tools, vol. 3, pp. 199–210 (2015)
24. Carruth, D.W.: Virtual reality for education and workforce training. In: 2017 15th International Conference on Emerging eLearning Technologies and Applications (ICETA), pp. 1–6 (2017). https://doi.org/10.1109/ICETA.2017.8102472
25. Chakravarty, S., Chava, R.V.S.P., Fox, E.A.: Dialog acts classification for question-answer corpora. In: ASAIL@ICAIL (2019). https://api.semanticscholar.org/CorpusID:195693905
26. Chiang, C., et al.: Examining the effectiveness of 3D virtual reality training on problem-solving, self-efficacy, and teamwork among inexperienced volunteers helping with drug. jmir.org (2021). https://www.jmir.org/2021/11/e29862/
27. Corradini, A., Montanari, U., Rossi, F., Ehrig, H., Heckel, R., Löwe, M.: Algebraic Approaches to Graph Transformation-Part I: Basic Concepts and Double Pushout Approach, pp. 163–245. World Scientific (1997)
28. Cukurova, M., Kent, C., Luckin, R.: Artificial intelligence and multimodal data in the service of human decision-making: a case study in debate tutoring. Br. J. Edu. Technol. **50**, 3032–3046 (2019)
29. Gaffney, C., Dagger, D., Wade, V.: Evaluation of ACTSim: a composition tool for authoring adaptive soft skill simulations. In: Nejdl, W., Kay, J., Pu, P., Herder, E. (eds.) Adaptive Hypermedia and Adaptive Web-Based Systems, pp. 113–122. Springer, Heidelberg (2008)
30. Sedrakyan, G., Leony, D., Muñoz-Merino, P.J., Kloos, C.D., Verbert, K.: Evaluating student-facing learning dashboards of affective states. In: Lavoué, É., Drachsler, H., Verbert, K., Broisin, J., Pérez-Sanagustín, M. (eds.) Data Driven Approaches in Digital Education, pp. 224–237. Springer (2017)

31. Ding, D., Burger, F., Brinkman, W.-P., Neerincx, M.A.: Virtual reality negotiation training system with virtual cognitions. In: IVA 2012. LNCS (LNAI), vol. 7502, pp. 119–128. Springer, Cham (2017). https://doi.org/10.1007/978-3-319-67401-8_14
32. D'Mello, S., Graesser, A.: Dynamics of affective states during complex learning. Learn Instr. **22**, 145–157 (2012). https://doi.org/10.1016/j.learninstruc.2011.10.001
33. D'mello, S., Graesser, A.: Autotutor and affective autotutor: learning by talking with cognitively and emotionally intelligent computers that talk back. ACM Trans. Interact. Intell. Syst. **2** (2013). https://doi.org/10.1145/2395123.2395128
34. D'mello, S.K., Kory, J.: A review and meta-analysis of multimodal affect detection systems. ACM Comput. Surv. (CSUR) **47**, 1–36 (2015)
35. Dong, J., Zhuang, D., Huang, Y., Fu, J.: Advances in multi-sensor data fusion: algorithms and applications. Sensors **9**, 7771–7784 (2009). https://doi.org/10.3390/s91007771
36. Drewes, F., Kreowski, H.J., Habel, A.: Hyperedge replacement graph grammars, pp. 95–162 (1997). https://doi.org/10.1142/9789812384720_0002
37. Ehrig, H., Ehrig, K., Prange, U., Taentzer, G.: Implementation of typed attributed graph transformation by AGG. In: Fundamentals of Algebraic Graph Transformation, pp. 305–323 (2006)
38. Ehrig, H., et al.: Algebraic Approaches to Graph Transformation-Part II: Single Pushout Approach and Comparison with Double Pushout Approach, pp. 247–312. World Scientific (1997)
39. Ehrig, H., Prange, U., Taentzer, G.: Fundamental theory for typed attributed graph transformation. In: Ehrig, H., Engels, G., Parisi-Presicce, F., Rozenberg, G. (eds.) ICGT 2004. LNCS, vol. 3256, pp. 161–177. Springer, Heidelberg (2004). https://doi.org/10.1007/978-3-540-30203-2_13
40. Engelfriet, J., Rozenberg, G.: Node Replacement Graph Grammars, pp. 1–94 (1997). https://doi.org/10.1142/9789812384720_0001
41. Fernández, M., Kirchner, H., Pinaud, B., Vallet, J.: Labelled graph rewriting meets social networks. In: Lucanu, D. (ed.) Rewriting Logic and Its Applications, pp. 1–25. Springer, Cham (2016)
42. Forsberg, B.N., Lundström, A., Gulliksen, J.: VR for HR - a case study of human resource development professionals using virtual reality for social skills training in the workplace. LNCS (including subseries Lecture Notes in Artificial Intelligence and Lecture Notes in Bioinformatics), vol. 14145, pp. 231–251 (2023). https://doi.org/10.1007/978-3-031-42293-5_17
43. Furuno, T., et al.: 3D bow and posture measurements for virtual reality customer service training system. joig.net. https://doi.org/10.18178/joig.9.4.152-156. https://www.joig.net/uploadfile/2021/1124/20211124053356215.pdf
44. Gao, W., Wu, H., Siddiqui, M.K., Baig, A.Q.: Study of biological networks using graph theory. Saudi J. Biol. Sci. **25**, 1212–1219 (2018). https://doi.org/10.1016/J.SJBS.2017.11.022
45. Giannakos, M.: Theory and Intermediate-Level Knowledge in Multimodal Learning Analytics (2023)
46. Giannakos, M.N., Sharma, K., Pappas, I.O., Kostakos, V., Velloso, E.: Multimodal data as a means to understand the learning experience. Int. J. Inf. Manage. **48**, 108–119 (2019). https://doi.org/10.1016/j.ijinfomgt.2019.02.003
47. Gillies, M., Pan, X.: Virtual reality for social skills training. In: Proceedings of the Virtual and Augmented Reality to Enhance Learning and Teaching in Higher Education Conference 2018, pp. 83–92. No. 8. IM Publications Open (2018)

48. Graesser, A.C., D'Mello, S., Hu, X., Cai, Z., Olney, A., Morgan, B.: Autotutor. In: Applied Natural Language Processing: Identification, Investigation and Resolution, pp. 169–187. IGI Global (2012)
49. Gratch, J., DeVault, D., Lucas, G.: The benefits of virtual humans for teaching negotiation. In: Traum, D., Swartout, W., Khooshabeh, P., Kopp, S., Scherer, S., Leuski, A. (eds.) IVA 2016. LNCS (LNAI), vol. 10011, pp. 283–294. Springer, Cham (2016). https://doi.org/10.1007/978-3-319-47665-0_25
50. Grawemeyer, B., Mavrikis, M., Holmes, W., Gutierrez-Santos, S., Wiedmann, M., Rummel, N.: Affecting Off-Task Behaviour: How Affect-Aware Feedback Can Improve Student Learning, pp. 104–113. Association for Computing Machinery (2016). https://doi.org/10.1145/2883851.2883936
51. Gritta, M., Lampouras, G., Iacobacci, I.: Conversation graph: data augmentation, training, and evaluation for non-deterministic dialogue management. Trans. Assoc. Comput. Linguist. **9**, 36–52 (2021). https://doi.org/10.1162/tacl_a_00352
52. Hammer, D., Wildavsky, A.: The open-ended, semistructured interview: an (almost) operational guide. Craftways, 57–101 (2018). https://doi.org/10.4324/9780203794517-5
53. Hara, T., Maeda, H., Komatsubara, S., Taniqawa, T., Hirose, M.: Autonomous avatar for customer service training VR system. In: Proceedings - 2024 IEEE Conference on Virtual Reality and 3D User Interfaces Abstracts and Workshops, VRW 2024, pp. 835–836 (2024). https://doi.org/10.1109/VRW62533.2024.00214
54. Harmer, R., Cornec, Y.S.L., Legare, S., Oshurko, E.: Bio-curation for cellular signalling: the kami project. IEEE/ACM Trans. Comput. Biol. Bioinform. **16**, 1562–1573 (2017). https://doi.org/10.1109/TCBB.2019.2906164
55. Harmer, R., Oshurko, E.: Knowledge representation and update in hierarchies of graphs. J. J. Log. Algebr. Methods Program. **114**, 100559 (2020). https://doi.org/10.1016/J.JLAMP.2020.100559
56. Heckel, R., Taentzer, G.: Graph transformation for software engineers. Graph Transf. Softw. Eng. (2020). https://doi.org/10.1007/978-3-030-43916-3
57. Heimerl, A., Baur, T., André, E.: A transparent framework towards the context-sensitive recognition of conversational engagement (2016)
58. Henderson, N., Rowe, J., Paquette, L., Baker, R.S., Lester, J.: Improving affect detection in game-based learning with multimodal data fusion. In: Bittencourt, I.I., Cukurova, M., Muldner, K., Luckin, R., Millán, E. (eds.) AIED 2020. LNCS (LNAI), vol. 12163, pp. 228–239. Springer, Cham (2020). https://doi.org/10.1007/978-3-030-52237-7_19
59. Humm, L.B., Olsen, D., Morris, B., Fleming, M., Smith, M.: Simulated job interview improves skills for adults with serious mental illnesses. Stud. Health Technol. Inform. **199**, 50 (2014)
60. Hussain, M.S., Alzoubi, O., Calvo, R.A., D'mello, S.K.: Affect detection from multichannel physiology during learning sessions with autotutor (2011)
61. Ingraham, J., Garg, V.K., Barzilay, R., Jaakkola, T.: Generative models for graph-based protein design. Adv. Neural Inf. Process. Syst. **32** (2019)
62. Sui, Q.: Distance collaboration in virtual reality learning environments: a comprehensive study of educational innovation and social interaction, pp. 19–24. dl.acm.org (2024). https://doi.org/10.1145/3678392.3678416. https://dl.acm.org/doi/abs/10.1145/3678392.3678416
63. Ip, H.H., et al.: Enhance emotional and social adaptation skills for children with autism spectrum disorder: a virtual reality enabled approach. Comput. Educ. **117**, 1–15 (2018). https://doi.org/10.1016/J.COMPEDU.2017.09.010

64. Janssens, D., Rozenberg, G.: Graph grammars with neighbourhood-controlled embedding. Theoret. Comput. Sci. **21**(1), 55–74 (1982)
65. Johnson, M.T., Troy, A.H., Tate, K.M., Allen, T.T., Tate, A.M., Chapman, S.B.: Improving classroom communication: the effects of virtual social training on communication and assertion skills in middle school students. Front. Educ. **6** (2021)
66. Jonell, P., Fallgren, P., Doğan, F.I., Lopes, J., Wennberg, U., Skantze, G.: Crowdsourcing a self-evolving dialog graph. In: Proceedings of the 1st International Conference on Conversational User Interfaces. Association for Computing Machinery (2019). https://doi.org/10.1145/3342775.3342790
67. Kamei, A., Harriott, W.: Social emotional learning in virtual settings: intervention strategies. iejee.com **13**, 365–371 (2021). https://doi.org/10.26822/iejee.2021.196
68. Ke, F., Moon, J.: Virtual collaborative gaming as social skills training for high-functioning autistic children. Br. J. Educ. Technol. **49**, 728–741 (2018). https://doi.org/10.1111/BJET.12626
69. Ke, F., Moon, J., Sokolikj, Z.: Virtual reality-based social skills training for children with autism spectrum disorder. J. Spec. Educ. Technol. **37**(1), 49–62 (2022). https://doi.org/10.1177/0162643420945603
70. König, B., Nolte, D., Padberg, J., Rensink, A.: A tutorial on graph transformation. LNCS(including subseries Lecture Notes in Artificial Intelligence and Lecture Notes in Bioinformatics), vol. 10800, pp. 83–104 (2018)
71. Lahat, D., Adali, T., Jutten, C.: Multimodal data fusion: an overview of methods, challenges, and prospects. Proc. IEEE **103**, 1449–1477 (2015). https://doi.org/10.1109/JPROC.2015.2460697
72. Le, T.D., Imai, H., Negishi, Y., Kawachiya, K.: TFLMS: large model support in tensorflow by graph rewriting (2018). https://doi.org/10.48550/arXiv.1807.02037
73. Lee, J.H., Ostwald, M.J., Gu, N.: A justified plan graph (JPG) grammar approach to identifying spatial design patterns in an architectural style. Environ. Plan. B **45**(1), 67–89 (2018). https://doi.org/10.1177/0265813516665618
74. Liaw, S.Y., Ooi, S.W., Rusli, K.D.B., Lau, T.C., Tam, W.W.S., Chua, W.L.: Nurse-physician communication team training in virtual reality versus live simulations: Randomized controlled trial on team communication and teamwork attitudes. J. Med. Internet Res. **22**, e17279 (2020). https://doi.org/10.2196/17279
75. Liu, H., Fernando, O.N.N., Rajapakse, J.C.: Predicting affective states of programming using keyboard data and mouse behaviors. In: 2018 15th International Conference on Control, Automation, Robotics and Vision (ICARCV), pp. 1408–1413 (2018). https://doi.org/10.1109/ICARCV.2018.8581248
76. Malchanau, A.: Cognitive architecture of multimodal multidimensional dialogue management. Ph.D. thesis, Saarland University, Saarbrücken, Germany (2019). https://publikationen.sulb.uni-saarland.de/handle/20.500.11880/27372
77. Maras, K., Norris, J.E., Nicholson, J., Heasman, B., Remington, A., Crane, L.: Ameliorating the disadvantage for autistic job seekers: an initial evaluation of adapted employment interview questions. Autism **25**, 1060–1075 (2020)
78. Marchi, S., Targi, N., Liston, P.M., Parlangeli, O.: The possible role of empathy and emotions in virtual negotiation. Ergonomics **63**(3), 263–273 (2020). https://doi.org/10.1080/00140139.2019.1685678
79. Merrell, P.: Example-based procedural modeling using graph grammars. ACM Trans. Graph. **42** (2023). https://doi.org/10.1145/3592119
80. Milewski, B.: Category theory for programmers. Blurb (2018). https://bartoszmilewski.com/2014/10/28/category-theory-for-programmers-the-preface/

81. Moustafa, F., Steed, A.: A longitudinal study of small group interaction in social virtual reality, vol. 10. dl.acm.org (2018). https://doi.org/10.1145/3281505.3281527
82. Norris, J.E., Crane, L., Maras, K.: Interviewing autistic adults: adaptations to support recall in police, employment, and healthcare interviews. Autism **24**, 1506–1520 (2020)
83. Philip, J., Jiang, Y., Akdere, M.: Virtual reality technology for workplace training: the case for developing cultural intelligence. journals.sagepub.com **23**, 557–583 (2023). https://doi.org/10.1177/14705958231208297
84. Picard, R.W.: Affective Computing. MIT Press (2000)
85. Pittermann, J., Pittermann, A., Minker, W.: Emotion recognition and adaptation in spoken dialogue systems. Int. J. Speech Technol. **13**, 49–60 (2010). https://doi.org/10.1007/s10772-010-9068-y
86. Plump, D.: Essentials of term graph rewriting. Electron. Notes Theor. Comput. Sci. **51**, 277–289 (2002). https://doi.org/10.1016/S1571-0661(04)80210-X
87. Rosenfield, N.S., Lamkin, K., Re, J., Day, K., Boyd, L., Linstead, E.: A virtual reality system for practicing conversation skills for children with autism. Multimodal Technol. Interact. **3**, 28 (2019). https://doi.org/10.3390/MTI3020028
88. Rozenberg, G.: Handbook of Graph Grammars and Computing by Graph Transformation. World Scientific Publishing Co., Inc. (1997)
89. Sharma, K., Giannakos, M.: Multimodal data capabilities for learning: what can multimodal data tell us about learning? Br. J. Edu. Technol. **51**, 1450–1484 (2020)
90. Smith, M.J., et al.: Virtual reality job interview training and 6-month employment outcomes for individuals with schizophrenia seeking employment. Schizophrenia Res. **166**, 86–91 (2015). https://doi.org/10.1016/j.schres.2015.05.022
91. Smith, M.J., et al.: Virtual reality job interview training in adults with autism spectrum disorder. J. Autism Dev. Disord. **44**(10), 2450–2463 (2014). https://doi.org/10.1007/s10803-014-2113-y
92. Smith, M.J., et al.: Mechanism of action for obtaining job offers with virtual reality job interview training. Psychiatr. Serv. **68**, 747–750 (2017)
93. Spaulding, S., Gordon, G., Breazeal, C.: Affect-aware student models for robot tutors. In: Proceedings of the 2016 International Conference on Autonomous Agents & Multiagent Systems,pp. 864–872. International Foundation for Autonomous Agents and Multiagent Systems (2016)
94. Stanica, I., Dascalu, M.I., Bodea, C.N., Moldoveanu, A.D.B.: VR job interview simulator: where virtual reality meets artificial intelligence for education. In: 2018 Zooming Innovation in Consumer Technologies Conference, ZINC 2018, pp. 9–12 (2018). https://doi.org/10.1109/ZINC.2018.8448645
95. Stolcke, A., et al.: Dialogue act modeling for automatic tagging and recognition of conversational speech. Comput. Linguist. **26**, 339–373 (2000). http://arxiv.org/abs/cs/0006023
96. Suzuki, T., Ochiai, T., Hoshino, J.: Scenario-based customer service VR training system using second language. ieeexplore.ieee.org (2021). https://ieeexplore.ieee.org/abstract/document/9515961/
97. Tan, C.S.S., Schöning, J., Luyten, K., Coninx, K.: Informing intelligent user interfaces by inferring affective states from body postures in ubiquitous computing environments. In: Proceedings of the 2013 International Conference on Intelligent User Interfaces, pp. 235–246. Association for Computing Machinery (2013). https://doi.org/10.1145/2449396.2449427

98. Tao, J., Tan, T.: Affective computing: a review. In: Tao, J., Tan, T., Picard, R.W. (eds.) ACII 2005. LNCS, vol. 3784, pp. 981–995. Springer, Heidelberg (2005). https://doi.org/10.1007/11573548_125
99. Taylor, J.L., DaWalt, L.S., Marvin, A.R., Law, J.K., Lipkin, P.: Sex differences in employment and supports for adults with autism spectrum disorder. Autism **23**, 1711–1719 (2019)
100. Urma, R.G., Mycroft, A.: Source-code queries with graph databases-with application to programming language usage and evolution. Sci. Comput. Program. **97**, 127–134 (2015). https://doi.org/10.1016/J.SCICO.2013.11.010
101. Villani, D., Repetto, C., Cipresso, P., Riva, G.: May i experience more presence in doing the same thing in virtual reality than in reality? An answer from a simulated job interview. Interact. Comput. **24**, 265–272 (2012). https://doi.org/10.1016/j.intcom.2012.04.008
102. Voss, C., Petzold, F., Rudolph, S.: Graph transformation in engineering design: an overview of the last decade. AI EDAM **37**, e5 (2023). https://doi.org/10.1017/S089006042200018X
103. Walker, M., Passonneau, R.: Date: a dialogue act tagging scheme for evaluation of spoken dialogue systems (2001). https://www.aclweb.org/anthology/H01-1015
104. Wang, G.H., Zhang, J., Fu, G.S.: Predicting student behaviors and performance in online learning using decision tree. In: 2018 Seventh International Conference of Educational Innovation through Technology (EITT), pp. 214–219 (2018). https://doi.org/10.1109/EITT.2018.00050
105. Wanna, A., Coward, S., Drane, T., Constantinides, G.A., Ercegovac, M.D.: Multiplier optimization via e-graph rewriting. In: 2023 57th Asilomar Conference on Signals, Systems, and Computers, pp. 1528–1533 (2023). https://doi.org/10.1109/IEEECONF59524.2023.10476812
106. Wu, T.-H., Wu, F., Liang, C.-J., Li, Y.-F., Tseng, C.-M., Kang, S.-C.: A virtual reality tool for training in global engineering collaboration. Univ. Access Inf. Soc. (10), 1–13 (2017). https://doi.org/10.1007/s10209-017-0594-0
107. Xin-feng, Z., Jian-dong, W., Bin, L., Jun-wu, Z., Jun, W.: Methods to tackle state explosion problem in model checking. In: 2009 Third International Symposium on Intelligent Information Technology Application, vol. 2, pp. 329–331 (2009). https://doi.org/10.1109/IITA.2009.50
108. Yuan, S.N.V., Ip, H.H.S.: Using virtual reality to train emotional and social skills in children with autism spectrum disorder. Lond. J. Prim. Care **10**, 110–112 (2018). https://doi.org/10.1080/17571472.2018.1483000
109. Zapata-Rivera, D., Jackson, T., Katz, I.: Authoring conversation-based assessment scenarios. In: Design Recommendations for Intelligent Tutoring Systems, vol. 3, pp. 169–178 (2015)

Author Index

J. Endrullis and M. Tichy (Eds.): ICGT 2025, LNCS 15720, p. 247, 2025.
https://doi.org/10.1007/978-3-031-94706-3

The manufacturer's authorised representative in the EU is Springer Nature Customer Service Centre GmbH, Europaplatz 3, 69115 Heidelberg, Germany. If you have any concerns regarding our products, please contact ProductSafety@springernature.com

Printed and bound by CPI Group (UK) Ltd, Croydon, CR0 4YY
30/09/2026
02225585-0002